Models of Action:
Mechanisms for Adaptive Behavior

Models of Action:
Mechanisms for Adaptive Behavior

Edited by

Clive D. L. Wynne
University of Western Australia

John E. R. Staddon
Duke University

LAWRENCE ERLBAUM ASSOCIATES, PUBLISHERS
1998   Mahwah, New Jersey   London

Lawrence Erlbaum Associates, Inc., Publishers
10 Industrial Avenue
Mahwah, New Jersey 07430

Cover Design by Kathryn Houghtaling Lacey

Library of Congress Cataloging-in-Publication-Data

Models of action : mechanisms for adaptive behavior / edited by
Clive D. L. Wynne and John E. R. Staddon.
        p. cm.
    Includes bibliographical references and index.
    ISBN 0-8058-1597-X (alk. paper)
    1. Adaptability (Psychology) 2. Adjustment (Psychology)
  I. Wynne, Clive D. L. II. Staddon, J. E. R.
    BF335.M63   1997
    155.2'4—dc21   97-32852
                                                         CIP

Books published by Lawrence Erlbaum Associates are printed on
acid-free paper, and their bindings are chosen for strength and
durability.

Printed in the United States of America
10 9 8 7 6 5 4 3 2 1

Contents

Contributors

Ben Goertzel, College of Staten Island and Intelli Genesis Corporation, Staten Island, NY, USA.

Stephen Grossberg, Department of Cognitive and Neural Systems and Center for Adaptive Systems, Boston University, 677 Beacon Street, Boston, MA 02215, USA.

James Kehoe, Department of Psychology, University of New South Wales, Sydney, NSW 2052, Australia.

Peter Killeen, Department of Psychology, Arizona State University, Tempe, AZ 85287, USA.

John R. Koza, Computer Science Department, Stanford University, Margaret Jacks Hall, Stanford, CA 94305 USA

Nestor Schmajuk, Department of Psychology: Experimental, Duke University, Durham, NC 27706, USA

John Staddon, Department of Psychology: Experimental, Duke University, Durham, NC 27706, USA

David Urry, Department of Psychology: Experimental, Duke University, Durham, NC 27706, USA

Clive Wynne, Department of Psychology, University of Western Australia, Nedlands, WA 6907, Australia.

Silvano Zanutto, Department of Psychology: Experimental, Duke University, Durham, NC 27706, USA.

Preface

How much do we really know about the mechanisms of adaptive behavior? There is no way to measure, but it seems likely that compared to other sciences, we know very little indeed.

Wherever humans live, scavengers take our leavings. Here, by the Swan River in Western Australia, the silver gulls (*Larus novaehollandiae*) live largely on human trash. Yet, Europeans have been in Western Australia for less than 150 years: How did these gulls come to recognize our garbage as edible? How do they know where and when to find it? We do not know. There are rats, mice, and cockroaches everywhere. We are infested by pests that we would like to be rid of, but they defeat our destructive efforts quite effectively—and we understand their clever behavior almost not at all.

Our limited grasp of how intelligent animals work becomes embarrassingly obvious when we actually try to build smart robots. R2D2 and C3PO are works of fiction and are likely to remain so for some time.

Rats and gulls have adapted and continue to adapt to the ever-changing human environment. In psychology, the task of understanding adaptive behavior falls mainly to learning theory. Learning theorists around the turn of the century, inspired by the idea of evolutionary continuity, borrowed the model systems approach from the rest of biology. In physiology, for example, it is usual to study liver, vein, or cell in whatever species is most convenient. Similarly, learning theorists analyzed adaptive behavior in handy species such as rats and pigeons. Assuming uniformity and regularity, they looked for simple mathematical laws that could apply to all animals, humans included.

Two interrelated trends changed the study of adaptive behavior through the 1960s. The cognitive revolution shifted research emphasis to human behavior, including *mental life*. Cognitive psychology appealed to the digital computer as a metaphor to model adaptive behavior and thought. At the same time, electronic computing devices—robots—were being built that could be viewed as having adaptive behavior of their own. These two enterprises—using

computers to model human problem solving, and making artificially intelligent
agents—became deeply intertwined.

Although it began around the same time as the others, a third approach to
modeling adaptive behavior has only recently come into its own. As digital
computers have improved, it has become possible to use them to model
biological information-processing systems. The first variant of the *quasibiological approach* to gain widespread currency is known variously as neural
network modeling, parallel distributed processing, or neurocomputing: Computers are programmed to emulate the behavior of networks of neuron-like
elements. These models necessarily capture only a small a part of the complexity of real neural circuits. Nevertheless, they often behave in ways far
more human- (or animal-) like than programs based on linear information-
processing models.

Still more recently, computers have been programmed to model the process
of evolution by natural selection. This genetic algorithm approach to problem
solving is still in its infancy but promises new advances in our understanding
of adaptive behavior, both as a biologically more realistic model of human and
animal behavior and as a tool for the development of artificial adaptive systems.
This book attempts to represent all these aspects of the quasibiological
approach to adaptive behavior.

The contributions of Killeen, Staddon and Zanutto, Wynne, and to some
extent Kehoe, are influenced by the learning theory tradition. Both Staddon
and Zanutto and Killeen take as their aim the development of a model of a
relatively basic behavioral phenomenon: In Staddon and Zanutto's case this is
feeding regulation. These authors are concerned with accounting for the
pattern of eating in meals and bouts observed in many species with a model
that is, on the one hand, as simple as possible, but that does not eschew
mention of hypothetical internal states.

Killeen applies himself to modeling the energizing and shaping impact of
reinforcers on behavior. He takes his theoretical inspiration from physics and
talks about a dynamics of reinforcement. In so doing he clarifies distinctions
that have long been ambiguous, such as the distinction between the energizing
and the associative effects of incentives on behavior.

Wynne takes on an ostensibly highly complex and cognitive performance,
namely that of transitive inferential reasoning, and explores the extent to
which simple traditional learning models can account for the available data.
His modeling approach is conservative, exploring the extent to which "good
old-fashioned" learning theories can account for performance on an apparently
complex task.

Kehoe's chapter can be viewed as a bridge from traditional learning theory
to quasibiological neural network models. His is also one of the chapters with
a more cognitive flavor. Kehoe's approach is to take an apparently very simple

problem—to what extent is responding to a compound stimulus controlled by responding to its elements?—and explore the simplest model necessary to account for the known facts. Following a discussion of various simple models in the learning theory tradition, he concludes with a distributed neural network model, which, as he points out, embodies many principles of older mathematical learning theories.

Like Kehoe, researchers Schmajuk, Urry, and Zanutto also offer a bridge from traditional associative learning theory to neural network models. Schmajuk et al. take avoidance and escape learning as their chosen task and develop a neural network model that combines aspects of both classical and operant conditioning to account for a "frightening complexity" of data.

Whereas the neural network models of Kehoe and of Schmajuk et al. are concerned with accounting for one set of phenomena in detail, Grossberg presents a model covering a wide range of phenomena including recognition memory, category formation, generalization, the timing of adaptive responses, and more. In addition to working with units that share some properties with neurons, the parts of Grossberg's model are explicitly identified with the circuitry of particular pieces of brain tissue. This means that the model can make predictions about the effects of lesions of parts of the brain.

Koza's approach is also biologically motivated, not by brain physiology, but by the evolutionary process. The busy work of chromosomes exchanging information, reproducing, and surviving inspires a system for solving a model problem. Koza designs computer programs that survive (or are culled) and reproduce according to processes modeled on biological evolutionary principles. The particular issue he tackles in this chapter is the problem of introducing structure into computer programs that have been developed by a process of synthetic genetic evolution. Typically the processes of crossover and mutation tend to destroy higher-level structure (like subroutines in standard computer programs). Koza takes a lesson from evolution and shows how it can be applied to improve programs derived from genetic operations by permitting the evolution of subroutines.

If one way of ordering the chapters in this volume is in terms of broad historical trends in the study of adaptive behavior to which they can be allied, another is in terms of simplicity versus complexity. Several authors (Killeen, Staddon and Zanutto, and Wynne are the clearest examples) quite explicitly argue that parsimony, searching for the simplest possible model for a set of data, is an important motivation for their work. At the other extreme, Goertzel urges us to break free from the confines of simple linear models and to embrace the modern mathematical tools enabling the consideration of far more complex systems. Using dynamical systems analysis, he finds structure in a range of complex-seeming phenomena at several different levels of study, from mood cycles to neural networks.

As well as the range of types of models for adaptive behavior covered, this volume offers the reader some insight into the thought processes behind the development of a particular model. So often in the journal literature, interesting theories appear as if from nowhere, without any background or context. This gives the impression that there is no history to an idea, and makes it difficult for the student to see how he or she could contribute to the further development of theory.

Our aim is to illustrate the range of thinking that goes into the study of adaptive behavior in the hope that readers (as well as our authors) will see new connections and new problems. We hope this volume contributes equally to an our understanding of everyday conundrums, like the feeding strategy of rats and gulls, and recondite research, like building smart robots that would make Isaac Asimov proud.

Before closing, it is our pleasure to acknowledge the kindness and patience of our contributors in responding to our requests. Clive Wynne thanks the University of Western Australia and the Australian Research Council for their support during the development of this volume; John Staddon similarly thanks Duke University, the National Science Foundation, and the National Institutes of Mental Health.

—*Clive D. L. Wynne*
—*John E. R. Staddon*

1

Learning the Language of Mind: Symbolic Dynamics for Modeling Adaptive Behavior

Ben Goertzel
College of Staten Island
Intelli Genesis Corp.

Symbolic dynamics is an analytical technique that reveals the linguistic structures hidden in the chaotic attractors of dynamical systems. I argue here that symbolic dynamics deserves a place at the center of cognitive and behavioral modeling.

The applicability of symbolic dynamics is explored on several different levels. After an exposition of the historical roots and basic concepts of symbolic dynamics, the relevance of these concepts to data analysis is explored by considering the problem of predicting mood fluctuations. Then, moving to the brain level, various techniques for constructing emergent grammar-producing neural networks are described. Next, the psynet model of psychological dynamics is outlined, and in this context it is argued that the recursively modular structure of the brain may perhaps reflect itself in emergent phrase-structure grammars. Finally, the role of symbolic dynamics as a unifier of various psychological modeling strategies is discussed, and in this light some new ideas on topological language spaces are introduced.

Language, which presumably evolved for the description of observable situations and personal feelings, has long since turned into something more powerful. It can be used to describe anything from the counterintuitive behavior of quarks to the imagined mating habits of space aliens. It is indispensable, not only to communication, but to many of our thought processes as well.

Even so, the languages that we use in everyday life are only a small subset of the realm of possible languages. In theoretical computer science, one finds the notion of a *formal language*—defined as a collection of entities (the alphabet) together with a collection of combination rules. The rules tell you which combinations of entities are correct and which are incorrect, or else which are more correct than others. The universe of formal languages contains all human languages, from Arabic to Yiddish, and an infinite realm of other languages as well, some extremely simple and some incomprehensibly complex.

What use are all these other languages, one might wonder, if no one can speak them? A principal use is in the theory of computer programming languages. What I argue here is that these formal languages are also essential for the modeling of mind and brain. The key to this relationship is the technique of *symbolic dynamics*.

Symbolic dynamics is a mathematical/computational tool for expressing patterns of change over time as formal languages. It appears in mathematics and physics as a tool for the study of nonlinear iterations and differential equations, and for the analysis of complex time-series data. It involves the transformation of a sequence of values (e.g., real vectors) representing the trajectory of some dynamical system, into a sequence of letters or symbols drawn from an abstract alphabet. The patterns in the system's behavior make themselves apparent as linguistic patterns in the corpus of abstract symbol combinations. The emergent symbol system does not entirely capture the underlying nonlinear dynamics, but it captures the most significant abstract patterns in this dynamic.[1]

I argue that symbolic dynamics deserves to play a prominent role in the study of complex adaptive behavior. It has the potential to clarify at least two important issues: the meaning of abstract models of mind/brain, and the relation between symbolic reasoning and nonlinear neurodynamics. Symbolic dynamics presents a new vision of behavioral science as the study of dynamical systems and the topology of formal language spaces.

The structure of the chapter is as follows. The following section traces the concept of symbolic dynamics back to Leibniz, and casts Leibniz's philosophical speculations in mathematical form with a statement called the "Chaos Language Hypothesis." The next section introduces some basic ideas from dynamical systems theory and symbolic dynamics, using the Baker map and

---

[1]Some authors in mathematics and physics interpret the term *symbolic dynamics* to refer exclusively to the construction of symbol series from mathematically natural partitions (e.g., partitions based on the critical points of an iteration function). Here, I interpret the word more generally to refer to any instance of the construction of symbol sequences from trajectories of dynamical systems.

its generalizations as an illustrative example. Then, I explore the possibility of tracking human moods using symbolic dynamics, using some recent empirical data as a launching point for speculative, illustrative examples. I next demonstrate that simple neural networks are capable of manifesting arbitrary formal grammars in their chaotic dynamics, and discuss two methods for training neural networks to display structured chaos: the genetic algorithm, which has proven successful in simple instances, and an untried technique called attractor pattern learning. Finally, I discuss the role of symbolic dynamics in validating qualitative models of adaptive behavior, and toward this end present a novel topology for spaces of formal languages.

PAST AND FUTURE OF SYMBOLIC DYNAMICS

Only in the past few decades has symbolic dynamics emerged as a useful mathematical tool. Its conceptual roots, however, go back at least to Leibniz (1969). Leibniz proposed a "universal character for 'objects of imagination'"; he argued that "a kind of alphabet of human thoughts can be worked out and that everything can be discovered and judged by a comparison of the letters of this alphabet and an analysis of the words made from them" (pp. 221–222). This systematization of knowledge, he claimed, would lead to an appreciation of subtle underlying regularities in the mind and world: "[T]here is some kind of relation or order in the characters which is also in things . . . there is in them a kind of complex mutual relation or order which fits things; whether we apply one character or another, the products will be the same" (p. 225). In modern language, the universal characteristic was intended to provide for the mathematical description of complex systems like minds, thoughts, and bodies, and also to lead to the recognition of *robust emergent properties* in these systems, properties common to wide classes of complex systems. These emergent properties were to appear as *linguistic regularities*.

Leibniz did not get very far with this idea. He developed the language of formal logic (what is now called *Boolean logic*), but, like the logic-oriented cognitive psychologists and AI theorists of the 1960s–1980s, he was unable to build from the simple formulas of propositional logic to the complex, self-organizing systems that make up the everyday world. Today, with dynamical systems theory and other aspects of complex systems science, we have come much closer to realizing his ambitions. (One might argue, however, that in the intervening centuries, Leibniz's work contributed to the partial fulfillment of his program. The formal logic that he developed eventually blossomed into modern logic, computer science, and formal language theory. It is computer power, above all, that has enabled us to understand what little we do about the structure and dynamics of complex systems.)

Leibniz and Abraham (1995) pointed out the remarkable similarity between Leibniz's Universal Characteristic and the modern idea of a unified theory of complex system behavior; they singled out the concepts of attractors, bifurcations, Lyapunov exponents, and fractal dimensions as being important elements of the emerging universal characteristic. It is symbolic dynamics, however, that provides by far the most explicit and striking parallel between Leibniz's ideas and modern complex systems science. Symbolic dynamics does precisely what Leibniz prognosticated: It constructs formal alphabets, leading to formal "words" and "sentences" that reveal the hidden regularities of dynamical systems. As yet it does not quite live up to Leibniz's lofty aspirations—but there is reason to be optimistic.

Leibniz's speculations resonate wonderfully with recent ideas about the possibility of finding unifying laws of complex systems science (Goerner, 1993; Goertzel, 1994). Many have hypothesized the existence of such laws, but no one has been particularly clear about what form such laws might take. Symbolic dynamics gives some concrete ideas in this direction. Leibniz's idea of the Universal Characteristic, translated into modern terminology, suggests that there may be a small number of archetypal attractor structures common to a wide variety of complex systems. Mathematically speaking, the approximation of these archetypes should reveal itself as a clustering of inferred formal languages in formal language space (a notion that can be formalized by defining, as in the final section, an appropriate topology on formal language space).

This train of thought leads to the following formal hypothesis:

Chaos Language Hypothesis: The formal languages implicit in the trajectories of natural (in particular, psychological and social) dynamical systems show a strong tendency to "cluster" in the space of formal languages.

And this hypothesis suggests the following three-stage research program:

1. By computational analysis of data obtained from empirical studies and mathematical models, try to isolate the archetypal formal languages underlying complex psychological and social systems.
2. Analyze these languages to gain an intuitive and mathematical understanding of their structure.
3. Correlate these languages, as far as possible, with other quantitative and qualitative characterizations of the systems involved.

I suggest that this research program may be particularly fruitful in the context of cognitive and behavioral modeling. The brain, complex as it is, has proved particularly resistant to detailed quantitative analysis. It may be that linguistic analysis is the only way to go.

DYNAMICS, CHAOS, AND LINGUISTIC PATTERN

Symbolic dynamics has emerged out of the branch of mathematics known as dynamical systems theory. In this section, I review some concepts from dynamical systems theory and explain how symbolic dynamics connects the mathematics of differential equations and discrete iterations with the mathematics of formal languages.

Dynamics, in the most general sense, is the study of how things change over time. Mathematical dynamical systems theory is concerned mainly with the behavior of systems whose change over time is governed by briefly stated equations. It also, however, gives us deep qualitative insight into the behavior of the more complex dynamical systems that we see in the real world.

The key concept of dynamical systems theory is the *attractor*. An attractor is, quite simply, a characteristic behavior of a system. The striking insight of dynamical systems theory is that, for many mathematical and real-world dynamical systems, the initial state of the system is almost irrelevant. No matter where the system starts from, it will eventually drift into one of a small set of characteristic behaviors, a small number of attractors.

Some systems have fixed-point attractors, meaning that they drift into certain equilibrium conditions and stay there. Some systems have periodic attractors, meaning that after an initial transient period, they lock into a cyclic pattern of oscillation between a certain number of fixed states. Finally, some systems have attractors that are neither fixed points nor limit cycles, and are hence called *strange attractors*. The most complex systems possess all three kinds of attractors, so that different initial conditions lead not only to different behaviors, but to different *types* of behavior. To complicate things further, strange attractors come supplied with *invariant measures*, indicating the frequency with which different regions of the attractor are visited. Some dynamical systems have very simple attractors hosting subtly structured invariant measures; an example of this is given shortly.[2]

The formal definition of strange attractor is a matter of some contention. Rather than giving a mathematical definition, I prefer to give a dictionary definition that captures the common usage of the word. A strange attractor of a dynamic, as I use the term, is a collection of states that is: (a) invariant under the dynamic, in the sense that if one's initial state is in the attractor, so will be all subsequent states; (b) attracting in the sense that states that are near to the attractor but not in it will tend to get nearer to the attractor as time progresses; and (c) not a fixed point or limit cycle.

---

[2]Strictly speaking, this classification of attractors applies only to deterministic dynamical systems; to generalize them to stochastic systems, however, one must merely "sprinkle liberally with probabilities." For instance, a fixed point of a stochastic iteration $x_{n+1} = f(x_n)$ might be defined as a point that is fixed with probability one; or a p-fixed point might be defined as one for which $P(f(x) = x) > p$.

A great deal of attention has been paid to the fact that some dynamical systems are chaotic, meaning that, despite being at bottom deterministic, they are capable of passing many statistical tests for randomness. They *look* random. Under some definitions of strange attractor, dynamics on a strange attractor are necessarily chaotic; under my very general definition, however, this need not be the case. The many specialized definitions of chaos are even more various than the definitions of strange attractor.

Symbolic dynamics, in essence, is a way of studying the structure of strange attractors using formal language theory. The basic idea of symbolic dynamics is to divide the state space of a dynamical system into a finite number of regions or *cells*, numbered 1 through N. One then charts each trajectory of the system as an itinerary of cells. If a trajectory goes from cell 14 to cell 3 to cell 21 to cell 9, and so forth, its symbol sequence will begin {14, 3, 21, 9, . . .}. The collection of *code sequences* obtained in this way can tell one a great deal about the nature of the dynamics. They may be viewed as the corpus of some unknown language, and subjected to mathematical, statistical, and linguistic analysis. The regularities in this corpus (e.g., 9 never follows 21, except when it precedes 12; 21, 43, and 55 always follow either 66, 12, or 7, unless they follow the sequence 33, 44) may be taken as grammatical rules. The probabilistic regularities in this corpus (66 follows 71 83% of the time, and follows 12 9% of the time) may be taken as rules in a *probabilistic grammar.* A standard tool for proving a mapping chaotic is to show that it is topologically conjugate to a chaotic mapping acting on code sequences.

A simple example illuminates how the method is used in the study of mathematical iterations. Consider the Baker map

$$x_{n+1} = 2x_n \bmod 1 \quad (1)$$

Using the two cells [0,1/2] and [1/2,1], one finds that the dynamics of the Baker map lead to the map *sigma* on code sequences:

$$sigma(a_1 a_2 a_3 ...) = a_2 a_3 a_4 ... \quad (2)$$

This is called the shift map: It takes a code sequence and simply truncates it, removing the first symbol. This encodes the behavior of the Baker map in a particularly elegant way. It reveals why the map is chaotic: It throws away a whole symbol full of information at each time step.

The attractor of the Baker map is the whole interval [0,1], and the invariant measure is the uniform measure, which weights each point equally. In this case, symbolic dynamics does not lead to an interesting formal language. Instead, it gives rise to a language in which all possible combinations of letters are considered permissible. Every finite "word" of 0s and 1s, no matter how long, is permissible and will eventually occur in the dynamics of the Baker map. This is pure, unstructured chaos—the mathematical equivalent of the tower of Babel.

A related mathematical example that is much more subtle and interesting is the generalized Baker map

$$x_{n+1} = Phi(x_n) = (beta\ x_n + alpha) \bmod 1 \quad (3)$$

$$1 < beta < 2$$

A detailed study of this map is quite rewarding. A good special case to think about is $beta = 3/2$, $alpha = 0$.

The map Phi also has the whole interval [0,1] for an attractor, but its invariant measure is more subtly structured. In other words, no matter what value of x_0 one begins with, the values of x_i will wander all over the interval [0,1], but there will be a pattern to their distribution. If one charts the probabilities with which the x_i fall into different subintervals, one will obtain an approximation to the invariant measure of the map. The measure is invariant because these probabilities are the same for almost any initial value. As a fairly simple argument shows, the motion of the iterates of Phi through the invariant measure are chaotic.[3]

If one does symbolic dynamics on the generalized Baker map, using the natural cells [0,1/beta] and (1/beta,1], one finds that each point gives rise to a code sequence that is its expansion in base beta. Expansions in noninteger bases being somewhat subtler than binary expansions, one runs into certain complications: the code sequence space in question no longer contains all binary sequences, but only those special binary sequences that correspond to expansions in base beta. But the analysis essentially works the same; the iteration can be proved chaotic by reference to the chaoticity of a shift map on this strange code space (Goertzel, Bowman, & Baker, 1993).

Taking the case $beta = 3/2$, let us consider some sample expansions. We have, for instance,

$$.100... = 2/3$$
$$.0100... = 4/9$$
$$.00100... = 8/27$$
$$.10100... = 2/3 + 8/27 = 26/27$$

All these are unproblematic. Things get trickier, however, when one looks at something like

$$.1100... = 2/3 + 4/9 = 10/9.$$

---

[3]There is a sense, it should be noted, in which the distinction between attractors and invariant measures is really not conceptually important. If one constructs the "Frobenius-Perron operator" *Psi* corresponding to *Phi*, an operator that tells how *Phi* maps probability distributions onto probability distributions, then the invariant measure re-emerges as an attractor of *Psi*.

How can a decimal, a number with zero on the left hand side of the decimal point, be greater than one? Clearly something is wrong here. What is wrong is, specifically, that one is not dealing with an integer base!

Because a double-one is illegal in the leading position, it is also illegal everywhere else. After all, it is equally unsettling to look at

$$.01100... = 10/27 > 2/3 = .100...$$

Double ones are not the only forbidden sequences. Consider, for instance,

$$.1010100... = 2/3 + 8/27 + 32/243 = 266/243$$

The pattern 10101 is therefore disallowed in any base 3/2 expansion. On the other hand, a simple computation shows that

$$.100100100...$$

is just fine—skipping two zeros does the trick!

With a little creativity, one can construct a huge variety of disallowed sequences. This gives rise to the question: Precisely how many sequences are allowed? As the sequence length n tends to infinity, what percentage of the 2^n possible sequences are permitted? There is only one answer which makes any intuitive sense: approximately $(3/2)^n$ sequences should be allowed. In the general case, approximately $beta^n$ sequences should be allowed. Leo Flatto (personal communication, 1993) proved this result using number-theoretic techniques; however, his proof works only for the case $alpha = 0$. Working independently, Harold Bowman and I (Goertzel et al., 1993) proved the same theorem for the case of general $alpha$, in a way that relates naturally with the dynamics of the map.

The generalized Baker map illustrates a phenomenon that has escaped the notice of many commentators on chaos theory: Chaos, in itself, does not preclude the presence of emergent order. Chaotic dynamics has the potential to give rise to emergent statistical/algorithmic order of great subtlety and beauty. This potential for emergence is not always manifested: some real-world chaotic systems are, like the Baker map, devoid of regularity. But others are rife with emergent order. And one way to study this order is—as Leibniz intuited three centuries ago and symbolic dynamics now makes explicit—to view it as *linguistic form*.

One caveat must be noted. In these simple mathematical examples, I have assumed that the correct partition of state space was known in advance. Otherwise, the analysis would not have proceeded so smoothly. In general, however, there is no way to determine the best way to divide the state space of a dynamical system into cells. One must proceed by trial and error. One method that seems promising is to begin with a very large alphabet, and then group the letters of this alphabet into categories, resembling grammatical categories. The category labels may then be taken as letters for a new symbol sequence, which hopefully captures the essential structure of the system. This approach has been developed by myself and Gwen Goertzel (Goertzel & Goertzel, 1996), and is called the *Chaos Language Algorithm* (CLA). In its simplest version, the CLA is a three-stage process, consisting of

1. Discretization of trajectories of a dynamical system by symbolic dynamics. One divides the potential state space of the system into a finite number of disjoint regions and assigns each region a code symbol, thus mapping trajectories of the system into a series of code symbols.

2. Tagging of symbolic trajectory sequences using a self-organizing tagging algorithm. One takes the symbols derived in the first step and assigns each one to a certain category based on the idea that symbols that play similar roles in the trajectory should be in the same category. The tag of a code symbol is a number indicating the category to which it belongs. The code sequence is thus transformed into a tag sequence.

3. Inference of a grammar from the tag sequence produced in the previous step, and iterative improvement of the tagging to maximize grammar quality.

Applying the CLA to the generalized Baker map with *beta* $= 3/2$, for instance, one may begin by dividing the unit interval into 10 equally sized nonoverlapping subintervals. A trajectory of the system is then encoded as a series of integers from 0 to 9, where the tag i represents the subinterval $[i/10, (i+1)/10]$. The CLA, with appropriately chosen parameters, is able to form the optimal categories

Category 0:
0 1 2 3 4 5 6

Category 1:
7 8 9

These categories give rise to natural grammatical rules, such as the permissibility of 00, 01, and 10, but the impermissibility of 11.

SYMBOLIC DYNAMICS OF MOOD FLUCTUATIONS

The focus of this chapter is theoretical; but the question of symbolic dynamics as a tool for empirical data analysis is too important to overlook. The possibilities of symbolic dynamics in this regard are both diverse and profound. To take a single example, Paul Rapp (Theiler & Rapp, 1996) argued in a recent paper that the best way to understand the observed dynamics of the brain (e.g., EEG and ERP) is to discretize the complex time series and study the emergent symbolic dynamics. For another example, Goertzel and Goertzel (1996) used symbolic dynamics to study semantic and grammatical changes in written language. Some of my current research involves using symbolic dynamics methods to compare nonlinear psychophysical models with human psychophysical data. The possible applications of these ideas are as diverse as the mind itself.

Farther toward the neural level, symbolic dynamics may also provide a new way of thinking about the pioneering work of Freeman (1991) on the dynamics of olfactory perception. Freeman proposed a strange attractor with wings, each wing corresponding to a certain recognizable smell. The process of recognizing a smell corresponds to chaotic dynamics spread throughout the attractor; the process of having recognized a smell corresponds to periodic dynamics restricted to a single wing of the attractor. If a symbol were to be assigned to each wing, then this convergence from uncertainty to certainty would be made apparent as a *linguistic pattern*. Greater and greater knowledge would correspond to more and more restricted languages, the extreme cases being an initial language in which nearly all symbol combinations are permissible, and a final language consisting solely or mainly of repetitions of the same symbol.

But, though it is important to appreciate the broad applicability of these ideas, the role of symbolic dynamics in data analysis is best illustrated by detailed consideration of a single example. I (Goertzel, 1996) described an exploratory application of symbolic dynamics to the analysis of human mood fluctuations, using data collected by Allen Combs (1993). Here, rather than delving into the intricacies of this empirical research project, I use this work as a starting point for a more speculative and illustrative discussion of the role of symbolic dynamics in mood fluctuation.

The key point to remember, philosophically, is that language is just a way of combining things in time. The state space of a system at a given time may be very complicated, characterized by a huge number of different quantities. One may, as an observer, decide to categorize the state space, to divide it into a finite number of categories. Then the system's changing states over time can be represented as a series of categories. First the system was in Category 1, then it was in Category 2, then it was in Category 1 again, then Category 4, and so on.

For instance, suppose the system in question is a human being. Obviously, the condition of a human being is characterized by a huge number of different properties. But one can divide the various possible conditions of a person into a few classes, if one wishes. Using emotional labels, one can say that a person is happy or sad or ecstatic or wondrous, and so on. Using states of consciousness, one can use categories such as waking versus sleeping—or, ordinary waking state versus sleeping state versus just having woken-up state, versus stoned state versus tripping on LSD state, and so forth. Whatever categorization one has chosen, one can use this to track the change of the system—the person—over time. As the person changes over time, we can monitor what category the person is in at each time. The person's pattern of change over time is then represented as a certain series of category names, which is, formally speaking, interpretable as a series of letters or words in a language.

In other mood fluctuation experiments (Combs, 1993; Goertzel, 1996), the state space of a human being was divided into four categories based on emotional labels. Each subject marks two lines (Likert scales) at half-hour intervals (during waking hours), one for the dimension of happy versus sad; the other for the dimension of excited versus relaxed. This gives a state space that is a two-dimensional real vector. In order to apply symbolic dynamics, one needs to come up with categories that divide this state space into a finite number of regions, which will be the alphabet of the formal language. The simplest course is to consider four categories:

A—happy and excited
B—happy and relaxed
C—sad and excited
D—sad and relaxed

Given this coding, the data set produced by a person over a period of time would look like

AABCDBCADCCBBCDDACCCDCCDCAAABBC . . .

The idea of symbolic dynamics is that this list of letters is going to harbor implicit grammatical rules.

In practice, this experiment is fraught with difficulty, because of the difficulty of getting subjects to carry through with the recording over long periods of time, because of the high noise level of emotional self-assessment, and because of the irregularities in the data sequences introduced by the sleep–wake cycle. Here, however, we gleefully ignore these problems, and consider the experiment as we believe it would work if we had optimal data: data from subjects trained at self-observation, collected over a very long period of time. It should be kept in mind that our main goal at the moment is an understanding of the possible applications of symbolic dynamics, not an understanding of mood fluctuations.

What kinds of linguistic rules might arise in these time series? The simplest rules are *first-order constraints*, constraints on what tends to follow what (and indeed, these are the kinds of rules that have been observed in the real mood fluctuation data so far). One might find, for instance, that a person rarely moves from A to D or from D to A—that the pairs AD and DA rarely appear in the series. This leads to the probabilistic grammatical rule that an A tends to be followed by an A, B, or C. Or one might find that a certain person rarely goes from B to C or from C to B. These constraints would be pretty reasonable. Who goes from happy and excited all the way to sad and relaxed, without passing through intermediate stages?

Pushing a little further, one might find rules saying: "A has got to be followed by B or by C." Or, say: "A will nearly always be followed by B or by C." It could even happen, conceivably, that A, B, C, and D only occurred in a few combinations, say ABACAB, DABACA, BABA . . .—then one could look for patterns in the occurrence of these combinations, these emergent words.

These first-order transition rules may not seem like a very interesting kind of language, but they are only the beginning. For starters, there will be higher-order transition rules (i.e., rules involving sequences of length greater than two). An example would be: "If A is followed by C, then D is very likely to occur." In other words, ACD is probable, whereas ACA is unlikely. According to this rule, having been happy and excited, then sad and excited, one is not that likely to become happy and excited again; it is much more probable that one will get sad and relaxed.

In general, one finds that the simpler rules are more universal in nature, whereas the more complicated rules tend to be more individual. For instance, although ACD may be a very likely sequence for some people, for others it may be rare. A manic-depressive person might well tend to have excitement stay high while happiness and sadness oscillate back and forth, meaning, in symbolic dynamics terms, a lot of cycles ACACACA . . . , or perhaps ACDACDA . . . or ACDCACDCA . . . .

The patterns we have been discussing so far are all context-free rules. The next step up, in terms of complexity, is to context-sensitive rules, such as: "A follows B, but only in contexts where A occurs between C and D." Context-sensitive rules can always be summarized as collections of context-free transition rules, but they are more ordered than arbitrary collections of transition rules. They are higher-order patterns. For instance, one can list the following permissible sequences:

CABB
AABD
CABB
BACC

. . .

or else, using a context-sensitive rule, one can say something simple like: "XABY is only a rule when X is C and Y is D."

Examples of context-sensitive rules are transformation rules, such as one finds in the transformational grammar approach to human languages. The simplest transformation rules are movement rules, such as "Whenever you have ABCDB, you can replace it with BCDAB, and still have a permissible sequence." In this rule, the A is moving from the first to the penultimate position in the word, without disrupting the meaning.

One can, potentially, have very complicated situations here. For instance, suppose

ADCBD
ACBCC
ADBDD
ABDCB

are all permissible. This would lead one to believe the rule would be

AX_X

But instead, it might well happen that the rule

ABCDB

that fits the pattern is not a rule; instead one might have something like

BCDAB

as a rule instead. This situation, if it were to occur, would be a symbolic-dynamics analogue of what one finds in human linguistics, where

* I know the person was where

is ungrammatical even though

I know the person was here
I know the person was there
I know the person was in Topeka
I know the person was higher
. . .

are all permissible.

Based on the limited data analyzed so far, there is no reason to believe that such subtle patterns actually exist in human moods. On the other hand, there is no reason to rule it out. Mood fluctuations appear chaotic, but we know that chaos is just unpredictability in detail. Statistical and algorithmic predictions can still be made about chaotic systems; and symbolic dynamics is the foremost tool for making such predictions. Perhaps when we get to know a person, part of what we are doing is recognizing higher-order linguistic patterns in the time series of their behaviors.

THE SYMBOLIC POWER OF NEURAL NETWORKS

Symbolic dynamics is a general-purpose tool for understanding patterns of change over time—and thus, I have argued, it is pertinent to every type of dynamical behavioral model. In this section, I illustrate this point by considering the relationship between symbolic dynamics and *neural network* modeling.

There are many formal neural-network models. For simplicity, let us begin with the classic McCullough-Pitts model, in which an individual neuron is represented by a binary value, an element from the set {0,1}. Each neuron is understood to receive input to several other neurons, and send output to several other neurons. The output of a neuron at a given time is 1 if its total input exceeds a certain fixed threshold (which is different for different neurons), and zero otherwise.

The process by which this model is obtained from biological neural networks is itself, in part, an example of symbolic dynamics. A real neuron is a complex system, typically characterized by multiple real variables. In McCullough-Pitts models, however, this complexity is ignored by taking one variable, voltage, and classifying a neuron with the symbol 0 if the variable is below a certain threshold, or 1 otherwise. The state of a network with N neurons is then representable as a letter on an alphabet of size 2^N. The allowable transitions between one such letter and another define the dynamics of the neural network. The idea is that the dynamics on this *symbol space* are an adequate representation of the dynamics on the underlying multivariate real space.

McCullough-Pitts networks are very general and are not constrained to any particular variety of dynamics. As discrete systems, they cannot formally be chaotic, they must eventually repeat themselves; but they may become entrained in very, very long periodic orbits that emulate chaos over empirically meaningful time periods. Most neural-network models used in practice involve some form of fixed-point dynamics, but this reflects the needs of the humans designing and running the models more than it does the intrinsic nature of neural networks. A neural network with fixed-point dynamics is easy to monitor; its final state may be evaluated for adequacy or inadequacy. A complex, chaotic neural network is more difficult to study, and requires subtler methods of training (a conclusion that is hardly surprising, given the lengthy and subtle course of training required by our own huge biological neural networks).

Putting the practical issue of training aside, there are many ways to demonstrate that formal neural networks are *capable* of giving rise to arbitrarily complex symbolic dynamics. The simplest proof is indirect, and involves McCullough-Pitts networks. First, McCullough-Pitts networks are capable of simulating any finite automaton; this is a classical result that is shown, in essence, by constructing AND, OR, and NOT gates out of simple binary-valued threshold neurons. Thus, if one can construct a finite automaton displaying a given symbolic dynamics, one can realize this automaton using McCullough-Pitts neurons. Taking the abstraction one step farther, if one can construct a mathematical iteration on a bounded set, displaying a given symbolic dynamics, then one can approximate this iteration to any degree of accuracy by a finite automaton, and one has one's result. The universal symbolic power of neural networks is thus obtained as a simple consequence of the universal symbolic power of mathematical iterations.

Finally, it is easy to construct a dynamical system whose state space represents a given grammar. One wishes an iteration that displays chaotic dynamics leading to a given collection of letter occurrence probabilities p_i, and transition probabilities pij, where i and j come from $\{0,...,N-1\}$. First, one takes each of the N letters involved, and assigns to them, in alphabetical order, the N subintervals $(i/N,(i+1)/N)$, where i = $0,...,N-1$. Let I_i denote the subinterval assigned to letter i, where the length of I_i is determined by the desired probability p_i. Divide each subinterval I_i into M equally sized subintervals (subsubintervals) $I_{i,r}$, r = 1, . . . ,M. Here the number M should be very large; for instance, M might represent the size of the corpus from which the grammar was constructed, and should be divisible by N. The function f is defined to be linear over each subinterval $I_{i,r}$. Each letter j gets a number of subintervals $I_{i,r}$ proportional to the probability p_{ij} of the transition from i to j. Over each subinterval $I_{i,r}$ that is assigned to j, the function f is defined to be the linear function mapping the left endpoint of $I_{i,r}$ into the left endpoint of I_j, and the right endpoint of $I_{i,r}$ into the right endpoint of I_j.

Now, given this construction, suppose one constructs symbolic dynamics on the partition defined by the I_j, for all the words j. One finds that the probabilistic grammar obtained in this way is precisely the probabilistic grammar from which the dynamical system was constructed. For models incorporating k^{th} order transitions, where k exceeds 1, the standard trick suffices: instead of looking at intervals corresponding to words, look at intervals corresponding to k-tuples of words. If the probabilistic grammar in question is, say, a grammar of English, then the dynamical system "knows" English.[4] These piecewise linear maps can be approximated by finite-state automata, which can in turn be realized by McCullough-Pitts neural networks. Thus we have demonstrated the *theoretical* plausibility of realizing any symbol system whatsoever with neural networks—as an emergent structure, a natural consequence of complex, chaotic dynamics.

This construction is somewhat artificial: there would be no point to building a neural network this way. It is possible, however, to envision *training* neural networks to display given symbolic dynamics structures. This may even be done blindly: it seems that the genetic algorithm (Goldberg, 1988) may be used as a tool for evolving neural networks with given symbolic dynamics. Using networks of 5 to 10 neurons, each one following the discrete chaotic neuron iteration of Mayer-Kress (1994; this is not a McCullough-Pitts model but a more complex iteration that attempts biological realism), and a population size in the range 20–50, it is possible to evolve neural networks whose strange attractors display various patterns of permissible and impermissible words (e.g., the second-order pattern of the generalized Baker map, in which 11 and 101 are disallowed). These are only simple, preliminary experiments, but they indicate that evolutionary processes are at least plausible candidates for the training of neural networks to produce formal languages. Given the existence of a strong body of evidence for the evolutionary character of neurodynamics and learning (Edelman, 1988; Goertzel, 1993a), this is an encouraging datum.

Another route toward the training of neural networks with specified symbolic dynamics is a strategy that might be called *attractor pattern learning*. Attractor pattern learning applies just as well to continuous-valued neural networks as to McCullough-Pitts networks; for simplicity, however, I assume the latter, and further restrict attention to the case of deterministic, rather than probabilistic grammars. Suppose the state of the neural network at a given time is encoded as a binary vector of length J, where J is the number of neurons (e.g., 01110100...101, where the bit in the rth position indicates whether the rth neuron is on or off). One wishes to adjust the weights of the network so that the network will display chaotic dynamics leading to a given collection of letter occurrence probabilities p_i, and transition probabilities p_{ij}, where i and j come from $\{0,...,N-1\}$. The key is to associate each letter i with a certain scan pattern of the form $Q_i = q_{i1} \ldots q_{iJ}$, where each q_{ik} is either a 0, a 1, or a "don't care" symbol (#). For instance, in a six-neuron network, one might have $Q_2 = 01\#10\#$. Each such scan pattern corresponds to a certain *region* R_i of the state space of the network; for instance, $01\#10\#$ corresponds to that region in which the first and fifth neurons are inactive, but the second and fourth neurons are active.

---

[4]According to Harris's (1993) analysis of English, all English sentences can be obtained by taking the results of this dynamical system and transforming them by a handful of simple transformation rules.

Let y_i denote the number of # symbols in Q_i (the generality of the regions denoted by Q_i). In order to get the occurrence probabilities p_i right, it suffices to pick y_i/J as close to p_i as possible. This approximation becomes more and more accurate as the network size J becomes larger. Next, in order to get the transition probabilities p_{ij}, one must *train* the network appropriately. Consider first the case where the p_{ij} are constrained to be either 0 or 1 (i.e., the case of nonprobabilistic grammar). Then it suffices to train the network to map states in Q_i into states in Q_j. This can be accomplished by picking random state-pairs (s_i, s_j), s_i and s_j in Q_i and Q_j, respectively, and training the network weights to realize these pairs. The randomness of the pairs guarantees that, once training is done, the periodicity of the network's trajectory will almost certainly be very long (i.e., the probability of length exponential in J tends to 1 as J tends to infinity).

If the p_{ij} are allowed to assume general values, the same strategy can be used, but the R_i must be divided into subregions, in the worst case, J subregions apiece. In order to accomplish this cleanly, the network size J may need to be larger than would be necessary for a deterministic grammar. A similar strategy may be used to incorporate higher-order transition probabilities, p_{ijk}.

These algorithms result in neural networks manifesting specified grammars in their dynamics. Along related lines, Goertzel and Goertzel (1995) demonstrated that one may construct a formal neural network capable of *inferring* formal grammars from the trajectories of dynamical systems. Thus, not only are neural networks capable of *manifesting* grammars in their dynamics, they are also capable of *recognizing* grammars in each other's dynamics. In this view, for instance, one might want to model word production in terms of two neural networks: one that follows a strange attractor whose various regions correspond to different words, and another that "reads off" these attractor regions and sends appropriate messages to the motor regions of the brain, thus initiating the process of word production.

Practical implementation of neural networks with structured chaos should be interesting, and will doubtless lead to many surprises. The conceptual implications of such networks, however, are already becoming clear. One is led to view human language, not as a sophisticated special-purpose add-on to neurodynamics, but as an extension of something that neural networks are doing at a relativey low level. The brain emerges as a system continually engaged with trying to understand the language it speaks to itself.

THE PSYNET MODEL

Neural networks are popular psychological models, and justifiably so. Not only do they qualitatively model various aspects of brain function, they are also intriguing nonlinear dynamical systems in their own right. My own particular sympathy, however, is for models that focus on the intermediate, *mesoscopic* level, rather than the microscopic level of neurons or the macroscopic level of logical symbol systems—the mesoscopic level of neuronal groups or mental processes.

There are several ways of approaching the mesoscopic level. For instance, as in Gregson's nonlinear psychophysics, one may seek to model the behavior of mental processes using simple nonlinear iterations. One may, as in Edelman's (1988) Neural Darwinism, explicitly take large groups or clusters of neurons as the fundamental elements of one's connectionist analysis. Or one may, as Kampis (1991) did, eliminate both neurons and continuous mathematics altogether, and model the brain as a collection of component processes acting on each other and transforming each other.

My own research on the modeling of adaptive behavior has led to a comprehensive mathematical theory that I call the *psynet model*. In the remainder of this section, I outline the psynet model and develop some ideas regarding its possible connection with symbolic dynamics. (For a complete exposition of the psynet model, see Goertzel, 1994, 1996).

First, I define a *magician system* as an evolving collection of entities called *magicians* that, by acting on one another, have the power to cooperatively create new magicians. Certain magicians are paired with *antimagicians*, magicians that have the power to annihilate them. At each time step, the dynamics of a magician system consists of two stages. First the magicians in the current population act on one another, producing a provisional new population. Then the magician–antimagician pairs in the provisional new population annihilate one another. The survivors are the new population at the next time step.

Mental structures, I argue, are simply collections of magician systems that are *attractors* of this magician dynamic. The magician dynamic connects autopoietic systems theory (Varela, 1978), the study of self-producing systems, with ordinary dynamical systems theory. In *Chaotic Logic* (Goertzel, 1994) I argued in detail that thoughts, beliefs and feelings are attractive magician systems, and proposed a master attractor called the dual network, a hierarchy of nested attractors that incorporates hierarchical perception/control and heterarchical memory in a coordinated way. Combs, in *Radiance of Being* (1995), argued extensively that *states of consciousness* should be understood as attractive magician systems.

Let S denote a set, to be called the space of magicians. Then S^*, the space of all finite sets composed of elements of S, with repeated elements allowed, is the space of *magician systems*. Using the convention of the previous paragraph, one may write

$$\text{System}_{t+1} = A[\text{System}_t] \quad (4)$$

where $System_t$ is an element of S^* denoting the magician population at time t, and A is the action operator, a function mapping magician populations into probability distributions of magician populations.

Let us assume, for simplicity's sake, that all magician interactions are ternary, involving one magician acting on another to create a third. In this instance the machinery of magician operations may be described by a binary algebraic operation *, so that where a, b, and c are elements of S, a*b = c is read "a acts on b to create c." The case of binary, quaternary, and other interactions may be treated in a similar way or, somewhat artificially, may be constructed as a corollary of the ternary case.

The action operator may be decomposed as

$$A[X] = F[\ R[X]\ ] \quad (5)$$

where R is the raw potentiality operator and F is a filtering operator. R is formally given by

$$R[System_t] = R[\{a_1, a_2, \ldots, a_{n(t)}\}]$$
$$= \{\ a_i * a_j \mid i,j = 1, \ldots, n(t)\} \quad (6)$$

Its purpose is to construct the raw potentiality of the magician system $System_t$, the set of all possible magician combinations that ensue from it. The role of the filtering operator, on the other hand, is to select from the raw potentiality those combinations that are to be allowed to continue to the next time step. This selection may be all-or-none, or it may be probabilistic. To define the filtering operator formally, let P^* denote the a space of all probability distributions on the space magician systems $S\cdot$. Then, F is a function that maps $S^* \times S^*$ into P^*.

Magician systems, thus defined, have no intrinsic geometry: all the magicians exist in the same zero-dimensional space. In chemical terms, one is dealing with a well-stirred system. But one may also consider graphical magician systems, magician systems that are specialized to some given graph G. Each magician is assigned a location on the graph as one of its defining properties, and magicians are only allowed to interact if they reside at the same or adjacent nodes. This does not require any reformulation of the previous fundamental equations, but can be incorporated in the filtering operator by setting F so that the magician combination $a_i * a_j$ has zero probability unless a_i and a_j occupy appropriate relative locations. The dual network, a structure postulated as a prerequisite of self- and reality-theories, is a magician system that resides on a special kind of graph.

This kind of system may at first sound like an absolute, formless chaos. But this glib perspective ignores something essential—the phenomena, well known for decades among European systems theorists (Kampis, 1991; Varela, 1978), of mutual intercreation and autopoiesis. Systems of magicians can interproduce. For instance, a_1 can produce a_2, while a_2 produces a_1. Or a_1 and a_2 can combine to produce a_3, while a_1 and a_3 combine to produce a_2, and a_1 and a_3 combine to produce a_2. The number of possible systems of this sort is truly incomprehensible. But the point is that, if a system of magicians is mutually interproducing in this way, then it is likely to *survive* the continual flux of magician interaction dynamics. Even though each magician will quickly perish, it will just as quickly be re-created by its co-conspirators. Autopoiesis creates self-perpetuating order amidst flux.

Some intercreative systems of magicians might be unstable; they might fall apart as soon as some external magicians start to interfere with them. Others will be robust; they will survive in spite of external perturbations. These are the ones that might be called autopoietic systems, in the sense of (Varela, 1978). Using a different vocabulary (Goertzel, 1994; 1996), these robust magician systems are called *structural conspiracies*. This leads to the natural hypothesis that thoughts, feelings and beliefs are structural conspiracies. They are stable systems of interproducing pattern/processes.

—Where **a** is in M, A **a** lies in M with probability p.
—Where N is also in S^*, and the distance $d(N,M)$ does not exceed x, there is probability q that $d(A(N), A(M))$ does not exceed $d(N,M)$.

These properties may be stated more rigorously in a number of ways. For instance, the references to probability may be made precise in a straightforward way using the Lebesgue measure. And the metric $d(N,M)$ may be defined in terms of the topology of the vector space $V(S^*)$, where the mapping V associates with each element **a** of S^* a vector **v** of integers, where v_i indicates the number of copies of magician a_i in **a**. Setting $V(M) = \{V(\mathbf{a}), \mathbf{a}$ in $M\}$, and setting $V(N)$ similarly, one may define $d(N,M) = d'(A(N), A(M))$, where d' is, say, the Hausdorff metric (Barnsley, 1988). Setting $p = q = 1$ yields a deterministic definition, but in general, it seems quite possible that natural systems may possess autopoietic attractors only under the more general stochastic definition.

Autopoiesis is not the end of the story. The next step is the intriguing possibility that, in psychological systems, there may be a global order to these autopoietic attractors, these structural conspiracies. Goertzel (1994) argued that these structures must spontaneously self-organize into larger autopoietic superstructures—and, in particular, into a special attracting structure called the *dual network*.

The dual network, as its name suggests, is a network of magicians that is simultaneously structured in two ways. The first kind of structure is *hierarchical*. Simple structures build up to form more complex structures, which build up to form yet more-complex structures, and so forth; and the more complex structures explicitly or implicitly govern the formation of their component structures. The second kind of structure is *heterarchical*: different structures connect to those other structures related to them by a sufficient number of pattern/processes. Psychologically speaking (Goertzel, 1993b, 1994), the hierarchical network may be identified with command-structured perception/control, and the heterarchical network may be identified with associatively structured memory.

Mathematically, the formal definition of the dual network is somewhat involved (Goertzel, 1994). A simplistic dual network, useful for guiding thought, though psychologically unrealistic, is a magician population living on a graph, each node of which is connected to certain heterarchical neighbor nodes and certain hierarchical child nodes. In terms of neural networks, one may think of a dual network as a special case of Alexander's (1995) notion of a fractal or recursively modular neural network: a multilevel neural network consisting of small neural networks (level 1), joined into meta-networks (level 2) whose nodes are neural networks, joined into meta-meta-networks (level 3), and so on. The dual network fits into Alexander's architecture but specifies a certain topography and functionality beyond the simple fact of self-similar structure.[5]

A *psynet*, then, is a magician system that has evolved into a dual network attractor. The core claim of the psynet model is that intelligent systems are psynets. This does not imply that all psynets are highly intelligent systems; for instance, Goertzel (1994) described a simplistic implementation of the psynet model that runs on an ordinary PC, and certainly does not deserve the label "intelligent." What makes the difference between intelligent and unintelligent psynets is above all, or so the model claims, *size*. It is interesting to ask how symbolic dynamics might arise within the psynet model. The dual network structure, in particular, would seem to have a relationship with the structure of emergent grammars. The remainder of this section explores this possible connection. The arguments are qualitative, rather than mathematical: the hope is that they will be suggestive, and stimulate further work. The goal is to explore, in a speculative way, what sorts of insights symbolic dynamics might eventually be expected to give into the dynamics of the *whole brain*. What might the language of the brain be like?

Goertzel (1995b) presented a heuristic called the *Structure-Dynamics Principle*, which states that, in many cases, the dynamic patterns revealed through symbolic dynamics overlap with the structural patterns observed in a system at a given time. Applying this to the self-similar structure of the dual network, we obtain the hypothesis that the brain may display self-similar

---

[5]This purported fractality of the brain's neural network probably has something to do with the oft-observed fractal dimensionality of EEGs, but the relationship is not yet entirely clear. As Alexander (1995) stressed, there are two sources of fractal dimension in the brain: recursively modular network structure and chaotic dynamics within particular modules.

patterns in its *dynamical behavior*—that the fractal structure of its pathways may be represented by a fractal structure in its emergent linguistic order.

What would it mean to display self-similar dynamical behavior patterns? Suppose one replaces the series of symbols with a series of phrase symbols, each one denoting a certain set of adjacent symbols within the original symbol sequence. Suppose the grammatical rules governing the phrase symbol sequence are about the same as the ones governing the original symbol sequence. And, suppose that one then groups the phrase symbols into metaphrase symbols, each one a set of adjacent phrase symbols. What if just about the same grammatical rules come out *again?* Then one has a self-similar linguistic structure: the grammatical structure of the whole is the same as the grammatical structure of the parts. Grammatical structure is scale-independent. This is nothing other than what linguists call *phrase structure.* In natural language, we encounter only a few nesting levels, so the fractality of phrase structure is not generally apparent. But the self-similar, recursively modular structure is still there, and is still significant.

The hypothesis suggested by the Structure-Dynamics Principle is, then, that the complex, autopoietic dynamics of dual networks tend to give rise to phrase structure grammars. Now, I present a tentative argument for the plausibility of this hypothesis. In a recursively modular network such as the dual network, the attractor states at level $k+1$ network are dependent on the activity of its component level k networks, and hence on the attractor states of its level k networks. *If* it happens that attractor states of level $k+1$ networks are triggered by specific *sequences* of attractor states of level k networks, then one has a system in which level $k+1$ network behavior corresponds to phrases of level k network behavior. In a system such as this, the possibility exists for the structure of level $k+1$ behavior to be the same as the structure of level k behavior, thus yielding a phrase-structure grammar. This is, I hypothesize, precisely the dynamics that enables the dual network to be an autopoietic attractor of mental process dynamics. Phrase-structure grammars are a formal instantiation of the idea of *self-similarity of mind.*

This is obviously a speculation. However, it is not entirely philosophical; it is an eminently *falsifiable* hypotheses. As we extend the tool of symbolic dynamics to more and more complex systems, we will be able to test hypotheses of this nature. We will be able to, in a very concrete sense, learn the language of chaos and complexity.

MENTAL MODELING AND TOPOLOGICAL LINGUISTICS

Let us now return to Leibniz and his concept of a universal characteristic. In terms of behavioral and cognitive modeling, the idea of symbolic dynamics as a universal characteristic has a very concrete meaning. The modeling of mental and neural structures and behaviors is a very subtle matter. There are no generally accepted criteria for evaluating models, for comparing one model to another. Where an exact match with human data is obtained, copious parameter setting is usually involved, thus raising serious questions of informativeness. The situation is even worse where, as with most models (including my own), the agreement with human behavior is only qualitative. In practice, a researcher's judgment of the quality of a given model is determined largely by *a priori* conceptual presuppositions. This situation cannot be completely remedied, a certain amount of value-ladenness being intrinsic to science, but it can be improved. Symbolic dynamics has the potential to serve as an important new tool for validating and comparing abstract psychological models. Using symbolic dynamics and related ideas, it becomes possible to formalize the notion of a qualitative model, and to compare, in a systematic way, the varying qualities of different qualitative models.

An important example of competing psychological models of uncertain status is the ongoing symbol-processing versus connectionism debate. The importance of symbol manipulation for certain aspects of human behavior is obvious, as is the fact that symbolic reasoning must somehow emerge out of subsymbolic neurodynamics. The bone of contention is whether the best path to understanding the mind proceeds via the study of symbol manipulation, via the study of neurodynamics, or via some kind of intermediate level, such as the self-organizing mental/neural process systems posited in the psynet model. Symbolic dynamics, as well as providing a means for the comparison of different theories, has a lot to say about the substance of this particular dilemma. It explains how symbol systems emerge from nonlinear dynamical systems—be they neural networks, evolutionary production systems, models of autopoietic self-organization, or whatever.

Symbolic dynamics gives one a concrete understanding of how a complex nonlinear system like the brain can give rise to a complex symbol system such as English. It makes it easy to see why current-neural network models fail to emulate the brain's linguistic capabilities. These neural network models are generally based on simple fixed-point dynamics, or more rarely, periodic dynamics. But symbolic dynamics clearly tells us that complex emergent linguistic structures only emerge from complex chaotic dynamical systems. We build nonchaotic neural network models because they are easier to control and understand; but the price for this decision is paid in the currency of emergent linguistic richness.

In the end, what symbolic dynamics suggests is that the choice of an underlying cognitive or behavioral model is not that crucial. Neural networks have the advantage of a crude resemblance to brain structure, but logic-based models simulate certain aspects of human behavior far better than the simplistic neural networks simulable on modern computers. Neither current neural-network models nor logic-based models capture the self-organizing creativity of the mind; for this, the best approach seems to be models based on process dynamics and autopoiesis (Goertzel, 1994). The important thing, in every case, is the overall emergent structure of the model's dynamics. The symbolic/subsymbolic distinction is in large part a distraction, along with the concern with manipulating parameters of models to obtain exact agreement with human data. We should be much more concerned with our failure to construct working computational models mirroring the incredible dynamical intricacy of mental/neural systems.

We cannot quite rest comfortably with this conclusion. If the linguistic level is the important one—if the key question is not whether a model correctly simulates all the details of the brain or mind, but whether it gets the emergent linguistic structures right—then we are left with the auxiliary question of what exactly it means to "get the emergent linguistic structures right." How do we measure the distance between different linguistic structures? In the next few paragraphs, I sketch some ideas in this regard. We have now reached the point, I believe, where symbolic dynamics hints at the need for new branches of mathematics: a topology of language, and perhaps even a calculus of language.

One natural approach to measuring distance between languages suggests itself: perhaps the distance between two languages is the amount of difficulty encountered in translating between one and the other. Suppose one has two languages, L_1 and L_2, each represented as an alphabet $\{0,....,N_k-1\}$, a collection W_k of n-tuples on the alphabet, and a collection of transition probabilities $\{p_{k,alpha}, alpha \text{ in } W_k\}$. A translation method *tau* from L_1 to L_2 is a mapping from W_1 to W_2, which is, in some sense, meaning-preserving. The problem then becomes to define the preservation of meaning in a rigorous way. In *Chaotic Logic* (Goertzel, 1994), the meaning of a linguistic entity x is defined as the collection of patterns (algorithmic regularities) in which that entity is involved. In other words, it consists of all patterns emergent in the set of pairs $\{(x,W)\}$, where W ranges over all states of the world that are roughly contemporaneous with the usage of W. This complete definition of meaning is mathematically unwieldy, however, and it is fortunate that in many cases an approximation will suffice. One can consider only the syntactic meaning of a word x, the collection of patterns that relate it to other words. A significant portion of this set is composed of repeated patterns (i.e., transition probabilities). Thus, an approximation of the meaning of a word i in one of our languages L_k is the collection of transition probabilities $p_{k,alpha}$, where the n-tuple *alpha* contains i.

Looking only at syntactic meaning, then, the goal of a translation *tau* from L_1 to L_2 is twofold: first, to make $p_{2,tau(x)}$ and $p_{1,x}$ as close as possible, for every x in W_1; and second, to make *tau*(x) as small as possible for each x (e.g., all else equal, it is better to map a word in L_1 into a word in L_2 rather than a pair of words in L_2). The error of a translation may be gauged by its failure at fulfilling these requirements; and the asymmetric distance between L_1 and L_2 may, as a first approximation, be understood as the error of the lowest-error translation from L_1 to L_2. The distance between the two languages is then most conveniently taken as the minimum of the two asymmetric distances (one from L_1 to L_2, the other from L_2 to L_1). It is not hard to see that this is a pseudometric (it is symmetric and obeys the triangle inequality), which becomes a metric if one considers two isomorphic languages to be identical.

Given such a formalism, the question of the qualitative similarity of a psychological model to a real system such as the brain may be explored in a rigorous way. One need merely analyze the dynamics of both systems by symbolic dynamics and compare the two languages obtained from the two different systems. Furthermore, having a distance measure on the space of formal languages, the notion of iterations on language space may be explored. The evolution of neural systems may be considered as an evolution of emergent linguistic systems. The passage from microscopic to mesoscopic to macroscopic may be understood as a progressive series of languages, becoming more and more revealing and restrictive as the focus becomes coarser and coarser.

CONCLUSION

What makes language so remarkable is that it is more than just a tool for communication; it is a tool for understanding. Symbolic dynamics enlarges on this point, instructing us that the structure of the universe may itself be understood as a linguistic structure. This is a deep idea that science is only beginning to absorb.

I believe that the formalism of symbolic dynamics is a necessary part of the modeling of complex adaptive behavior. It is natural for cognitive and behavioral modeling, in the same way that differential equations are natural for the study of astronomical systems and formal automata are natural for the study of programming languages. The sometimes opaque appearance of the mathematics should not be allowed to obscure the fundamental simplicity of the ideas. On a philosophical level, symbolic dynamics and topological language spaces tell you exactly what everyone involved in cognitive and behavioral modeling already knows: it is the simulation of qualitative behavior that is most important.

Symbolic dynamics gives a way of formalizing qualitative behavior, which is exactly what Leibniz was looking for in 1679. Leibniz had the right idea: to look at formal languages. He also saw the connection with geometry and dynamics. What he was missing was mainly the concept of structured chaos, and secondarily the concept of a formal language.

I have stressed that, in the symbolic dynamics view, the important thing is not the details of one's model, but the emergent linguistic structure to which it gives rise. Because of this, I have said, symbolic dynamics has a great potential for unifying models that live on different levels: microscopic, mesoscopic, and macroscopic. Neural network models display complex dynamics, but these dynamics can be approximately summarized as formal languages. Logic-based models generally display uninteresting dynamics, but this is not necessarily the case. Classifier systems (Holland, Holyoak, Nisbett, Thagard, 1986) are rule-based production systems with complex evolutionary dynamics, and their dynamics can be approximately summarized as formal languages. Mesoscopic models such as Gregson's (1989) nonlinear psychophysics or the psynet model also lead to dynamics that can be approximately summarized as formal languages.

Logic-oriented psychological theorists (Fodor, 1985) have argued vehemently for the necessity for a language of thought. Symbolic dynamics shows that a language of thought is easy to come by, if one assumes structured chaos in the underlying dynamical systems. In fact, the question is not one of a language of thought, but rather one of multiple languages emergent from multiple scales of complex dynamics. The challenge is to understand what kinds of dynamical systems give rise to the specific formal languages observed in human behavior.

At present, most work with symbolic dynamics and behavioral modeling has been exploratory and foundational, and the present chapter is no exception. We are still struggling to find the best ways to match the abstract concepts of formal language theory with the particularities of behavioral phenomena. However, the mathematical, philosophical, and psychological depth of the ideas is inarguable, and work is progressing at a rapid pace. I exhort anyone concerned with the modeling of complex adaptive behavior to explore the possibility of incorporating symbolic dynamics analysis in their work, whether as an aspect of a cognitive or behavioral model, a method for comparing different models, or a tool for analyzing empirical data.

REFERENCES

Combs, A. (1993). A process phenomenology of mind. In F. Abraham and R. Gilgen.
    (Eds.), *Chaos Theory in Psychology* (pp. xx–xx). Westport, CT: Greenwood Press.
Combs, A. (1995). *Radiance of being.* Edinburgh: Floris Press.
Fodor, J. (1985). *The modular mind.* Cambridge, MA: MIT Press.
Freeman, W. (1991). The physics of perception. *Scientific American, 34–41.*
Goerner, S. (1993). *The evolving ecological universe.* New York: Gordon and Breach.
Goertzel, B. (1993a). *The structure of intelligence.* New York: Springer-Verlag.
Goertzel, B. (1993b). *The evolving mind.* New York: Gordon and Breach.
Goertzel, B. (1994). *Chaotic logic.* New York: Plenum Press.

Goertzel, B. (1996). *From complexity to creativity.* New York: Plenum Press.

Goertzel, B., Bowman, H., & Baker, R. (1993). Dynamics of the Radix Expansion Map, *Journal of Mathematics and Mathematical Sciences, 17,* 143–148.

Goertzel, B., & Goertzel, G. (1995). The Markovian language network. In *Proceedings of ANZIIS-95, the Australia-New Zealand Conference on Intelligent Information Systems.* Published by IEEE on CD-ROM.

Goertzel, B., & Goertzel, G. (1996). The chaos language algorithm: A technique for inferring the grammatical structure overlying chaotic dynamics, applied to the generalized baker map.

Gregson, R. A. M. (1989). *Nonlinear psychophysics.* New York: Springer-Verlag.

Holland, J. H., Holyoak, K. J., Nisbett, R. E., & Thagard, P. R. (1986). *Induction: Processes of inference, learning, and discovery.* Boston, MA: MIT Press.

Leibniz, G. (1969). *On the universal characteristic.* In L. E. Loemker (Ed. & Trans.), *Gottfriend Wilhelm Leibniz: Philosophical papers and letters* (pp. xx–xx). Chicago: University of Chicago Press. (Original work published 1679)

Leibniz, G., & Abraham, F. D. (1995). The Leibniz-Abraham correspondence. In F. Abraham & R. Gilyen (Eds.), *Chaos theory in psychology* (pp. 3–7). Westport, CT: Greenwood Press.

Theiler, J., & Rapp, P. E. (1996). Re-examination of the evidence for low-dimensional, nonlinear structure in the human electroencephalogram. *Electroencephalography & Clinical Neurophysiology, 98,* 213–222.

2

Neural Substrates of Adaptively Timed Reinforcement, Recognition, and Motor Learning

Stephen Grossberg
Boston University

The concepts of *declarative memory* and *procedural memory* have been used to distinguish two distinct types of learning. A neural network architecture is described here that suggests how such memory processes work together as recognition learning, reinforcement learning, and sensory-motor learning all take place together during adaptive behaviors. To coordinate these processes, the hippocampal formation and cerebellum each contain circuits that learn to time their outputs adaptively. Within the architecture, hippocampal timing helps to maintain attention on motivationally salient goal objects during variable task-related delays, and cerebellar timing controls the release of conditioned responses. This property is part of the model's description of how conditionable cognitive–emotional interactions focus attention on motivationally valued cues, and how this process breaks down due to hippocampal ablation. The architecture also suggests how the hippocampal mechanisms that help rapidly to draw attention to salient cues could prematurely release motor commands if the release of these commands was not adaptively timed by the cerebellum.

The model hippocampal system modulates cortical recognition learning without actually encoding the representational information that the cortex encodes. These properties avoid the difficulties faced by several models that propose a direct hippocampal role in recognition learning. Learning within the

29

model hippocampal system controls adaptive timing and spatial orientation. Model properties hereby clarify how hippocampal ablations cause amnesic symptoms and difficulties with tasks that combine task delays, novelty detection, and attention toward goal objects amid distractions. When these model recognition, reinforcement, sensory-motor, and timing processes work together, they suggest how the brain can accomplish conditioning of multiple sensory events to delayed rewards, as during serial compound conditioning.

The chapter is divided into two parts: An intuitive introduction to the architecture's main design principles and mechanisms and a summary of mathematical equations and simulations that illustrate how the architecture achieves adaptively timed conditioning and attention.

HOW DO PROCESSES OF RECOGNITION, REWARD, AND ACTION INTERACT?

A central problem in the behavioral and cognitive neurosciences concerns how humans and other animals learn to recognize objects, to predict and attend to their rewarding or punishing consequences, and to perform appropriately timed actions capable of realizing or avoiding these consequences. Multiple brain regions participate in these processes, including the inferotemporal cortex, amygdala, hippocampal formation, and cerebellum. The complexity of these processes has led to the development of neural models that might shed light on their cellular and network properties. A neural architecture is described herein to suggest why both the hippocampus and the cerebellum contain circuits that are specialized for adaptive timing. Although the two timing circuits may share cellular and circuit properties, the architecture predicts that they carry out distinct functional roles during the learning and memory processes that subserve recognition and movement tasks.

These distinct roles are used to clarify several of the conceptual dichotomies that have been useful in research about normal and amnesic learning and memory. One such dichotomy concerns the distinctions between declarative memory and procedural memory, knowing that and knowing how, memory and habit, or memory with record and memory without record (Bruner, 1969; Mishkin, 1982, 1993; Ryle, 1949; Squire & Cohen, 1984). The amnesic patient HM exemplified this distinction by learning and remembering motor skills such as the assembly of the Tower of Hanoi without being able to recall having done so (Bruner, 1969; Cohen & Squire, 1980; Mishkin, 1982; Ryle, 1949; Scoville & Milner, 1957; Squire & Cohen, 1984). HM's surgical lesion included extensive parts of the hippocampal formation and amygdala. Subsequent animal studies have shown that damage to the hippocampal formation (Ammon's horn, dentate gyrus, subiculum, fornix) and the parahippocampal region (entorhinal, perirhinal, and parahippocampal cortices) can reproduce analogous amnesic symptoms (Mishkin, 1978; Squire & Zola-Morgan, 1991). These results implicate this aggregate hippocampal system in the processes that regulate declarative memory, or "knowing that." Such processes support a competence for learning recognition categories and being able to access them flexibly in a task-specific way (Eichenbaum, Otto, & Cohen, 1994).

A parallel line of research has implicated the cerebellum in the processing of procedural memory, or knowing how. The cerebellum is an essential circuit for conditioning discrete adaptive responses during eye movements, arm movements, nictitating membrane movements, and jaw movements (Ebner & Bloedel, 1981; Gilbert & Thach, 1977; Ito, 1984; Lisberger, 1988; Optican & Robinson, 1980; Thompson, 1988; Thompson et al., 1984, 1987). Models of cerebellar learning have been developed over the years to help explain these motor conditioning data (Albus, 1971; Bullock, Fiala, & Grossberg, 1994; Fujita, 1982a, 1982b; Grossberg, 1969a; Grossberg & Kuperstein, 1986; Ito, 1984; Lisberger, 1988; Marr, 1969).

A third line of research on learning and memory concerns cognitive–emotional interactions, including how a conditioned stimulus (CS) such as a tone or light, when paired with an unconditioned stimulus (US) such as a shock, can learn to generate conditioned responses (CR), such as fear or limb withdrawal, that were originally elicited only by the US. Such learning is optimal at a range of positive interstimulus intervals (ISI) that are characteristic of the animal and the task, and is greatly attenuated at zero ISI and long ISIs (Smith, 1968). Although the amygdala has been identified as a primary site in the expression of emotion and stimulus–reward association (Aggleton, 1993), the hippocampal formation has also been implicated in the processing of cognitive–emotional interactions. In particular, Thompson et al. (1987) distinguished two types of learning that go on during conditioning of the rabbit nictitating membrane response (NMR): "conditioned fear" learning linked to the hippocampus and "learning of the discrete adaptive response" within the cerebellum (p. 82). In addition, removal of the hippocampal formation greatly attenuates attentional blocking (Rickert, Bennett, Lane, & French, 1978; Schmajuk, Spear, & Isaacson, 1983; Solomon, 1977). Blocking is the process whereby conditioning of a cue CS_1 to a US prevents a second cue CS_2 from being conditioned to US when it is later presented before US as part of a simultaneous $CS_1 + CS_2$ stimulus compound. Much experimental and theoretical work has suggested that CS_2 loses its ability to be conditioned to US because it is an irrelevant cue that predicts no more about the US than does CS_1 when presented alone (Grossberg, 1975, 1982; Kamin, 1969). Blocking enables a learning subject to attend selectively to relevant cues.

The present chapter synthesizes, into a single neural architecture, models that have been developed to explain data from each of these three areas. This synthesis clarifies how the various models work together to control behavior. In particular, it suggests why both the cerebellum and the hippocampal system may need adaptive timing circuits for their normal functioning. We suggest that the hippocampal mechanisms that help to draw attention rapidly to salient cues could prematurely release motor commands were these commands not adaptively timed by the cerebellum. To reach such conclusions as efficiently as possible, this part of the chapter provides just enough information about the component models to understand how they can work together to explain key data. Mathematical equations of some key model processes are described later in the chapter. Others are developed in detail in other articles that are cited later.

The chapter is devoted to an exposition of just one theory as no other theory of which we are aware has explained such a large database or articulated the design principles that support this explanatory range. Some comparisons with other models are found in Grossberg and Merrill (1996).

Multiple Roles for the Hippocampal System?

Why should a single, albeit complex, brain region like the hippocampal system be involved in so many processes: recognition learning, reinforcement learning, and motivated attention? A clue is provided by neural data and models about how each of these processes works. In particular, both recognition learning and reinforcement learning are regulated by a matching process whereby bottom-up stimuli from the outside world are matched against top-down learned expectations to determine whether attentive learning or memory search will occur. The unblocking paradigm illustrates this matching process for the case of reinforcement learning (Kamin, 1969). The unblocking paradigm is a variant of the blocking paradigm in which the US changes intensity in the two learning episodes. Thus if CS_1 is followed by one US intensity (US_1), and the compound stimulus $CS_1 + CS_2$ is followed by a different US intensity (US_2), then CS_2 can become conditioned to the US, unlike in the blocking paradigm, and with an emotional valence that depends on the sign of the difference $US_1 - US_2$ between US_1 and US_2 (Kamin, 1969). The mismatch between the actual intensity US_2 and the expected intensity US_1 triggers a memory search that attentionally unblocks the representation of CS_2 that is stored in short term memory, and enables it to learn to predict the change in US intensity (Grossberg, 1975). This memory search helps to focus attention on that subset of sensory cues that predicts motivationally salient outcomes in a given context, and to block those that do not.

Recognition learning is accomplished by interactions between inferotemporal cortex (IT) and hippocampal formation, among other brain areas (Desimone, 1991; Desimone & Ungerleider, 1989; Eichenbaum, Otto, & Cohen, 1994; Gochin, Miller, Gross, & Gerstein, 1991; Harries & Perrett, 1991;

Mishkin, 1978, 1982; Mishkin & Appenzeller, 1987; Perrett, Mistlin, & Chitty, 1987; Schwartz, Desimone, Albright, & Gross, 1983; Squire & Zola-Morgan, 1991). These interactions include the matching process that modulates the course of recognition learning in IT cortex and the course of reinforcement learning in thalamo-cortico-amygdala circuits. I next analyze some models of how these recognition and reinforcement learning circuits interact with motor learning circuits. It is shown that the behavioral success of this interaction requires both types of circuits to be adaptively timed.

Self-Organizing Feature Maps and Adaptive Resonance

The first type of model results from an analysis of how humans and animals rapidly learn to categorize and name events and their contexts in real time. These Adaptive Resonance Theory (ART) models have been used to help explain and predict a large body of cognitive and neural data about recognition learning, recall, attention, priming, and memory search (Carpenter & Grossberg, 1991, 1993; Grossberg, 1982b, 1987, 1988a). ART systems realize this synthesis by incorporating mechanisms that solve a fundamental problem about learning and memory that is called the *stability–plasticity dilemma.* An adequate self-organizing recognition system must be capable of plasticity in order to learn rapidly about significant new events, yet its memory must also remain stable in response to irrelevant or often-repeated events. Thus we can learn to recognize many new faces without risking the unselective forgetting of our parents' faces. In ART, interactions between an attentional subsystem and an orienting subsystem, or novelty detector, self-stabilize the learning process as the network becomes familiar with an environment by categorizing the information within it in a way that leads to behavioral success (Grossberg, 1980).

Learning takes place in the attentional subsystem. Its processes include activation of short-term memory (STM) traces, incorporation through learning of STM information into longer-lasting, long-term memory (LTM) traces, and interactions between pathways that carry specific information with nonspecific pathways that modulate the specific pathways. These interactions between specific STM and LTM processes and nonspecific modulatory processes regulate the stability–plasticity balance during normal learning, as follows.

The attentional subsystem undergoes both bottom-up learning and top-down learning between processing levels such as those denoted by F_1 and F_2 in Fig. 2.1. Level F_1 contains a network of nodes, or cell populations, each of which is activated by a particular combination of sensory features. Level F_2 contains a network of nodes that represent recognition codes, or categories, which are selectively activated by the activation patterns across F_1. Each F_1 node sends output signals to a subset of F_2 nodes. Each F_2 node thus receives

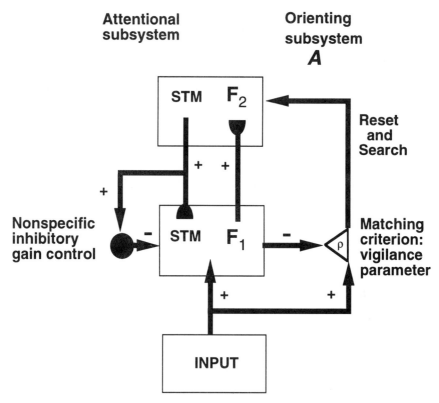

FIG. 2.1. An example of a model ART circuit in which attentional and orienting circuits interact. Level F_1 encodes a distributed representation of an event by a short-term memory (STM) activation pattern across a network of feature detectors. Level F_2 encodes the event using a compressed STM representation of the F_1 pattern. Learning of these recognition codes occurs at the long term memory (LTM) traces within the bottom-up and top-down pathways between levels F_1 and F_2. The top-down pathways read out learned expectations whose prototypes are matched against bottom-up input patterns at F_1. The size of mismatches in response to novel events are evaluated relative to the vigilance parameter ρ of the orienting subsystem A. A large enough mismatch resets the recognition code that is active in STM at F_2 and initiates a memory search for a more appropriate recognition code. Output from subsystem A can also trigger an orienting response.

inputs from many F_1 nodes. The thick pathway from F_1 to F_2 in Fig. 2.1 represents the array of diverging and converging pathways, for simplicity.

Learning takes place at the synapses denoted by semicircular endings in the $F_1 \rightarrow F_2$ pathways. Pathways that end in arrowheads do not undergo learning. This bottom-up learning enables F_2 nodes to become selectively tuned to particular combinations of activation patterns across F_1 by changing their LTM traces.

Why is bottom-up learning insufficient in a system that can autonomously solve the stability–plasticity dilemma? This analysis was carried out in that part of the ART model that combines bottom-up associative learning and lateral inhibition for purposes of learned categorization. This type of model is often called a self-organizing feature map, competitive learning, or learned vector quantization. In such a model, as shown in Fig. 2.2A, an input pattern registers itself as a pattern of activity, or STM, across the feature detectors of level F_1. Each F_1 output signal is multiplied or gated, by the adaptive weight, or LTM trace, in its respective pathway. All these LTM-gated inputs are added up at their target F_2 nodes. Competitive interactions, mediated by lateral inhibition within F_2, contrast-enhance this input pattern. Even if many F_2 nodes may receive inputs from F_1, lateral inhibition acts to cause a much smaller set of F_2 nodes to store their activation in STM.

It is useful to think of all the STM signals that converge on an F_2 node as an STM pattern, or vector. Likewise, all the LTM traces that multiply these signals on their way to a prescribed F_2 node form an LTM vector. The operation of adding up the LTM-gated signals at each F_2 node is called the inner product, or *dot product*, of the two vectors. It measures how similar the two vectors are, and increases as a function of their similarity. The LTM traces thereby *filter* the STM signal pattern and generate larger inputs to those F_2 nodes whose LTM patterns are most similar to the STM pattern.

As noted above, the lateral inhibition among F_2 nodes selects just a few of the more active F_2 nodes for STM storage. This contrast-enhancing operation enables many input patterns at F_1 that share similar input features to be classified by a small set of F_2 nodes. The F_2 nodes hereby become category nodes that are capable of classifying the inputs to F_1.

In a self-organizing feature map, only the F_2 nodes that win the contrast-enhancing competition and store their activity in STM can influence the learning process. STM activity at the winning F_2 nodes selectively opens a learning gate at the LTM traces that abut these nodes. These LTM traces can then approach, or track, the input signals in their pathways, a process called steepest descent. This learning law is thus often called gated steepest descent, or instar learning. In its simplest form, this learning law can be expressed by the equation

$$\tfrac{d}{dt} w_{ij} = f(x_j)(-w_{ij} + S_i), \qquad (1)$$

where $d/dt\, w_{ij}$ is the time rate of change of the LTM trace, or adaptive weight, w_{ij} from the ith F_1 node to the jth F_2 node, $f(x_j)$ is the learning gating signal that becomes positive only if the postsynaptic activity, or potential, x_j of the jth F_2 node becomes sufficiently large, and S_i is the ith bottom-up signal. This learning rule was introduced into neural network models in Grossberg (1969a) and is the learning rule that was used to introduce ART (Grossberg, 1976b). As it is

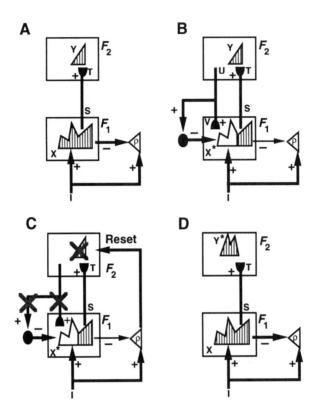

FIG. 2.2. ART search for a recognition code: (A) The input pattern **I** is instated across the feature detectors at level F_1 as a short-term memory (STM) activity pattern **X**. Input **I** also nonspecifically activates the orienting subsystem A; see Fig. 2.1. STM pattern **X** is represented by the hatched pattern across F_1. Pattern **X** both inhibits A and generates the output pattern **S**. Pattern **S** is multiplied by long-term memory (LTM) traces and added at F_2 nodes to form the input pattern **T**, which activates the STM pattern **Y** across the recognition categories coded at level F_2. (B) Pattern **Y** generates the top-down output pattern **U**, which is multiplied by top-down LTM traces and added at F_1 nodes to form the prototype pattern **V** that encodes the learned expectation of the active F_2 nodes. If **V** mismatches **I** at F_1, then a new STM activity pattern X^* is generated at F_1. X^* is represented by the hatched pattern. It includes the features of **I** that are confirmed by **V**. Inactivated nodes corresponding to unconfirmed features of **X** are unhatched. The reduction in total STM activity which occurs when **X** is transformed into X^* causes a decrease in the total inhibition from F_1 to A. (C) If inhibition decreases sufficiently, A releases a nonspecific arousal wave to F_2, which resets the STM pattern **Y** at F_2. (D) After **Y** is inhibited, its top-down prototype signal is eliminated, and **X** can be reinstated at F_1. Enduring traces of the prior reset lead **X** to activate a different STM pattern Y^* at F_2. If the top-down prototype due to Y^* also mismatches **I** at F_1, then the search for an appropriate F_2 code continues until a more appropriate F_2 representation is selected. Then an attentive resonance develops and learning of the attended data is initiated. [Reprinted with permission from Grossberg & Merrill (1996).]

36

tracking the signals in its pathway, such an LTM trace w_{ij} can either increase (if the signal S_i is large) or decrease (if the signal S_i is small). It thus combines Hebbian and antiHebbian learning properties in a way that has been used to model neurophysiological data about hippocampal LTP and LTD (Artola & Singer, 1993; Levy, 1985; Levy & Desmond, 1985) and adaptive tuning of cortical feature detectors during the visual critical period (Rauschecker & Singer, 1979; Singer, 1983).

In particular, as Table 2.1 shows, significant postsynaptic activity, mediated by the gating signal $f(x_j)$, is needed to cause any change in w_{ij}. If this modulatory gate opens, then w_{ij} may increase or decrease, depending on the relative size of S_i. Because S_i, in turn, may influence the amount of postsynaptic activity x_j via the presynaptic signal $S_i w_{ij}$, various secondary effects can occur that are beyond the scope of this discussion (see Carpenter & Grossberg, 1990). It is perhaps worth noting, however, that an early prediction (Grossberg, 1968b, 1969b, 1974) suggested that synaptic learning would be mediated by a postsynaptic process of protein synthesis and receptor sensitization that controls a coordinated presynaptic process of transmitter production. The postsynaptic signal process was predicted to be triggered by an inward Ca^{++} current that is antagonistic to Mg^{++}. Coordinated presynaptic and postsynaptic changes were predicted to depend on the inward Ca^{++} current in synergy with an inward Na^+ current and an outward K^+ current. Similar concepts have been used, in greatly elaborated form, to explain recent data about LTP and LTD; for example, see Artola and Singer (1993) and Kuno (1995). Gated steepest descent learning may thus be viewed as a first approximation to a much more complex cascade of biochemical events.

The net effect of such learning is to train the LTM vectors of the winning F_2 category nodes to become more similar to the STM patterns that they filter.

TABLE 2.1
Instar Learning Rule

	S_i	w_{ij}	x_j		
	Case 1	Case 2		Case 3	Case 4
State of S_i	+	−		+	−
State of x_j	+	+		−	−
State of w_{ij}	⇧	⇩		⇔	⇔

+ = active
− = inactive
⇧ = increase
⇩ = decrease
⇔ = no change

As a result, the winning F_2 categories sharpen their tuning curves to respond more selectively to the STM patterns that they have experienced.

Self-organizing feature map models were introduced and computationally characterized in Malsburg (1973) and Grossberg (1976a, 1978). These models were subsequently applied and further developed by many authors, notably Kohonen (1984). They exhibit many useful properties, especially if not too many input patterns, or clusters of input patterns, perturb level F_1 relative to the number of categorizing nodes in level F_2. Grossberg (1976a) proved under these sparse environmental conditions that category learning is stable, with LTM traces that track the statistics of the environment, are self-normalizing, and oscillate a minimum number of times. Also, the F_2 category selection rule, like a Bayesian classifier, tends to minimize error.

It was also proved, however, that under more general environmental conditions, learning becomes unstable and subject to catastrophic forgetting. Such a model could forget the faces of your parents while learning a new face. This memory instability is due to basic properties of associative learning and lateral inhibition. Although a gradual switching off of plasticity can partially overcome the problem, such a mechanism cannot work in a learning system whose plasticity is maintained throughout adulthood. These results put into sharp focus the problem of how the brain dynamically self-stabilizes its memory while remaining open to new experiences throughout life, a topic that has attracted increasing interest (Kandel & O'Dell, 1992). An analysis of this instability, together with data about categorization, conditioning, and atten-tion, led to the introduction of ART models that self-stabilize the memory of self-organizing feature maps in response to an arbitrary stream of input patterns (Grossberg, 1976b).

The Link Between Top-Down Matching, Hypothesis Testing, and Stable Learning

In an ART model, learning does not occur when some winning F_2 activities are stored in STM. Instead, activation of F_2 nodes may be interpreted as making a hypothesis about an input at F_1. When F_2 is activated, it quickly generates an output pattern that is transmitted along the top-down adaptive pathways from F_2 to F_1. These top-down signals are multiplied in their respective pathways by LTM traces at the semicircular synaptic knobs of Fig. 2.2B. The LTM-gated signals from all the active F_2 nodes are added to generate the total top-down feedback pattern from F_2 to F_1. This pattern plays the role of a learned expectation. Activation of this expectation may be interpreted as "testing the hypothesis," or "reading out the prototype," of the active F_2 category. As shown in Fig. 2.2B, ART networks are designed to match the "expected prototype" of the category against the bottom-up input pattern, or exemplar, to F_1. Nodes that are activated by this exemplar are suppressed if they do not correspond to large LTM traces in the top-down prototype pattern. The resultant F_1 pattern encodes the cluster of input features that the network deems relevant to the hypothesis based upon its past experience. This resultant activity pattern, called X^* in Fig. 2.2B, encodes the pattern of features to which the network "pays attention."

If the expectation is close enough to the input exemplar, then a state of *resonance* develops as the attentional focus takes hold. The pattern X^* of attended features reactivates the F_2 category Y which, in turn, reactivates X^*. The network locks into a resonant state through a positive feedback loop that dynamically links, or binds, X^* with Y. Damasio (1989) used the term *convergence zones* to describe such a resonant process. Such resonances are capable of binding spatially distributed features into synchronous and coherent states, both in cortico-cortical and thalamocortical feedback networks (Grossberg, 1976b; Grossberg & Somers, 1991).

Neurophysiological data that are consistent with the prediction that ART-like resonances exist between LGN and V1 have recently been reported (Sillito, Jones, Gerstein, & West, 1994). In particular, it was suggested in Grossberg (1980) that top-down corticogeniculate feedback would selectively amplify monocular LGN activations that are consistent with the oriented binocular cortical cells that activate the feedback, while inhibiting LGN cells that are not. In addition, top-down feedback by itself, as in all ART systems, was suggested not to be fully able to activate LGN cells. In support of this prediction, Sillito et al. (1994) reported that

> cortically induced correlation of relay cell activity produces coherent firing in those groups of relay cells with receptive field alignments appropriate to signal the particular orientation of the moving contour to the cortex. This increases the gain of the input for feature-linked events detected by the cortex. ... The cortico-thalamic input is only strong enough to exert an effect on those LGN cells that are additionally polarized by their retinal input ... the feedback circuit searches for correlations that support the "hypothesis" represented by a particular pattern of cortical activity. (pp. 479–482)

Gove, Grossberg, and Mingolla (1995) showed how this type of corticogeniculate feedback and resonance can be used as part of a larger model of cortical visual processing to simulate data about brightness perception and illusory contours.

Similar ART matching and resonance rules have been used to explain and predict behavioral and brain data from other task domains. For example, Carpenter and Grossberg (1993) used ART matching and resonance rules to explain data about visual object recognition and medial temporal amnesia. Govindarajan, Grossberg, Wyse, and Cohen (1994) used ART matching and resonance rules to simulate auditory psychophysical data about acoustic source segregation when multiple sources harmonically overlap, as during a cocktail party. Grossberg, Boardman, and Cohen (1994) used ART matching and resonance rules to simulate psychophysical data about variable-rate speech categorization. Grossberg and Stone (1986a) used such rules to explain data about lexical priming and decision making. Grossberg, Roberts, Aguilar, and Bullock (1997

) used ART matching and resonance rules to explain neural data about multimodal control of saccadic eye movements. Why should similar matching and resonance rules be used in so many brain systems?

ART shows how these matching and resonance rules can be used to help solve the noise-saturation dilemma in any brain system that dynamically adjusts and maintains its parameters to cope with changing environmental conditions throughout life. The matched resonant state, rather than bottom-up activation, is predicted to drive the learning process. The resonant state persists long enough, at a high enough activity level, to activate the slower learning process; hence the term *adaptive resonance* theory. ART systems learn prototypes, rather than exemplars, because the attended feature vector X^*, rather than the input exemplar itself, is learned. Both the bottom-up LTM traces that tune the category nodes and the top-down LTM traces that filter the learned expectation learn to correlate activation of F_2 nodes with the set of all *attended* X^* vectors that they have ever experienced. These attended STM vectors assign less STM activity to features in the input vector I that mismatch the learned top-down prototype V than to features that match V. Bottom-up activations that are not supported by large top-down LTM traces are hereby suppressed, and hence cannot destabilize the learning process.

Prototype Learning and Exemplar Learning
Using Vigilance Control

A similar type of matching by similarity across arrays of features has been used to quantitatively fit categorization data from human subjects (Estes, 1994). Models of this type need to assume that every input exemplar that a subject has ever processed is stored. Such models face formidable problems of memory storage and retrieval, and have not yet been shown capable of real-time autonomous categorization of complex databases. ART models computationally elaborate the idea that humans learn prototypes (Posner & Keele, 1968, 1970), which save greatly on memory resources by allowing many exemplars to be represented by a single category prototype. ART models have also been used for real-time autonomous categorization of complex databases (e.g., Asfour, Carpenter, & Grossberg, 1995; Asfour, Carpenter, & Lesher, 1993; Bachelder, Waxman, & Seibert, 1993; Baloch & Waxman, 1991; Bradski & Grossberg, 1994; Carpenter, Grosberg, Markuzon, Reynolds, & Rosen, 1992; Carpenter, Grossberg, & Reynolds, 1991, 1995; Carpenter & Ross,

1993; Carpenter & Tan, 1993; Caudell, Smith, Escobedo, & Anderson, 1994; Dubrawski & Crowley, 1994; Gjerdingen, 1990; Goodman et al., 1992; Ham & Han, 1993; Harvey, 1993; Kasperkiewicz, Racz, & Dubrawski, 1995; Keyvan, Durg, & Rabelo, 1993; Metha, Vij, & Rabelo, 1993; Moya, Koch, & Hostetler, 1993; Seibert & Waxman, 1992; Suzuki, Abe, & Ono, 1994; Suzuki, 1995; Wienke, Xie, & Hopke, 1994).

Given that ART systems learn prototypes, how can they also learn to recognize unique experiences, such as a particular view of a friend's face? The prototypes that are learned by ART systems accomplish this by realizing a qualitatively different concept of prototype than that offered by previous models. In particular, ART prototypes form in a way that is designed to conjointly maximize category generalization while minimizing predictive error (Carpenter, Grossberg, & Reynolds, 1991; Carpenter et al., 1992). As a result, ART prototypes can automatically learn individual exemplars when environmental conditions require highly selective discriminations to be made. How the matching process achieves this is discussed below.

First, let us consider what happens if the mismatch between bottom-up and top-down information is too great for a resonance to develop. Then the F_2 category is quickly reset and a memory search for a better category is initiated. This combination of top-down matching, attention focusing, and memory search is what stabilizes ART learning and memory in an arbitrary input environment. The attentional focusing by top-down matching prevents inputs that represent irrelevant features at F_1 from eroding the memory of previously learned LTM prototypes. In addition, the memory search resets F_2 categories so quickly when their prototype V mismatches the input vector I that the more slowly varying LTM traces do not have an opportunity to correlate the attended F_1 activity vector X^* with them. Conversely, the resonant event, when it does occur, maintains and amplifies the matched STM activities for long enough and at high enough amplitudes for learning to occur in the LTM traces.

Whether or not a resonance occurs depends on the level of mismatch, or novelty, that the network is prepared to tolerate. Novelty is measured by how well a given exemplar matches the prototype that its presentation evokes. The criterion of an acceptable match is defined by an internally controlled parameter called vigilance (Carpenter & Grossberg, 1987a, 1987b, 1991). The vigilance parameter is computed in the orienting subsystem A; see Fig. 2.1. Vigilance weighs how similar an input exemplar I must be to a top-down prototype V for resonance to occur. Resonance occurs if $\rho |I| - |X^*| \leq 0$. This inequality says that the F_1 attentional focus X^* inhibits A more than the input I excites it. If A remains quiet, then an $F_1 \leftrightarrow F_2$ resonance can develop and category learning can ensue.

Either a larger value of ρ or a smaller match ratio $|X^*||I|^{-1}$ makes it harder to satisfy the resonance inequality. When ρ grows so large or $|X^*||I|^{-1}$ is so small that $\rho|I| - |X^*| > 0$, then A generates an arousal burst, or novelty wave, that resets the STM pattern across F_2 before new learning can occur, and initiates a bout of hypothesis testing, or memory search. During search, the orienting subsystem interacts with the attentional subsystem (Figs. 2.2C and 2.2D) to rapidly reset mismatched categories and to select better F_2 representations with which to categorize novel events at F_1, without risking unselective forgetting of previous knowledge. Search may select a familiar category if its prototype is similar enough to the input to satisfy the resonance criterion. The prototype may then be refined by attentional focusing before learning occurs. If the input is too different from any previously learned prototype, then an uncommitted population of F_2 cells is selected and learning of a new category is initiated.

Because vigilance can vary across learning trials, recognition categories capable of encoding widely differing degrees of generalization or abstraction can be learned by a single ART system. Low vigilance leads to broad generalization and abstract prototypes. High vigilance leads to narrow generalization and to prototypes that represent fewer input exemplars, even a single exemplar. Thus a single ART system may be used, say, to learn abstract prototypes with which to recognize abstract categories of faces and dogs, as well as exemplar prototypes with which to recognize individual faces and dogs. A single system can learn both, as the need arises, by increasing vigilance just enough to activate A if a previous categorization leads to a predictive error (Carpenter et al., 1992; Carpenter, Grossberg, & Reynolds, 1991).

Corticohippocampal Interactions and Medial Temporal Amnesia

As sequences of inputs are practiced over learning trials, the search process eventually converges on stable categories. It has been mathematically proved (Carpenter & Grossberg, 1987a, 1991) that familiar inputs directly access the category whose prototype provides the globally best match, without requiring a search. This property helps to explain how we can recognize familiar objects so quickly, even though we may know about many other things. Unfamiliar inputs continue to engage the orienting subsystem to trigger memory searches for better categories until they become familiar. New categories can continue to form until the memory capacity is fully utilized. Memory capacity is determined by the number of category nodes in F_2, which can be chosen to be as large as one desires without degrading system performance.

The process whereby search is automatically disengaged is a form of memory consolidation that emerges from network interactions. This type of *emergent consolidation* does not preclude structural consolidation at individual cells, because the amplified and prolonged activities that subserve a resonance may be a trigger for learning-dependent cellular processes, such as protein synthesis and transmitter production.

The attentional subsystem of ART has been used to model aspects of inferotemporal (IT) cortex, and the orienting subsystem models part of the hippocampal system. The interpretation of ART dynamics in terms of IT cortex led Miller, Li, and Desimone (1991) to successfully test the prediction that cells in monkey IT cortex are reset after each trial in a working memory task. To illustrate the implications of an ART interpretation of IT–hippocampal interactions, Carpenter and Grossberg (1993) described how a lesion of the ART model's orienting subsystem creates a formal memory disorder with symptoms much like the medial temporal amnesia that is caused in animals and patient HM after hippocampal system lesions. In particular, such a lesion *in vivo* causes unlimited anterograde amnesia; limited retrograde amnesia; failure of consolidation; tendency to learn the first event in a series; abnormal reactions to novelty, including perseverative reactions; normal priming; and normal information processing of familiar events (Cohen, 1984; Graf, Squire, & Mandler, 1984; Lynch, McGaugh, & Weinberger, 1984; Squire & Butters, 1984; Squire & Cohen, 1984; Warrington & Weiskrantz, 1974; Zola-Morgan & Squire, 1990).

Unlimited anterograde amnesia occurs because the network cannot carry out the memory search to learn a new recognition code. Limited retrograde amnesia occurs because familiar events can directly access correct recognition codes. Before events become familiar, memory consolidation occurs that utilizes the orienting subsystem (Fig. 2.1C). This failure of consolidation does not necessarily prevent learning per se. Instead, learning influences the first recognition category activated by bottom-up processing, much as "amnesics are particularly strongly wedded to the first response they learn" (Gray, 1982, p. 253). Perseverative reactions can occur because the orienting subsystem cannot reset sensory representations or top-down expectations that may be persistently mismatched by bottom-up cues. The inability to search memory prevents ART from discovering more appropriate stimulus combinations to attend. Normal priming occurs because it is mediated by the attentional subsystem.

Similar behavioral problems have been identified in hippocampectomized monkeys. Gaffan (1985) noted that fornix transection "impairs ability to change an established habit in a different set of circumstances that is similar to the first and therefore liable to be confused with it" (p. 94). In ART, a defective orienting subsystem prevents the memory search whereby different representations could be learned for similar events. Pribram (1986) called such a process a "competence for recombinant context-sensitive processing" (p.

362). These ART mechanisms illustrate how memory consolidation and novelty detection may be mediated by the same neural structures (Zola-Morgan & Squire, 1990), why hippocampectomized rats have difficulty orienting to novel cues (O'Keefe & Nadel, 1978), and why there is a progressive reduction in novelty-related hippocampal potentials as learning proceeds in normal rats (Deadwyler, West, & Lynch, 1979; Deadwyler, West, & Robinson, 1981). In ART, the orienting system is automatically disengaged as events become familiar during the memory consolidation process.

ART properties hereby provide an alternative to the popular hypothesis that the hippocampal formation somehow temporarily stores recognition codes from all sensory modalities before the temporal cortex can more permanently do so (Eichenbaum, Otto, & Cohen, 1994; Marr, 1971; McClelland, McNaughton, & O'Reilly, 1994; Milner, 1989). This hypothesis faces formidable obstacles as soon as one seriously tries to model how such a process could work. For example, how could the hippocampal system rapidly store all the information that one can recall after seeing an exciting movie? McClelland, McNaughton, and O'Reilly (1994) admitted that their model cannot do this. In fact, not only is fast learning impossible, but "the sequential acquisition of new data can lead to *catastrophic interferences* with what has previously been learned." Only if learning is slow and carefully interleaved on sufficiently small and regular databases can it occur at all in this type of model. Such a model fails to solve the stability–plasticity dilemma. Grossberg and Merrill (1996) provided a comparative analysis of the ART corticohippocampal model of medial temporal amnesia with alternative amnesia models, both in terms of their explanatory power and their plausibility as neural mechanisms.

A Prediction About How Corticohippocampal Interactions Control the Specificity of Learned Prototypes

The ART conception of temporal–hippocampal interactions suggests the following prediction. Level F_2 properties may be compared with properties of cell activations in inferotemporal cortex (IT) during recognition learning in monkeys. The ability of F_2 nodes to learn categories with different levels of generalization clarifies how some IT cells can exhibit high specificity, such as selectivity to views of particular faces, whereas other cells respond to broader features of the animal's environment (Desimone & Ungerleider, 1989; Gochin et al., 1991; Harries & Perrett, 1991; Mishkin, 1982; Mishkin & Appenzeller, 1987; Perrett, Mistlin, & Chitty, 1987; Schwartz, Desimone, Albright, & Gross, 1983; Seibert & Waxman, 1991). Moreover, when monkeys are exposed to easy and difficult discriminations (Spitzer, Desimone, & Moran, 1988), "in the difficult condition the animals adopted a stricter internal criterion for discriminating matching from nonmatching stimuli ... the animals' internal representations of the stimuli were better separated, independent of the criterion used to discriminate them ... increased effort appears to cause enhancement of the responses and sharpened selectivity for attended stimuli" (pp. 339–340). These are also properties of model cells in F_2 due to the role of vigilance control. ART prototypes represent smaller sets of exemplars at higher vigilance levels, so a stricter matching criterion is learned. These exemplars match their finer prototypes better than do exemplars that match a coarser prototype. This better match more strongly activates the corresponding F_2 nodes.

This property suggests that operations that make the novelty-related potentials of the hippocampus more sensitive to input changes may trigger the formation of more selective inferotemporal recognition categories. Can such a correlation between IT discrimination and hippocampal potentials be recorded, say, when monkeys learn easy and difficult discriminations? Conversely, operations that progressively block the expression of hippocampal novelty potentials are suggested to cause learning of coarser recognition categories, with amnesic symptoms as a limiting case.

The conclusion that no learning occurs in the ART orienting system does not force the theory to deny that some types of learning do occur in the hippocampal system. The model suggests that these learning processes are involved in adaptively timed modulation of reinforcement learning and aspects of spatial orientation, as discussed next.

A Framework for Temporal Learning

Before providing this discussion, it is appropriate to comment on how an ART-based system could rapidly learn the information in a movie. Such an analysis requires that the processes whereby individual events are recognized and recalled are supplemented by processes involved in the learning and recognition of temporally ordered sequences, or lists, of events. There are many levels on which this class of problems could be approached, and it seems fair to say that no available theory proposes a complete explanation of this competence. On the other hand, the critique of alternative models has been made on the level of their inability to rapidly and stably learn large amounts of information, notably temporally ordered information. This is not a problem in an ART-based system.

A framework for accomplishing this was described in Grossberg (1978) using a combination of ART category learning, working memories, temporal associative learning networks, and predictive feedback within the system. A great deal of work has since been done to carry this program further. For example, ART-based architectures, called VIEWNET systems, are capable of rapidly and stably learning to recognize 3-D objects by categorizing their 2-D views and learning to associate their 2-D view categories with 3-D object categories that are invariant under changes of familiar 2-D view (Bradski & Grossberg, 1994, 1995). Properties of these 2-D view and 3-D object category nodes may be compared with neural responses from distinct cell populations in monkey inferotemporal cortex (Logothetis, Pauls, Buelthoff, & Poggio, 1994).

The 3-D object categories may, in turn, be stored in a working memory (Baddeley, 1986; Grossberg, 1978) that can encode both object representations and their temporal order in STM. This type of working memory is designed so that its contents may rapidly and stably be learned and categorized by another ART network, whose active nodes are said to code list categories. This list categorization process has been proved to retain its stability even as new information continues to be stored in the working memory through time (Bradski, Carpenter, & Grossberg, 1992, 1994; Cohen & Grossberg, 1986, 1987; Grossberg, 1978; Grossberg & Stone, 1986a). Interactions between such a working memory and its list categories have been used to explain data from experiments about the sequential performance of stored motor commands (Boardman & Bullock, 1991; Grossberg & Kuperstein, 1989), about errors in serial item and order recall due to rapid visual attention shifts (Grossberg & Stone, 1986a), about errors and reaction times during lexical priming and episodic memory experiments (Grossberg & Stone, 1986b), and about data concerning word superiority, phonemic restoration, and backward effects on speech perception (Cohen & Grossberg, 1986; Grossberg, 1986). Such a working memory design thus seems to be used in several modalities. This is plausible when one realizes that the design embodies a few simple principles that enable its temporally evolving STM patterns to be stably categorized in LTM.

Frontal cortex provides a likely neural substrate for such a working memory (Goldman-Rakic, 1994). Here, information from multiple sensory modalities converges and may interact with subcortical reward mechanisms to sustain an attentional focus on salient goals (Gaffan, 1994; Knight, 1994). Can ART systems learn multimodal list categories and focus attention on predictively successful ones?

Multimodal information distributed across a working memory may indeed be integrated into ART categories (Asfour, 1994; Asfour, Carpenter, Grossberg, & Lesher, 1993). Such an ART system, called Fusion ARTMAP, is designed to solve the credit assignment problem of selectively resetting those input channels that are causing predictive errors. In addition, ART models of cognitive–emotional interactions have been described to suggest how attention may be selectively allocated to event categories that have high salience due to prior reinforcement and how less salient events may be attentionally blocked (Grossberg, 1975, 1982a, 1984; Grossberg & Levine, 1987; Grossberg &

Merrill, 1992); also see below. They have also been used to explain and predict cognitive data about human decision making under risk as a manifestation of cognitive-emotional neural mechanisms (Grossberg & Gutowski, 1987), and to shed some light on how these cognitive-emotional interactions may break down during mental depression (Grossberg, 1972b, 1984).

The motivationally modulated list categories may, in turn, be recurrently linked together by an associative learning network that helps to predict the categories most likely to occur in a given temporal context. Such networks have been used to model the position-dependent error gradients and learning rates that are observed during human verbal learning and to predict how this process breaks down in schizophrenic subjects (Grossberg, 1969c, 1982b; Grossberg & Pepe, 1970, 1971). Finally, the attended list categories may be used to predict the next images that are expected by the system, a one-to-many process called *outstar learning* (Grossberg, 1968a, 1978, 1980). One possible anatomical substrate of this type of predictive learning is fronto-temporal projections (Gaffan, 1994).

Taken together, these architectural elements may be called a *resonant avalanche*. This name acknowledges the role of resonance in stabilizing the learning process, and of the avalanche of temporal associations in predicting the events that the system next expects to experience. (For a summary of avalanches at different levels of complexity, see Grossberg, 1978.) Although the theory of resonant avalanches has not yet been completely developed, there are enough mathematical, computational, and data simulation results available to conclude that ART systems escape the critique of various other memory models that was proposed previously.

Adaptively Timed Cognitive–Emotional and Sensory-Motor Interactions

Let us now return to the question of what sorts of learning are predicted to occur in the hippocampal system by an ART-based model. As in our remarks about fronto-temporal interactions, this discussion includes an analysis of issues concerning reinforcement and temporal processing. The model fronto-temporal interactions that were reviewed previously concern a type of *macro*-timing that integrates information across a series of events. The model fronto-temporal-hippocampal interactions now discussed consider a type of *micro*-timing that calibrates how long motivated attention may be allocated to a single predicted event.

Some authors (e.g., Eichenbaum, Otto, & Cohen, 1994) have dichotomized the representational properties of hippocampal memory processing—namely, those relating to recognition learning and memory—as being orthogonal functional properties from hippocampal temporal processing properties. It is unclear why a single brain structure should combine properties if they are indeed orthogonal. The adaptive timing model described next suggests how these representational and temporal processes may be linked. The timing model is part of a larger model system that controls how cognitive–emotional and sensory-motor interactions are coordinated, including how classical and instrumental conditioning are adaptively timed and modulated by cognitive recognition processes (Baloch & Waxman, 1991; Grossberg, 1971, 1972b, 1975, 1982a, 1987; Grossberg & Levine, 1987; Grossberg & Merrill, 1992; Grossberg & Schmajuk, 1987).

This cognitive–emotional model suggests that (at least) three types of internal representation interact during conditioning: sensory representations S, drive representations D, and motor representations M (Fig. 2.3). The S representations are categorical thalamo-cortical representations of external events, including the object recognition categories that are learned by IT cortex and linked to frontal cortex via fronto-temporal interactions. The D representations include hypothalamic and amygdala circuits, at which homeostatic and reinforcing cues converge to generate emotional reactions and motivational decisions. The M representations include cerebellar circuits that control discrete adaptive responses. Three types of learning take place among these representations: $S \rightarrow D$ conditioned reinforcer learning that converts a CS into a reinforcer by pairing activation of its sensory representation S with activation of the drive representation D that receives input from a salient US or other conditioned reinforcer CS; $D \rightarrow S$ incentive motivational learning whereby an activated drive representation D may learn to prime the sensory representations S of all cues, including CSs, that have consistently been activated when it has; and $S \rightarrow M$ habit, or motor, learning whereby the sensory-motor maps, vectors, and gains that are involved in motor control may be adaptively calibrated.

These processes contribute to the modulation of declarative memory by motivational feedback and to the learning and performance of procedural memory. Thus, learned $S \rightarrow D \rightarrow S$ positive feedback quickly draws attention to motivationally salient cues and blocks activation of less salient cues via lateral inhibition among the S categories. $D \rightarrow S$ motivational feedback also energizes the release of discrete adaptive $S \rightarrow M$ responses. Based on a theoretical analysis, the final common path of the drive representations D, at or after the stage at which motivational decisions are made, was predicted to intersect or be modulated by the hippocampal formation (Grossberg, 1975, 1982b). In support of this prediction, Thompson et al. (1984, 1987) have shown that emotional conditioning (as in the $S \rightarrow D$ circuit) influences hippocampal sites, whereas motor conditioning (as in the $S \rightarrow M$ circuit) occurs within the cerebellum. In addition, hippocampal ablation attenuates blocking (Rickert, Bennett, Lane, & French, 1978; Schmajuk, Spear, & Isaacson, 1983; Solomon, 1977).

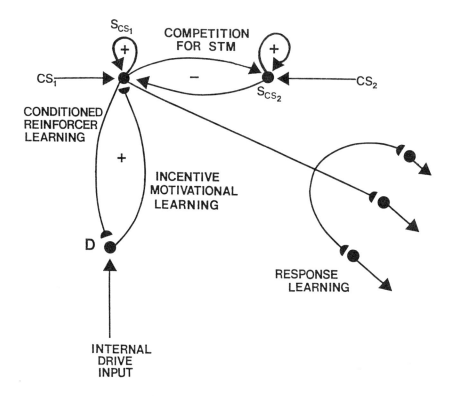

FIG. 2.3. Schematic conditioning circuit: Conditioned stimuli (CS_i) activate sensory categories (S_{CS_i}) that compete among themselves for limited capacity short-term memory activation and storage, as at level F_2 in an ART circuit. The activated S_{CS_i} representations elicit trainable signals to drive representations D and motor command representations M, denoted *response learning*. Learning from a sensory representation S_{CS_i} to a drive representation D is called *conditioned reinforcer learning*. Learning from D to a S_{CS_i} is called *incentive motivational learning*. Signals from D to S_{CS_i} are elicited when the combination of conditioned sensory plus internal drive inputs is sufficiently large. Sensory representations that win the competition in response to the balance of external inputs and internal motivational signals can activate motor command pathways.

Blocking fails in the model when D → S feedback is impaired, as follows. In the complete model, when the S population activities that categorize conditioned reinforcers are amplified by strong conditioned S → D → S attentional feedback, they can block activation of other S populations via S → S lateral inhibition. When D → S feedback is removed, amplification and its blocking effect are eliminated. See Grossberg and Levine (1987) for blocking simulations. These model properties clarify how damage to the hippocampal system that involves both its drive-modulatory and orienting functions can result in either impaired or abnormally strong utilization of contextual cues (due to a failure of S → D → S feedback to enhance salient cues), and a failure of flexible reset and memory search for appropriate cues to attend (due to a failure of the orienting subsystem A to trigger these events).

Why should a single brain region, like the hippocampal system, modulate both recognition learning and reinforcement learning? We suggest that this is so in part because the same adaptive timing and orienting processes modulate both types of learning (Grossberg & Merrill, 1992; Grossberg & Schmajuk, 1989). This linkage clarifies how the hippocampal system may mediate tasks like delayed nonmatch to sample (DNMS) wherein both temporal delays and novelty-sensitive recognition processes are involved (Gaffan, 1974; Mishkin & Delacour, 1975). The proposed adaptive timing and orienting properties of the hippocampal system are envisaged to cooperate in the following way. As shown in Figs. 2.3 and 2.4, S → D → S feedback can rapidly focus attention on motivationally salient cues, as inhibition from D to the orienting subsystem inhibits orienting reactions that would otherwise occur in response to irrelevant situational cues. The inhibition from D to the orienting subsystem helps to model competition between consummatory and orienting behaviors (Staddon, 1983).

Another process is, however, needed to prevent the premature reset of attention by potentially distracting irrelevant cues during variable task-specific delays. For example, suppose that an animal inspects a food box right after a signal occurs that has regularly predicted food delivery in 6 seconds. When the animal inspects the food box, it perceives the nonoccurrence of food during the subsequent 6 seconds. These nonoccurrences disconfirm the animal's sensory expectation that food will appear in the magazine. Because the perceptual processing cycle that processes this sensory information occurs at a much faster rate than 6 seconds, it can compute this sensory disconfirmation many times before the 6 second delay has elapsed. Why is not the mismatch between the learned expectation of food and the percept of no-food treated like a predictive failure? Why, as often occurs when a previously rewarded cue is no longer rewarded, does the mismatch not trigger reset of attention, frustration, forgetting, and exploratory behavior? Were this to happen, humans and animals would restlessly explore their environments without being able to wait for delayed rewards.

The central issue is: What spares the animal from erroneously reacting to these *expected nonoccurrences* of food during the first 6 seconds as predictive failures? Why does the animal not immediately become so frustrated by the nonoccurrence of food that it shifts its attentional focus and releases exploratory behavior aimed at finding food somewhere else? Alternatively, if the animal does wait, but food does not appear after the 6 seconds have elapsed,

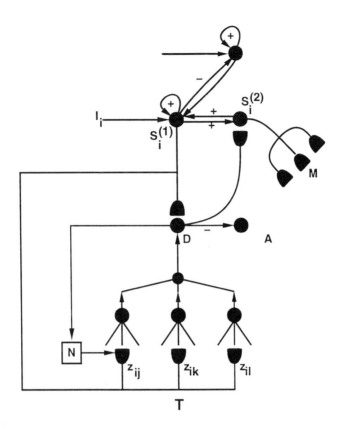

FIG. 2.4. A spectrally timed conditioning model with feedback pathways $D \to S^{(2)} \to S^{(1)}$ that are capable of focusing attention in an adaptively timed fashion on reinforcing events. The sensory representations S of Fig. 2.3 are here broken into two successive levels $S^{(1)}$ and $S^{(2)}$. Levels $S^{(1)}$ and $S^{(2)}$ interact via reciprocal excitatory pathways. The excitatory pathways $S^{(1)} \to D$ and $D \to S^{(2)}$ are, as in Fig. 2.3, adaptive. Representations in $S^{(2)}$ can, however, fire only if they receive convergent signals from $S^{(1)}$ and D. Then they deliver positive feedback to $S^{(1)}$ and bias the competition to focus attention on their respective features and to attentionally block inhibited features. Prior to conditioning, a CS can only be stored in STM at $S^{(1)}$ and can subliminally prime $S^{(2)}$ and D representations without supraliminally firing these representations. After conditioning, the CS can trigger strong conditioned $S^{(1)} \to D \to S^{(2)} \to S^{(1)}$ feedback and rapidly draw attention to itself as it activates the emotional representations and motivational pathways controlled by D. Representation D can also inhibit the orienting subsystem as it focuses attention upon motivationally valued sensory events. The sensory representations $S^{(1)}$ send parallel pathways to a spectral timing circuit T whose adaptive weights z sample the Now Print, or teaching signal, N, that is transiently activated by changes in the activity of the drive representation D. After conditioning of T takes place, adaptively timed readout from T can maintain attention on task-relevant cues for a learned duration via the $T \to D \to S$ feedback pathway. Timed signals also inhibit the orienting subsystem via the $T \to D \to A$ pathway and thereby help to prevent distracting events from interfering with planned consummatory acts. [Reprinted with permission from Grossberg & Merrill (1996).]

why does the animal then react to the *unexpected nonoccurrence* of food by becoming frustrated, shifting its attention, and releasing exploratory behavior?

Grossberg and Schmajuk (1989) and Grossberg and Merrill (1992) argued that a primary role of the timing mechanism is to inhibit, or *gate*, the process whereby a disconfirmed expectation would otherwise negatively reinforce previous consummatory behavior, shift attention, and release exploratory behavior. The process of registering sensory mismatches or matches is not itself inhibited; if the food happened to appear earlier than expected, the animal could still perceive it and eat. Instead, the effects of these sensory mismatches on reinforcement, attention, and exploration are inhibited.

Spectral Timing in the Hippocampus and Deficits Due to its Removal

In order to realize this property, we suggested that a *spectral timing* circuit $S \rightarrow T$ operates in parallel with the fast $S \rightarrow D \rightarrow S$ emotional conditioning circuit (Fig. 2.4) to maintain attention on salient cues during variable task-specific delays. Different populations of cells in T can be conditioned to respond selectively to different ISI intervals. The total population output sums the output from all cells in the spectrum. Remarkably, this population response accurately models the ISI, even though no single cell does (Fig. 2.5). Learned $S \rightarrow T$ timing *maintains* inhibition of the orienting subsystem and, in the example noted previously, enables attention to be maintained on motivationally salient, goal-related cues within the 6-second delay. If food does not occur even after 6 seconds or more have elapsed, then the adaptive timing circuit becomes quiet, and subsequent ART mismatches can trigger attentional reset, frustration, forgetting, and exploration in a manner modeled in Grossberg (1987).

Grossberg and Merrill (1992) predicted that this spectral timing circuit T exists in the hippocampal dentate–CA3 region to explain neurophysiological data showing that hippocampal–CA3 pyramidal cell firing often mirrors the temporal delays observed in the conditioned nictitating membrane response (Berger, Berry, & Thompson, 1986). We suggested that subsets of hippocampal dentate cells respond at different rates to generate the spectral representation that controls the adaptively timed population response at CA3 pyramidal cells. Nowak and Berger (1992) reported experimental evidence that is consistent with this prediction.

If the hippocampal system is removed, should animals and humans always have problems with DNMS and related tasks that involve stimulus delays? In the model, when the timing circuit T is removed, attention may more easily be distracted from goal objects during task-related delays. On the other hand,

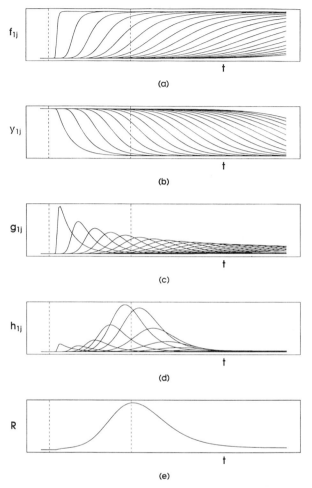

FIG. 2.5. A computer simulation of spectral timing: (a) In response to a CS input I_i in Fig. 2.4, a spectrum of population activities x_{ij} react at different rates and generate signals $f_{ij} = f(x_{ij})$; (b) each signal causes a transmitter y_{ij} in its pathway to become inactivated, or habituate, at a different rate; (c) the transmitters y_{ij} multiply, or gate, the signals f_{ij} to generate net signals $g_{ij} = f_{ij}y_{ij}$ that sample overlapping time intervals; (d) the sampling signals g_{ij} and the US, expressed via the teaching signal N, conjointly activate adaptive weights, or LTM traces, z_{ij}, which generate adaptively gated output signals $h_{ij} = g_{ij}z_{ij}$; (e) although individual signals h_{ij} do not satisfactorily time the ISI, the population sum $R = \Sigma_j h_{ij}$ of the adaptive signals does accurately time the ISI (dotted vertical lines). Parameters and signal functions: $g = 0.2$, $a_y = 1.0$, $b_y = 125.0$, $a_z = 1.0$, $d = 0.0$, $e = 0.02$, $a_E = 240.0$, $a_A = 1.2$, $b_A = 120.0$, $g_A = 12.0$, $a_D = 120.0$, $b_D = 120.0$, $g_D = 0.0$, $f_D(S) = [S - 0.05]^+$, $a_C = 0.5$, $b_C = 25.0$, $f_C(D) = [D - 0.05]^+$, $f_A(A) = [A - 0.1]^+$, $F_X(A) = [A - 0.7]^+$, $r_j = 10.125 / (0.0125 + j)$; and the intensities of the CS and US inputs I_i in (1) equal 2. The abscissa variable t is time. [Reprinted with permission from Grossberg & Merrill (1992).]

53

if the orienting subsystem is also removed, then flexible reset of attention in response to novel events is impaired, thereby eliminating a key mechanism whereby a distracting event could undermine performance. If the attentional system remains intact, then direct activation of individual recognition codes in response to a familiar event is still possible, and the matching process per se can partially update short-term memory. However, the network can no longer flexibly search for the proper configuration of targets to attend, especially in the presence of complex spatial layouts that include distracting cues. The lack of timed control over variable delays can thus harm behavior more when it is necessary to shift attention among different sets of cues. Gaffan (1992) described analogous data from hippocampectomized monkeys.

Both DNMS performance at brief delays and single-pair object discrimination learning with brief intertrial intervals are spared in hippocampal subjects (Eichenbaum, Otto, & Cohen, 1994). In the model, this is also true because the fast $S \rightarrow D \rightarrow S$ attentional circuit remains intact. Long interstimulus delays, say of a day, also spare the performance of animals in some conditions (Mishkin, Malamut, & Bachevalier, 1984). These results have led some investigators to claim that the hippocampal system subserves a memory store of *intermediate* duration (Eichenbaum, Otto, & Cohen, 1994). As already noted, how the hippocampal system could create such a representation before it is transferred to the appropriate neocortical representations across several modalities has never been explained, and faces serious conceptual difficulties.

The ART model does not need to posit any such hippocampal memory store. At short delays, the fast feedback $S \rightarrow D \rightarrow S$ system helps to focus attention on motivationally salient objects and to initiate attentional blocking. The failure of blocking at intermediate delays due to removal of the $S \rightarrow T$ circuit leads to abnormally strong utilization of contextual cues. This processing failure causes little problem at long delays because potentially disruptive cues, being so widely separated in time, decay before they can compete for attention. These properties can be inferred from the model simulations of blocking by Grossberg and Levine (1987). It has not, to our knowledge, yet been tested whether the spectral timing circuit that is proposed to exist in dentate–CA3 plays the role described previously in the DNMS paradigm.

Spectrally Timed Gain Control in the Cerebellum

Why is adaptive timing also needed in the motor conditioning circuit? This need is clarified by the fact that the $S \rightarrow D \rightarrow S$ circuit focuses attention quickly on motivationally salient cues and can thereby just as quickly activate the motor circuit (Fig. 2.3). Without adaptive timing within the motor circuit itself, the conditioned response could be prematurely released. Thus, the clear survival advantage of attending quickly to motivationally important sensory events could disrupt the properly timed execution of responses contingent on these events. The model suggests that this problem does not occur during normal behaviors because the hippocampal dentate–CA3 circuit and the cerebellar motor circuit are both adaptively timed. These distinct timing functions have been dissociated through ablation (Ebner & Bloedel, 1981; Gilbert & Thach, 1977; Optican & Robinson, 1980; Thompson, 1988; Thompson et al., 1984, 1987) and ISI shift experiments during which the peak time of the hippocampal trace can change before the peak time of the discrete adaptive response (Hoehler & Thompson, 1980). The model suggests that orienting responses may be inhibited by the hippocampal dentate–CA3 timing circuit during the same time intervals when conditioned responses are disinhibited by the cerebellar timing circuit. This coordinated action extends the classical idea that consummatory and orienting responses are mutually inhibitory.

Recent experiments on conditioning the rabbit NMR suggest that response learning occurs within a subcortical cerebellar pathway, whereas response timing occurs within the cerebellar cortex (Perrett, Ruiz, & Mauk, 1993). If the cortical timing circuit is ablated, then motor responses are, indeed, prematurely released. These experimental results are consistent with the classical hypothesis that a fast cerebellar motor pathway—here interpreted to be subcortical (Lisberger, 1988)—can learn a conditioned gain appropriate to the response using climbing fiber inputs as a teaching signal (Albus, 1971; Fujita, 1982a, 1982b; Grossberg, 1969a; Grossberg & Kuperstein, 1986; Marr, 1969).

It is also hypothesized that adaptive timing is learned by a spectral timing circuit in which parallel fiber–Purkinje cell cortical synapses use climbing fiber inputs as a teaching signal (Fig. 2.6). In this conception, cortical learning opens a timed gate by removing Purkinje cell inhibition from subcortical sites. As the timed gate opens, the subcortical motor pathway can read out its learned gain with the correctly timed ISI between CS and US. Learned suppression of Purkinje cell output may be accomplished by conditioned long term depression, or LTD (Hoehler & Thompson, 1980; Ito, 1984). Eight key data properties have been simulated by this model (Bullock, Fiala, & Grossberg, 1994; Fiala, Grossberg, & Bullock, 1996):

> Model Purkinje cell activity decreases in the interval following the onset of the CS; model nuclear cell responses match CR topography; CR peak amplitude occurs at the US onset; a discrete CR peak shift occurs with a change in ISI between CS and US; mixed training at two different ISIs produces a double-peaked CR; peak CR acquisition and response rates depend unimodally on the ISI; CR onset latency decreases during training; and maladaptively-timed, small-amplitude CRs result from ablation of cerebellar cortex.

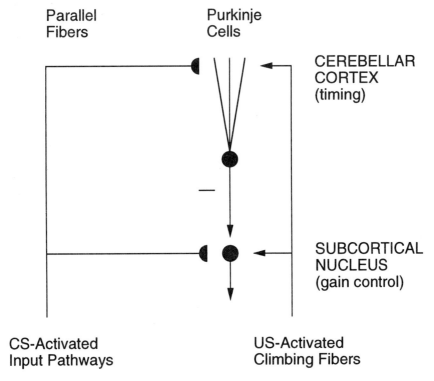

Parallel   Purkinje
Fibers   Cells

CEREBELLAR
CORTEX
(timing)

SUBCORTICAL
NUCLEUS
(gain control)

CS-Activated   US-Activated
Input Pathways   Climbing Fibers

FIG. 2.6. A model of adaptively timed cerebellar conditioning: US-activated climbing fibers provide a teaching signal that causes adaptively timed long-term depression at parallel fiber–Purkinje cell synapses, thereby disinhibiting the inhibitory effect of tonic Purkinje cell outputs on cerebellar nuclear cells. The climbing fibers also control learning of adaptive gains along subcortical pathways through the nuclear cells. The net effect of learning is to open an adaptively timed Purkinje gate that enables learned gains to be expressed at the correct time. [Reprinted with permission from Grossberg & Merrill (1996).]

Some striking cellular and circuit homologs exist between these model cerebellum and hippocampal timing mechanisms. Both control an inhibitory gate that modulates another learning process, and both occur on dendrites whose summed output across a spectrum of rate-sensitive cell sites determines the collective timed response. These similarities suggest the prediction that both the hippocampal dentate cell and cerebellar Purkinje cell dendrites may undergo similar biophysical events during conditioning.

Cooperative Hippocampal and Cerebellar Timing
During Serial Compound Conditioning

How do the hippocampal and cerebellar timing circuits cooperate during timed behaviors? We illustrate such cooperation by explaining paradoxical data about serial compound conditioning, during which a sequence CS1–CS2–US of two CSs precedes a US (Kehoe & Morrow; 1984; Kehoe, Gibbs, Garcia, & Gormenzano, 1979; Kehoe, Marshall-Goodall, & Gormenzano, 1987). Robust serial compound conditioning to CS1 can occur even if primary CS1–US conditioning at the same ISI, in the absence of CS2, is ineffective. This happens, for example, if the CS1–CS2 ISI = 2,400 msec and the CS2–US ISI = 400 msec (Kehoe & Morrow, 1984). How does the occurrence of CS2 enable CS1 to bridge the 2,800 msec ISI before US occurs?

We suggest that CS2 can reactivate the sensory representation S1 of CS1 via the drive representation D along the feedback pathway $CS2 \rightarrow S2 \rightarrow D \rightarrow S1$, and thereby restart the $S1 \rightarrow T$ and $S1 \rightarrow M$ timing circuits. In particular, on the first learning trial, the activity of S1 does not persist until US occurs, but the activity of S2 does. As a result, $S2 \rightarrow D$ and $D \rightarrow S2$ conditioning start to occur. On later learning trials, S1 is active when CS2 occurs. Thus S1 is active when S2 activates D. S1 can hereby also learn to activate D, and D can be reciprocally conditioned to both S1 and S2 via the $D \rightarrow S1$ and $D \rightarrow S2$ feedback pathways. In this way, activation of D by CS2 reactivates S1 and restarts its timing circuits, so that they are active when the US occurs. As a result, $S1 \rightarrow M$ conditioning of the NMR is possible, but is released earlier than the 2,800 msec ISI between CS1 and US.

This explanation clarifies why, if the ISI between CS1 and CS2 is short enough, then CS2 elicits less NMR conditioning than it does when it is conditioned to the US at the same ISI without the occurrence of CS1 (Kehoe et al., 1979). If the CS1–CS2 delay is short enough, S1 can partially block S2 because $S1 \rightarrow D \rightarrow S1$ feedback is still strong when CS2 occurs. Conversely, if the total CS1–US ISI is increased, then CS2 can elicit more NMR conditioning than it would in the absence of CS1. Here, S1's activity subsides by the time S2 occurs, but it primes D with residual activity that can amplify $S2 \rightarrow D \rightarrow S2$ and $S2 \rightarrow T$ conditioning when CS2 and US occur. Kehoe, Horne, Macrae, & Horne, (1993) have shown that a spectral timing model can, indeed, be used to simulate key properties of serial compound conditioning data.

START: A UNIFIED MODEL OF ADAPTIVE TIMING
AND CONDITIONED REINFORCER LEARNING

The hippocampal adaptive timing model depicted in Fig. 2.4 is mathematically defined here. It combines Spectral Timing mechanisms with mechanisms from Adaptive Resonance Theory. Hence it is called the START model (Grossberg

& Merrill, 1992). The START model builds on a previous model of reinforcement learning whose processing stages have been compared with behavioral and neural data in a series of previous articles. Here we provide just enough review and exposition to define the model and to compare its emergent properties with illustrative data.

Although the model is helpful for the explanation of both classical and operant conditioning data, here each conditionable sensory event is called a conditioned stimulus, or CS. The ith sensory event is denoted by CS_i. Event CS_i activates a population of cells that is called the ith sensory representation S_i (Fig. 2.3). Another population of cells, called a drive representation D, receives a combination of sensory, reinforcement, and homeostatic (or drive) stimuli. Reinforcement learning, emotional reactions, and motivational decisions are controlled by D (Grossberg, 1971, 1972a, 1982b). In particular, a reinforcing event, such as an unconditioned stimulus, or US, is capable of activating D.

Various authors have invoked representations analogous to drive representations. Bower and his colleagues have called them emotion nodes (Bower, 1981; Bower, Gilligan, & Monteiro, 1981), and Barto, Sutton, and Anderson (1983) called them adaptive critic elements. During conditioning, presentation of a CS_i before a US causes activation of S_i followed by activation of D. Such pairing causes strengthening of the adaptive weight, or long-term memory trace, in the modifiable synapses from S_i to D. This learning event converts CS_i into a conditioned reinforcer. Conditioned reinforcers hereby acquire the power to activate D via the conditioning process.

In the START model, reinforcement learning in $S_i \rightarrow D$ pathways is supplemented by a parallel learning process that is concerned with adaptive timing. As shown in Fig. 2.4, both of these learning processes output to D, which in turn inhibits the population of cells in the orienting subsystem A. The orienting subsystem is a source of nonspecific arousal signals that are capable of initiating frustrative emotional reactions, attention shifts, and orienting responses. The inhibitory pathway from D to A is the gate that prevents these events from occurring.

Limited Capacity Short-Term Memory.   The sensory representations S_i compete for a limited capacity, or finite total amount, of activation. Winning populations are said to be stored in short-term memory, or STM. The competition is carried out by an on-center, off-surround interaction among the populations S_i. The property of STM storage is achieved by using recurrent, or feedback, pathways among the populations. A tendency to select winning populations is achieved by using membrane equations, or shunting interactions, to define each population's activation, and a proper choice of feedback signals between populations (Grossberg, 1973, 1982b). Expressed mathematically, each CS_i activates an STM representation S_i whose activity S_i obeys the shunting on-center, off-surround competitive feedback equation:

$$\tfrac{d}{dt}S_i = -\alpha_A S_i + \beta_A(1 - S_i)(I_i(t) + f_S(S_i)) - \gamma_A S_i \sum_{k \neq i} f_S(S_k). \quad (2)$$

In Equation 2, $I_i(t)$ is the input that is turned on by presentation of CS_i. Term $-\alpha_A S_i$ describes passive decay of activity S_i. Term $\beta_A(1 - S_i)(I_i(t) + f_S(S_i))$ describes the excitatory effect on S_i of the input $I_i(t)$ and the feedback signal $F_S(S_i)$ from population S_i to itself. Activity S_i can continue to grow until it reaches the excitatory saturation point, which is scaled to equal 1 in Equation 2. Term $-\gamma_A S_i \sum_{k \neq i} f_S(S_k)$ describes inhibition of S_i by competitive signals $F_S(S_k)$

from the off-surround of populations $k \neq i$. Figure 2.7 summarizes a computer simulation of how a brief CS_1 gives rise to a sustained STM activation S_1, which is partially inhibited by competition from S_{0S} activation in response to a US. The signal function f_S may be chosen to have any of several forms without qualitatively altering model properties. In this chapter, the simple threshold-linear, half-wave rectification function

$$f(w) = [w - \mu]^+ \equiv \max(w - \mu, 0) \quad (3)$$

is used, except in Equation 9, which uses a sigmoid, or S-shaped, signal function.

Drive Representation. The computer simulations reported herein use only a single drive representation D. Explanations of data arising from competing drive representations are discussed in Grossberg (1984, 1987). The activity D of the drive representation D obeys the equation

$$\tfrac{d}{dt}D = -\alpha_D D + \beta_D \sum_i f_D(S_i)C_i + \gamma_D R. \quad (4)$$

In Equation 4, term $-\alpha_D D$ describes the passive decay of activity D. Term $\beta_D \sum_i f_D(S_i)C_i$ describes the total excitatory effect of all the sensory repre-

sentations S_i on D. In this term, the signal function f_D is chosen as in Equation 3, and C_i is the adaptive weight, or long-term memory (LTM) trace, in the pathway from the sensory representation S_i of CS_i to the drive representation D. This LTM trace is denoted by C_i because its size measures how well S_i can activate D, and thus how CS_i $(i \geq 1)$ has become a conditioned reinforcer through learning. Because C_i multiplies $f_D(S_i)$, a large activation of S_i will have a negligible effect on D if C_i is small, and a large effect on D if C_i is large. Coefficient C_0 is set equal to a large value from the start because it enables the US to activate D via its sensory representation S_0. Term $\gamma_D R$ describes the

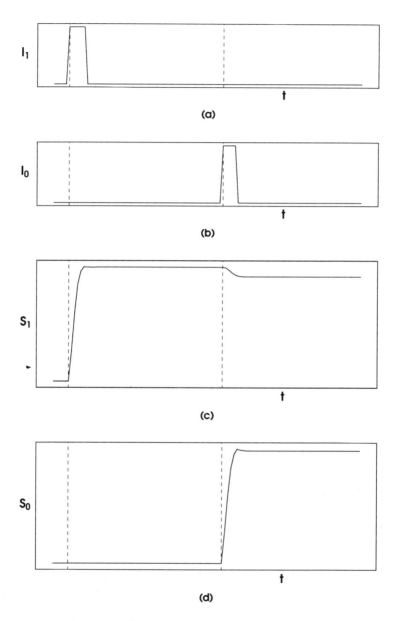

FIG. 2.7. In a START model, STM storage of a brief CS is achieved by positive feedback within the sensory representation S. CS attenuation by the US is dynamically controlled by the strength of recurrent inhibitory signals. (a) Input I_1 activated by CS_1; (b) Input I_0 activated by US; (c) STM activation of CS_1 sensory representation; (D) STM activation of US sensory representation. [Reprinted with permission from Grossberg & Merrill (1992).]

total output of the spectral timing circuit to D. Output R is defined in Equation 12.

Figure 2.8c summarizes a computer simulation in which the activity D responds to CS and US signals after 50 conditioning trials. Figures 2.8a and 2.8b summarize the corresponding STM traces S_1 of the CS and S_0 of the US, respectively.

Conditioned Reinforcement. The adaptive weight C_i that calibrates conditioned reinforcement obeys a gated learning law (Grossberg, 1969b):

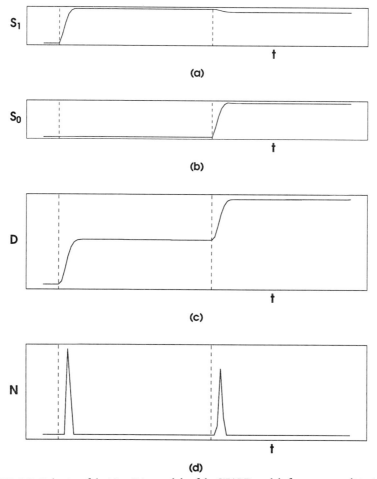

FIG. 2.8. Behavior of the Now Print module of the START model after many conditioning trials: (a) Activation of the sensory representation S_1 by the CS; (b) Activation of the sensory representation S_0 by the US; (c) The resultant activation D of the drive representation D; (d) The resultant Now Print signal, N. [Reprinted with permission from Grossberg & Merrill (1992).]

$$\tfrac{d}{dt}C_i = \alpha_C S_i(-C_i + \beta_C(1 - C_i)f_C(D)). \quad (5)$$

Learning by C_i is turned on and off by the signal S_i from S_i, which thus acts like a learning gate, or modulator. Once turned on, C_i performs a time-average of activity at the drive representation D via the signal $f_C(D)$, which is chosen as in Equation 3. Activity C_1 cannot exceed the finite value 1, due to the shunting term $1 - C_i$. The value of C_i can both increase and decrease during the course of learning. The remaining equations of the model describe the adaptive timing process.

Activation Spectrum.   The START model is said to control "spectral" timing because each drive representation D is associated with a population of cell sites whose members react at a spectrum of rates r_j. Neural populations whose elements are distributed along a temporal or spatial parameter are familiar throughout the nervous system. Two examples are populations of spinal cord cells that obey the size principle (Henneman, 1957, 1985), and the spatial frequency-tuned cells of the visual cortex (Jones & Keck, 1978; Musselwhite & Jeffreys, 1985; Parker & Salzen, 1977a, 1977b; Parker, Salzen, & Lishman, 1982a, 1982b; Plant, Zimmern, & Durden, 1983; Skrandies, 1984; Vassilev, Manahilov, & Mitov, 1983; Vassilev & Strashimirov, 1979; Williamson, Kaufman, & Brenner, 1978).

The spectral activities x_{ij} that are associated with drive representation D and activated by sensory representation S_i obey the equation:

$$\tfrac{d}{dt}x_{ij} = r_j(-x_{ij} + (1 - x_{ij})f_x(S_i)), \quad (6)$$

where f_x satisfies Equation 3. By Equations 2 and 6, presentation of CS_i to S_i via an input I_i generates an output signal $f_x(S_i)$ that activates the local potentials x_{ij} of all cell sites in the target population. The potentials x_{ij} respond at rates proportional to $r_j, j = 1, 2, \ldots, N$. These potentials activate the next processing stage via signals

$$f(x) = \frac{x_{ij}^8}{\delta_{ij}^8 + x_{ij}^8} \quad (7)$$

Signal $f(x_{ij})$ is a sigmoid function of activity x_{ij}. Figure 2.5a shows the activation spectrum $f(x_{ij}(t))$ that arises from presentation of CS_i to S_i via input I_i in Equation 2, using a choice of rate parameters r_j in Equation 6 that range from 10 (fast) to 0.0025 (slow).

Habituative Transmitter Spectrum.   Each spectral activation signal $f(x_{ij})$ interacts with a habituative chemical transmitter y_{ij} via the equation:

$$\tfrac{d}{dt} y_{ij} = \alpha_y (1 - y_{ij}) - \beta_y f(x_{ij}) y_{ij}. \quad (8)$$

According to Equation 8, the amount of neurotransmitter y_{ij} accumulates to a constant target level 1, via term $\alpha_y(1 - y_{ij})$, and is inactivated, or *habituates*, due to a mass action interaction with signal $f(x_{ij})$, via term $-\beta_y f(x_{ij}) y_{ij}$. The different rates r_j at which each x_{ij} is activated causes the corresponding y_{ij} to become habituated at different rates. The family of curves $y_{ij}(t)$, $j = 1, 2, \ldots,$ n, is called a habituation spectrum. The signal functions $f(x_{ij}(t))$ in Fig. 2.5a generate the habituation spectrum of $y_{ij}(t)$ curves in Fig. 2.5b.

Gated Signal Spectrum. Each signal $f(x_{ij})$ interacts with y_{ij} via mass action. This process is also called *gating* of $f(x_{ij})$ by y_{ij} to yield a net signal g_{ij} that is equal to $f(x_{ij}) y_{ij}$. Each gated signal $g_{ij}(t) \equiv f(x_{ij}(t)) y_{ij}(t)$ has a different rate of growth and decay, thereby generating the gated signal spectrum shown in Fig. 2.5c. In these curves, each function $g_{ij}(t)$ is a unimodal function of time, where function $g_{ij}(t)$ achieves its maximum value M_{ij} at time T_{ij}, T_{ij} is an increasing function of j, and M_{ij} is a decreasing function of j.

These laws for the dynamics of a chemical transmitter were described in Grossberg (1968b, 1969a). They capture the simplest first-order properties of a number of known transmitter regulating steps (Cooper, Bloom, & Roth, 1974), such as transmitter production (term α_y), feedback inhibition by an intermediate or final stage of production on a former stage (term $-\alpha_y y_j$), and mass action transmitter inactivation (term $-\beta_y f(x_j) y_j$). Alternatively, they can be described as the voltage drop across an RC circuit, or the current flow through an appropriately constructed transistor circuit. These properties are sufficient to explain the article's targeted data, so finer transmitter processes, such as transmitter mobilization effects, are not considered herein.

Spectral Learning Law. Learning of spectral timing obeys a gated steepest descent equation

$$\tfrac{d}{dt} z_{ij} = \alpha_z f(x_{ij}) y_{ij} (-z_{ij} + N), \quad (9)$$

where N is the Now Print signal that is defined in Equation 10. Each *long-term memory (LTM) trace* z_{ij} in Equation 9 is activated by its own sampling signal $g_{ij} = f(x_{ij}) y_{ij}$. The sampling signal g_{ij} turns on, or *gates*, the learning process, and causes z_{ij} to approach N during the sampling interval at a rate proportional to g_{ij}. The attraction of z_{ij} to N is called *steepest descent*. Thus Equation 9 is a variant of the gated steepest descent equation that was defined in Equation 1. Each z_{ij} changes by an amount that reflects the degree to which the curves $g_{ij}(t)$ and $N(t)$ have simultaneously large values through time. If g_{ij} is large when N is large, then z_{ij} increases in size. If g_{ij} is large when N is small, then

z_{ij} decreases in size. As in Equation 5, z_{ij} can either increase or decrease as a result of learning.

Now Print Signal. A transiently active Now Print signal, N modulates the learning process of Equation 9. The signal N may be activated either by a US or by a CS that has already become a conditioned reinforcer. Figure 2.4 shows that both the US and a conditioned reinforcer CS can activate the drive

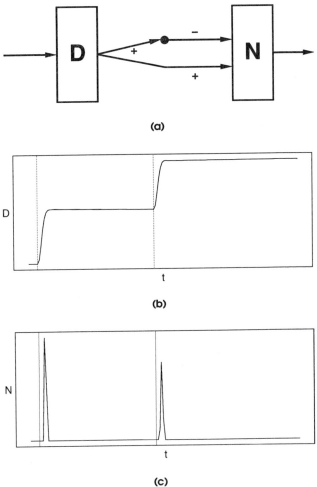

FIG. 2.9. Generation of a Now Print signal: (a) The output of a drive representation D is converted into a Now Print signal, N, by passing this output through a fast excitatory pathway and a slower inhibitory pathway, whose signals converge at N; (b) Simulation of the activity D of D in response to two successive inputs, with the first response larger; (c) Activity N of scales with the size of the increment in D. All parameters as in Fig. 2.5. [Reprinted with permission from Grossberg & Merrill (1992).]

representation D. Equation 4 describes this property mathematically. The Now Print signal N (for example, in Fig. 2.9c) is turned on by sufficiently large and rapid increments in the activity D of D (for example, in Fig. 2.9b). As Figs. 2.9b and 2.9c illustrate, the Now Print signal, N approximates the time derivative of the drive activity D. A neurophysiologically plausible way to achieve this property is to assume that the transient signal N is derived from the sustained activity D by the action of a slow inhibitory interneuron (see Fig. 2.9a). The transformation from sustained activity D to transient activity N can be realized mathematically by the function

$$N = [f_C(D) - E - \varepsilon]^+. \quad (10)$$

In Equation 10, E is the activity of an inhibitory interneuron that time-averages $f_C(D)$, as in equation

$$\tfrac{d}{dt} E = \alpha_E(-E + f_C(D)), \quad (11)$$

before inhibiting the direct excitatory signal $f_C(D)$. Equation 10 means that $N = 0$ if $f_C(D) - E \le \varepsilon$, and $N = f_C(D) - E - \varepsilon$ if $f_C(D) - E > \varepsilon$. By Equation 10, N responds rapidly to an increment in $f_C(D)$. By Equation 11, the inhibitory interneuron activity, E responds more slowly to $f_C(D)$. As E grows, it inhibits the influence of $f_C(D)$ on N, by Equation 10, thereby shutting N off. As noted below, an important property of N is that it increases in amplitude, but not significantly in duration, in response to larger inputs $f_C(D)$.

As noted previously, the time interval between CS onset and US onset is called the *interstimulus interval*, or ISI. Using the spectral learning law (Equations 9 to 11), the individual LTM traces differ in their ability to learn at different values of the ISI. This is the basis of the network's timing properties. Figure 2.10 illustrates how six different LTM traces z_j, $i = 1, \ldots$, 6, learn during this simulated learning experiment. The CS and US are paired during 4 learning trials, after which the CS is presented alone on a single performance trial. In this computer simulation, the CS input $I_{CS}(t)$ remained on for a duration of 0.05 time units on each learning trial. The US input $I_{US}(t)$ was presented after an ISI of 0.5 time units and remained on for 0.05 time units. The upper panel in each part of the figure depicts the gated signal function $g_{ij}(t)$ with r_j chosen at progressively slower rates. The middle panel plots the corresponding LTM trace $z_{ij}(t)$.

Doubly Gated Signal Spectrum. The lower panel plots the twice-gated signal $h_{ij}(t) = f(x_{ij}(t))y_{ij}(t)z_{ij}(t)$. Each twice-gated signal function, $h_{ij}(t)$ registers how well the timing of CS and US is learned and read-out by the ith processing channel. In Fig. 2.10b, where the once-gated signal $g_{ij}(t)$ peaks at approximately the ISI of 0.5 time units, the LTM trace $z_{ij}(t)$ shows the maximum learning. The twice-gated signal $h_{ij}(t)$ also shows a maximal en-

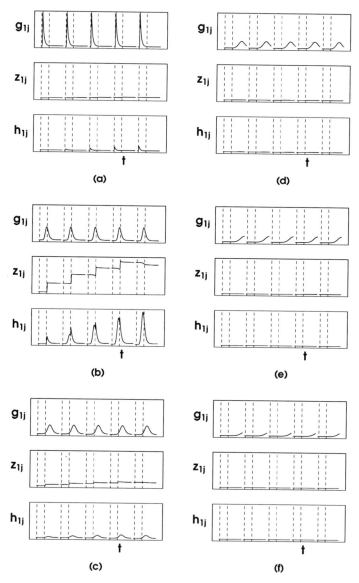

FIG. 2.10. Selective learning within different spectral populations at a fixed ISI = 0.5 time units. Each three-image panel from (a) to (f) represents the gated signal $g_{1j}(t)$ [top], long-term memory trace $z_{1j}(t)$ [middle], and doubly gated signal $h_{1j}(t) = g_{1j}(t)z_{1j}(t)$ [bottom], at a different value of j. In (a), $j = 1$; in (b), $j = 17$; in (c), $j = 33$; in (d), $j = 49$; in (e), $j = 65$; in (f), $j = 81$. The same parameters as in Fig. 2.5 were used. [Reprinted with permission from Grossberg & Merrill (1992).]

hancement due to learning, and exhibits a peak of activation at approximately 0.5 time units after onset of the CS on each trial. This behavior is also generated on the fifth trial, during which only the CS is presented.

Output Signal. The output of the network is the sum of the twice-gated signals, $h_{ij}(t)$ from all the spectral components corresponding to all the CS_i. Thus:

$$R = \sum_{i,j} f(x_{ij}) y_{ij} z_{ij}. \quad (12)$$

The output signal computes the cumulative learned reaction of the whole population to the input pattern. Figure 2.5e shows the function R derived from the pooled signals h_{ij} shown in Fig. 2.5d. A comparison of Figs. 2.5c, 2.5d, and 2.5e illustrate how the output $R(t)$ generates an accurately timed response from the cumulative partial learning of all the cell sites in the population spectrum. The once-gated signals $g_{ij}(t)$ in Fig. 2.5c are biased towards early times. The twice-gated signals $h_{ij}(t)$ in Fig. 2.5d are biased towards the ISI, but many signals peak at other times. The output $R(t)$ in Fig. 2.5e combines these partial views into a cumulative response that peaks at the ISI.

The Problem of Self-Printing During Adaptively Timed Secondary Conditioning

The START model of Grossberg and Merrill (1992) overcame four types of problems whose solution is needed to explain behavioral and neural data about adaptively timed conditioning. These are the problems of: (a) self-printing during adaptively timed secondary conditioning, (b) asymmetric effects of increasing CS or US intensity on timed responding, (c) different effects of US duration on timing than on reinforcement, and (d) combinatorial explosion of network pathways. These problems and their solution by the START model are reviewed later, together with supportive data. Problems (a), (c), and (d) were not solved by the Grossberg and Schmajuk (1989) model.

A major problem for any model of adaptive timing is to explain how adaptively timed secondary conditioning can occur. In primary conditioning, a conditioned stimulus CS_1 is paired with an unconditioned stimulus US until CS_1 becomes a conditioned reinforcer. In secondary conditioning, another conditioned stimulus CS_2 is paired with CS_1 until it, too, gains reinforcing properties. Various experiments have shown that the conditioned response to CS_2 can be adaptively timed (Gormezano & Kehoe, 1984; Kehoe, Marshall-Goodell, & Gormezano, 1987). Indeed, Gormezano and Kehoe (1984) claimed that, in their experimental paradigm, "first- and second-order conditioning follow the same laws" (p. 314), although they also acknowledged that some variables may differentially effect first-order and second-order conditioning in other paradigms.

Adaptively timed secondary conditioning could easily erase the effects of adaptively timed primary conditioning in the following way. For CS1 to act as a conditioned reinforcer, CS1 must gain control of the pathway along which the US activates its reinforcing properties. Suppose that CS1 activated its sensory representation S1 via an input (ICS1) pathway and that US expressed its reinforcing properties via an input (IUS) pathway. Also, suppose that conditioned reinforcer learning enabled CS1 to activate IUS. Thereafter, presentation of CS1 would simultaneously activate both the ICS1 pathway and the IUS pathway. This coactivation would create new learning trials for CS1 with a zero ISI. In other words, CS1 could self-print a spectrum with zero ISI due to CS1–CS1 pairing via the ICS and conditioned IUS pathway. Thus, as CS1 became a conditioned reinforcer, it could undermine the timing that it learned through CS1–US pairing during primary conditioning. Such self-printing could, for example, occur on secondary conditioning trials when a CS2 is followed by a conditioned reinforcer CS1.

Simulations of Secondary Conditioning

The START model overcomes the self-printing problem with its use of a transient Now Print signal N, as in Equation 10. During primary conditioning, onset of the US causes a brief output burst from N. During secondary conditioning, onset of the conditioned reinforcer CS1 also causes a brief output burst from N. However, the spectrum activated by CS1 takes some time to build up, so essentially all of its activities x_{ij} and sampling signals $f(x_{ij})y_{ij}$ are very small during the brief interval when N is large (compare Figs. 2.5c and 2.9c). By the spectral learning law Equation 9, negligible self-printing occurs. The main effect of the self-printing that does occur is to reduce every spectral LTM trace z_{ij} in Equation 2.9 by a fixed proportion of its value, thus scaling down the size of R(t) without changing the timing of its peak.

Figure 2.11a depicts the model output R(t) when the Now Print threshold $\hat{I}$ in Equation 10 is set to a high enough level to guarantee that no self-printing or secondary conditioning occur. Here, CS1 never activates a Now Print signal. Figure 2.11b shows the output when $\hat{I}$ is set lower, thus allowing secondary conditioning and some self-printing to occur. Correct timing still obtains.

Figure 2.12 shows how the model behaves during secondary conditioning. The left-hand half of each panel shows the output of the model in response to the primary conditioned stimulus CS1, and the right-hand half of each panel shows the model output in response to the secondary conditioned stimulus CS2. The peak time arising from the presentation of CS2 occurs

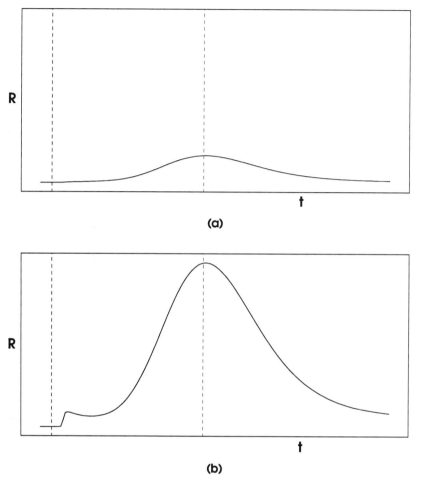

FIG. 2.11. The effect of self-printing on the output of the model. (a) A large threshold ∈ in the Now Print signal abolishes self-printing and secondary conditioning. It generates the lower output $R(t)$. (b) A smaller threshold allows secondary conditioning and self-printing without a loss of timing. It generates the larger output $R(t)$. [Reprinted with permission from Grossberg & Merrill (1992).]

near the expected time of arrival of CS1, rather than the expected time of the US. This property is consistent with the environment that a model or animal experiences, because the subject never sees CS2 paired with the primal US, but rather sees it paired as a predictor of CS1, which serves as a CR in this context.

Asymmetric CS and US Processing in Timing Control

Although CS1 can attain properties of a conditioned reinforcer through CS1–US pairing, this does not imply that all the functional properties of a conditioned reinforcer and an unconditioned stimulus are interchangeable. For example, increasing the intensity of a conditioned reinforcer CS1 can speed up the clock (Maricq, Roberts, & Church, 1981; Meck & Church, 1987; Wilkie, 1987),

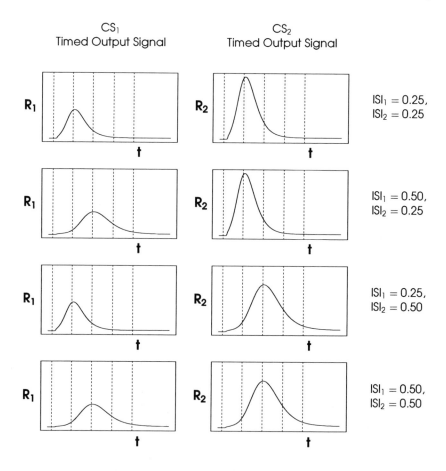

FIG. 2.12. START model output $R(t)$ during secondary conditioning with varying ISIs between the first and second CS, and between the second CS and the US. Notation ISI_1 below denotes the ISI between CS_1 and US, and ISI_2 denotes the ISI between CS_2 and CS_1. On each learning trial either CS_1–US or CS_2–CS_1 occur, but not CS_2–CS_1–US. The curves are drawn with CS_1–US pairings in the left column and CS_2–CS_1 pairings in the right column. The vertical bars occur at successive 0.25 time unit intervals: (a),(b) ISI_1 = .25, ISI_2 = .25; (c),(d) ISI_1 = .5, ISI_2 = .25; (e),(f) ISI_1 = .25, ISI_2 = .5; (g),(h) ISI_1 =

whereas increasing the intensity of a primary US can increase the amplitude of conditioned response, but does not change its timing (Smith, 1968).

The way that parametric changes of CS and US cause different effects on adaptive timing places constraints on possible mechanisms of how adaptive timing is learned during secondary conditioning. Although the CS acquires reinforcing properties of a US when it becomes a conditioned reinforcer, it may not acquire all of its timing properties. The proposed solution of the self-printing problem in Grossberg and Merrill (1992) suggests how different responses may be caused by an increase in CS intensity or US intensity. This explanation holds even if the CS_1 and US sensory representations S_1 and S_0, respectively, each send signals along the same types of pathways to the drive representation and the adaptive timing circuit.

An increase in CS_1 intensity causes an increase in the amplitude of input $I_1(t)$ in Equation 2. The larger input causes a larger peak amplitude of activity S_1 in Equation 2, and a larger signal $f_x(S_1)$ in Equation 6, where the rate with which a spectral activation x_{1J} reacts to signal $f_x(S_1)$ equals $r_j(1 + f_x(S_1))$. Thus an increase in CS_1 intensity speeds up the processing of *all* spectral activations x_{ij}. Because CS_1 is a conditioned reinforcer, some of its LTM traces z_{ij} are nonzero. Thus the total output R in Equation 12 peaks at an earlier time, and causes the total output D from D in Equation 4 to also peak at an earlier time.

In contrast, a primary reinforcer such as a US does not generate a significant output $R(t)$ from the spectral timing circuit, even if it is allowed to generate a large signal $f_x(S_0)$ to the adaptive timing circuit in Equation 6. This is true because a large US generates a signal $f_x(S_0)$ to the spectral activations in Equation 6 at the same time that it generates a large signal $f_D(S_0)$ to D in Equation 4 and a large Now Print signal N in Equation 9. Thus, a US creates the conditions of a "zero ISI experiment" for purposes of spectral learning. All the LTM traces z_{0j} in Equation 9 therefore remain very small in response to any number of US representations. An increase in US amplitude thus cannot cause speed-up of the output $R(t)$ in Equation 12, because this output remains approximately zero in response to any US intensity. In summary, the same mechanism that explains how the self-printing problem is avoided also explains why an increase in CS intensity, but not US intensity, speeds up the conditioned response.

The primary effect of an increase in US intensity is to increase the amplitude of the signal $f_D(S_0)$ in Equation 4 to the drive representation, D. This causes an increase in the amplitude of D and thus an increase in the amplitude of the conditioned response that is modulated by D. This explanation of how a US increases the amplitude of the conditioned response also holds if the US sends no signal $f_x(S_0)$ directly to the adaptive timing circuit. Grossberg and Schmajuk (1989) further discussed this issue.

Different Effects of US Duration on Emotional Conditioning and Adaptive Timing: Sustained and Transient Responses

The existence of a transient Now Print signal, N plays a central role in our explanations of how to avoid self-printing during secondary conditioning, and of different effects of CS and US intensity on learned timing. Another type of data lends support to the hypothesis that the activity, D and the Now Print signal, N both exist but respond to the US in different ways. These data show that an increase in US duration can significantly increase the strength of emotional conditioning (Ashton, Bitgood, & Moore, 1969; Boe, 1966; Borozci, Storms, & Broen, 1964; Church, Raymond, & Beauchamp, 1967; Keehn, 1963; Strouthes, 1965). How can a brief Now Print signal, N whose duration does not increase significantly with US duration coexist with emotional conditioning properties that do increase significantly with US duration?

An answer can be given using properties of drive representations, D. The activation D of a drive representation by a US does persist longer when the US duration is increased, and does thereby increase the strength of emotional conditioning at the S $\rightarrow$ D synapses that are modeled by Equations 4 and 5; see Grossberg (1972a, Section 4) and Grossberg (1982a) for further discussion of this property. This sustained activation, D of a drive representation gives rise to a transient Now Print signal, N at a different processing stage—a transient detector—that is downstream from D itself, as displayed in Figs. 2.4 and 2.9. Thus, D and N represent responses of sustained cells and transient cells—a distinction familiar from visual perception—which here instantiate different functional properties of emotional conditioning and conditioning of adaptive timing, respectively. The parametric data properties summarized previously illustrate that the processes of emotional conditioning and adaptive timing, although linked, are not the same.

The Problem of Combinatorial Explosion: Stimulus Versus Drive Spectra

According to any spectral timing theory, each CS_i activates a sensory representation S_i that broadcasts signals along many parallel pathways. This can lead to a combinatorial explosion of cell bodies if the spectra are incorrectly instantiated. For example, suppose that each pathway activated a different cell, and that each cell's activity computed a different x_{ij}, $j = 1, 2, \ldots, n$. Then there would exist as many copies of the spectral timing model as there are sensory representations in the brain. In addition, each spectrum contains 80 activities x_{ij} in our computer simulations. Such a model would require a huge number of cells to represent a different spectrum for every possible sensory representation. This is, in fact, the type of circuit used in the Grossberg–Schmajuk model.

In the START model, each *drive* representation, not every *sensory* representation, has its own spectral cells. Thus, the pathways from all sensory representations that correspond to any given drive representation share the same neurons. This modification greatly reduces the number of cells that are needed to achieve spectral timing of arbitrary conditionable CS–US combinations, as there are many fewer drive representations (e.g., for hunger, thirst, sex, etc.) than there are sensory representations. As in Fig. 2.4, each spectrum is computed in parallel with its drive representation. As the present simulations consider only one type of reinforcer, only one drive representation is depicted. In general, each CS sends an adaptive pathway to every drive representation to which it can be conditioned, as well as adaptive pathways sufficient to sample the corresponding spectral representation. The coordinates of each drive representation and its spectrum encode reinforcement and homeostatic variables. In contrast, the CS-activated pathways to these circuits carry signals that reflect the sensory features of the CSs.

Thus, the fact that different perceptual stimuli may elicit characteristic responses at the cells that represent adaptive timing does not, in itself, imply that these perceptual stimuli are encoded at those cells. Grossberg and Merrill (1992) suggested how hippocampal cells can form an adaptive timing circuit, and how dendrites of hippocampal pyramidal cells can represent a drive-based spectrum that avoids the combinatorial explosion in the dentate–CA3 circuit. This interpretation is used to suggest an explanation of anatomical, neurophysiological, and neuropharmacological data about this hippocampal circuit that are consistent with a functional role in adaptive timing. The Grossberg and Merrill (1992) article also provided computer simulations of data from experiments employing partial reinforcement (Roberts, 1981), ISI shifts (Coleman & Gormezano, 1971), time-averaging due to multiple CSs (Holder & Roberts, 1985), and multiple ISIs (Millenson, Kehoe, & Gormezano, 1977; Smith, 1968).

CONCLUSIONS

The neural model described herein suggests how the hippocampal system and cerebellum may cooperate to control adaptively timed recognition learning, motivated attention, and conditioned responding. The model clarifies how the hippocampal system may combine novelty-based modulation of recognition learning and reinforcement learning with a competence for adaptively timed attention and inhibition of orienting responses. In particular, it suggests how orienting responses may be inhibited by the hippocampal dentate–CA3 timing circuit during the same time interval during which goal-oriented conditioned responses are released by adaptively timed opening of the cerebellar Purkinje cell gate.

The model distinguishes between the micro-timing that is needed to determine how long motivated attention needs to be focused on a single predicted goal event and the macro-timing whereby attention is maintained during the planned performance of a sequence of actions leading to a goal. Both sorts of timing would appear to be at work during many behaviors. A partially developed theory of how they are coordinated clarifies some aspects of the complex pattern of connections that exists between the temporal cortex, frontal cortex, and hippocampal system.

Why the hippocampal system should play a role in spatial orientation is also consistent with this modeling framework. This link is established when one poses the question of how an animal can direct its goal-oriented attentive behaviors among sets of environmental landmarks that vary in their motivational salience. Such a perspective is consistent with the proposal that the hippocampal system can play a role as a cognitive map (Leonard & McNaughton, 1990; O'Keefe, 1990; O'Keefe & Nadel, 1978), suitably defined, without denying its relevance for the control of approach-avoidance behaviors (Amsel, 1993). How to computationally integrate the steering role of reinforcement and motivation into a self-organizing network for spatial orientation remains an open problem. Despite these theoretical gaps, the ART models that have already been developed put mechanistic flesh on the metaphorical bones of declarative memory and procedural memory by articulating new behavioral principles, neural mechanisms, and experimental explanations and predictions that can be used to clarify how a freely moving individual flexibly learns about and acts on valued goal objects in a timely fashion.

ACKNOWLEDGMENTS

Supported in part by the Defense Advanced Research Projects Agency (ONR N00014-92-J-4015) and the Office of Naval Research (ONR N00014-92-J-1309, ONR N00014-95-1-0657, and ONR N00014-95-1-0409). We wish to thank Cynthia E. Bradford and Diana Meyers for their valuable assistance in the preparation of the manuscript.

REFERENCES

Aggleton, J. P. (1993). The contribution of the amygdala to normal and abnormal emotional states. *Trends in Neurosciences, 16*, 328–333.

Albus, J. S. (1971). A theory of cerebellar function. *Mathematical Biosciences, 10,* 25–61.

Amsel, A. (1993) Hippocampal function in the rat: Cognitive mapping or vicarious trial and error. *Hippocampus, 3,* 251–256.

Artola, A., & Singer, W. (1993). Long-term depression of excitatory synaptic transmission and its relationship to long-term potentiation. *Trends in Neurosciences, 16,* 480–487.

Asfour, Y. R. (1994). *Neural networks for multi-sensor fusion and classification.* Unpublished doctoral thesis, Boston University.

Asfour, Y. R., Carpenter, G. A., & Grossberg, S. (1995). Landsat image segmentation using the fuzzy ARTMAP neural network. In *Proceedings of the world congress on neural networks, Washington, I,* 150–156.

Asfour, Y. R., Carpenter, G. A., Grossberg, S., & Lesher, G. (1993). Fusion ARTMAP: A neural network architecture for multi-channel data fusion and classification. In *Proceedings of the world congress on neural networks* (Vol. II, pp. 210–215). Hillsdale, NJ: Lawrence Erlbaum Associates.

Ashton, A. B., Bitgood, S. C., & Moore, J. W. (1969). Auditory differential conditioning of the rabbit nictitating membrane response: III. Effects of US shock intensity and duration. *Psychonomic Science, 15,* 127–128.

Bachelder, I. A., Waxman, A. M., & Seibert, M. (1993). A neural system for mobile robot visual place learning and recognition. In *Proceedings of the world congress on neural networks,* (Vol. I, pp. 512–517). Hillsdale, NJ: Lawrence Erlbaum Associates.

Baddeley, A. D. (1986). *Working memory.* Oxford, England: Clarendon.

Baloch, A. A., & Waxman, A. M. (1991). Visual learning, adaptive expectations, & behavioral conditioning of the mobile robot MAVIN. *Neural Networks, 4,* 271–302.

Barto, A. G., Sutton, R. S., & Anderson, C. W. (1983). Neuron-like adaptive elements that can solve difficult learning problems. *IEEE Transactions, SMC-13,* 834–846.

Berger, T. W., Berry, S. D., & Thompson, R. F. (1986). Role of the hippocampus in classical conditioning of aversive and appetitive behaviors. In R. L. Isaacson & K. H. Pribram (Eds.), *The hippocampus* (Vol. 4, pp. 203–239). New York: Plenum.

Boardman, I., & Bullock, D. (1991). A neural network model of serial order recall from short-term memory. In *Proceedings of the international joint conference on neural networks,* Seattle, (Vol. II, pp. 879–884). Piscataway, NJ: IEEE Service Center.

Boe, E. E. (1966). Effect of punishment duration and intensity on the extinction of an instrumental response. *Journal of Experimental Psychology, 72,* 125–131.

Borozci, G., Storms, L. H., & Broen, W. E. (1964). Response suppression and recovery of responding at different deprivation levels as functions of intensity and duration of punishment. *Journal of Comparative and Physiological Psychology, 58,* 456–459.

Bower, G. H. (1981). Mood and memory. *American Psychologist, 36,* 129–148.

Bower, G. H., Gilligan, S. G., & Monteiro, K. P. (1981). Selectivity of learning caused by adaptive states. *Journal of Experimental Psychology: General, 110,* 451–473.

Bradski, G., Carpenter, G. A., & Grossberg, S. (1992). Working memory networks for learning temporal order with application to 3–D visual object recognition. *Neural Computation, 4,* 270–286.

Bradski, G., Carpenter, G. A., & Grossberg, S. (1994). STORE working memory networks for storage and recall of arbitrary temporal sequences. *Biological Cybernetics, 71,* 469–480.

Bradski, G., & Grossberg, S. (1994). A neural architecture for 3–D object recognition from multiple 2–D views. In *Proceedings of the world congress on neural networks,* (Vol 4, pp. 211–219). Hillsdale, NJ: Lawrence Erlbaum Associates.

Bradski, G., & Grossberg, S. (1995). Fast learning VIEWNET architectures for recognizing 3–D objects from multiple 2–D views. *Neural Networks, 8,* 1053–1080.

Bruner, J. S. (1969). *The pathology of memory.* In G. A. Talland & N. C. Waugh (Eds.). New York: Academic Press.

Bullock, D., Fiala, J. C., & Grossberg, S. (1994). A neural model of timed response learning in the cerebellum. *Neural Networks, 7,* 1101–1114.

Carpenter, G. A., & Grossberg, S. (1987a). A massively parallel architecture for a self-organizing neural pattern recognition machine. *Computer Vision, Graphics, & Image Processing, 37,* 54–115.

Carpenter, G. A., & Grossberg, S. (1987b). ART 2: Stable self-organization of pattern recognition codes for analog input patterns. *Applied Optics, 26,* 4919–4930.

Carpenter, G. A., & Grossberg, S. (1990). ART 3: Hierarchical search using chemical transmitters in self-organizing pattern recognition architectures. *Neural Networks, 3,* 129–152.

Carpenter, G. A., & Grossberg, S. (Eds.). (1991). *Pattern recognition by self-organizing neural networks.* Cambridge, MA: MIT Press.

Carpenter, G. A., & Grossberg, S. (1992). Fuzzy ARTMAP: Supervised learning, recognition, & prediction by a self-organizing neural network. *IEEE Communications Magazine, 30,* 38–49.

Carpenter, G. A., & Grossberg, S. (1993). Normal and amnesic learning, recognition, and memory by a model of cortico-hippocampal interactions. *Trends in Neurosciences, 16,* 131–137.

Carpenter, G. A., Grossberg, S., Markuzon, N., Reynolds, J. H., & Rosen, D. B. (1992). Fuzzy ARTMAP: A neural network architecture for incremental supervised learning of analog multidimensional maps. *IEEE Transactions on Neural Networks, 3*(5), 698–713.

Carpenter, G. A., Grossberg, S., & Reynolds, J. H. (1991). ARTMAP: Supervised real-time learning and classification of nonstationary data by a self-organizing neural network. *Neural Networks, 4,* 565–588.

Carpenter, G. A., Grossberg, S., & Reynolds, J. H. (1995). A fuzzy ARTMAP nonparametric probability estimator for nonstationary pattern recognition. *IEEE Transactions on Neural Networks, 6,* 805–818.

Carpenter, G. A., & Ross, W. D. (1993). ART-EMAP: A neural network architecture for learning and prediction by evidence accumulation. In *Proceedings of the world congress on neural networks: Vol III* (pp. 649–656). Hillsdale, NJ: Lawrence Erlbaum Associates.

Carpenter, G. A., & Tan, A.-H. (1993). Rule extraction, fuzzy ARTMAP, & medical databases. In *Proceedings of the world congress on neural networks: Vol I* (pp. 501–506). Hillsdale, NJ: Lawrence Erlbaum Associates.

Caudell, T. P., Smith, S. D. G., Escobedo, R., & Anderson, M. (1994). NIRS: Large-scale ART-1 neural architectures for engineering design retrieval. *Neural Networks, 7,* 1339–1350.

Church, R. M., Raymond, G. A., & Beauchamp, R. D. (1967). Response suppression as a function of intensity and duration of punishment. *Journal of Comparative and Physiological Psychology, 63,* 39–44.

Cohen, M. A., & Grossberg, S. (1986). Neural dynamics of speech and language coding: Developmental programs, perceptual grouping, & competition for short term memory. *Human Neurobiology, 5,* 1–22.

Cohen, M. A., & Grossberg, S. (1987). Masking fields: A massively parallel neural architecture for learning, recognizing, & predicting multiple groupings of patterned data. *Applied Optics, 26,* 1866–1891.

Cohen, N. J. (1984). Preserved learning capacity in amnesia: Evidence for multiple memory systems. In L. Squire & N. Butters (Eds.), *The neuropsychology of memory* (pp. 83–103). New York: Guilford.

Cohen, N. J., & Squire, L. R. (1980). Preserved learning and retention of a pattern-analyzing skill in amnesia: Dissociation of knowing how and knowing that. *Science, 210,* 207–210.

Coleman, S. R., & Gormezano, I. (1971). Classical conditioning of the rabbit's (*Oryctocagus cuniculus*) nictitating membrane response under symmetrical CS–US interval shifts. *Journal of Comparative and Physiological Psychology, 77,* 447–455.

Cooper, J. R., Bloom, F. E., & Roth, R. H. (1974). *The biochemical basis of neuropharmacology* (2nd ed.). New York: Oxford University Press.

Damasio, A. R. (1989). The brain binds entities and events by multiregional activation from convergence zones. *Neural Computation, 1,* 123–132.

Deadwyler, S. A., West, M. O., & Lynch, G. (1979). Activity of dentate granule cells during learning: Differentiation of perforant path inputs. *Brain Research, 169,* 29–43.

Deadwyler, S. A., West, M. O., & Robinson, J. H. (1981). Entorhinal and septal inputs differentially control sensory-evoked responses in the rat dentate gyrus. *Science, 211,* 1181–1183.

Desimone, R. (1991). Face-selective cells in the temporal cortex of monkeys. *Journal of Cognitive Neuroscience, 3,* 1–8.

Desimone, R., & Ungerleider, L. G. (1989). Neural mechanisms of visual processing in monkeys. In F. Boller & J. Grafman (Eds.), *Handbook of neuropsychology* (Vol. 2, pp. 267–299). Amsterdam: Elsevier.

Dubrawski, A., & Crowley, J. L. (1994). Learning locomotion reflexes: A self-supervised neural system for a mobile robot. *Robotics and Autonomous Systems, 12,* 133–142.

Ebner, T. J., & Bloedel, J. R. (1981). Correlation between activity of Purkinje cells and its modification by natural peripheral stimuli. *Journal of Neurophysiology, 45,* 948–961.

Eichenbaum, H., Otto, T., & Cohen, N. J. (1994). Two functional components of the hippocampal memory system. *Behavioral and Brain Sciences, 17,* 449–472.

Estes, W. K. (1994). *Classification and cognition.* New York: Oxford University Press.

Fiala, J. C., Grossberg, S., & Bullock, D. (1996). Metabotropic glutamate receptor activation in cerebellar Purkinje cells as substrate for adaptive timing of the classically conditioned eye blink response. *Journal of Neuroscience, 16,* 3760–3774.

Fujita, M. (1982a). Adaptive filter model of the cerebellum. *Biological Cybernetics, 45,* 195–206.

Fujita, M. (1982b). Simulation of adaptive modification of the vestibulo-ocular reflex with an adaptive filter model of the cerebellum. *Biological Cybernetics, 45,* 207–214.

Gaffan, D. (1985). Hippocampus: Memory, habit, & voluntary movement. *Philosophical Transactions of the Royal Society of London, B308,* 87–99.

Gaffan, D. (1992). Amnesia for complex naturalistic scenes and for objects following fornix transection in the rhesus monkey. *European Journal of Neuroscience, 4,* 381–388.

Gaffan, D. (1994). Interaction of temporal lobe and frontal lobe in memory. In A. M. Thierry, J. Glowsinski, P. S. Goldman-Rakic, & Y. Christen (Eds.), *Motor and cognitive functions of the prefrontal cortex* (pp. 129–138). New York: Springer-Verlag.

Gilbert, P. F. C., & Thach, W. T. (1977). Purkinje cell activity during motor learning. *Brain Research, 128,* 309–328.

Gjerdingen, R. O. (1990). Categorization of musical patterns by self-organizing neuron-like networks. *Music Perception, 7,* 339–370.

Gochin, P. M., Miller, E. K., Gross, C. G., & Gerstein, G. L. (1991). Functional interactions among neurons in inferior temporal cortex of the awake macaque. *Experimental Brain Research, 84,* 505–516.

Goldman-Rakic, P. S. (1994). The issue of memory in the study of prefrontal function. In A. M. Thierry, J. Glowsinski, P. S. Goldman-Rakic, & Y. Christen (Eds.), *Motor and cognitive functions of the prefrontal cortex* (pp. 112–121). New York: Springer-Verlag.

Goodman, P., Kaburlasos, V., Egbert, D., Carpenter, G. A., Grossberg, S., Reynolds, J. H., Hammermeister, K., Marshall, G., & Grover, F. (1992). Fuzzy ARTMAP neural network prediction of heart surgery mortality. In G. A. Carpenter & S. Grossberg (Eds.), *Neural networks for learning, recognition, & control* (p. 48). Tyngsboro, MA: Wang Institute of Boston University.

Gormezano, I., & Kehoe, E. J. (1984). Associative transfer in classical conditioning to serial compounds. In M. L. Commons, R, J, Herrnstein, & A. R. Wagner (Eds.), *Quantitative analyses of behavior:* Vol. 3. Acquisition. (pp. 297–322). Cambridge, MA: Ballinger.

Gove, A., Grossberg, S., & Mingolla, E. (1995). Brightness perception, illusory contours, & corticogeniculate feedback. *Visual Nueroscience, 12,* 1027–1052.

Govindarajan, K. K., Grossberg, S., Wyse, L. L., & Cohen, M. A. (1994). A neural network model of auditory scene analysis and source segregation. *Technical Report CAS/CNS-TR-94-039,* Boston, MA: Boston University.

Graf, P., Squire, L. R., & Mandler, G. (1984). The information that amnesic patients do not forget. *Journal of Experimental Psychology: Learning, Memory, & Cognition, 10,* 164–178.

Gray, J. A. (1982). *The neuropsychology of anxiety: An enquiry into the functions of the septo-hippocampal system.* New York: Oxford University Press.

Grossberg, S. (1968a). Some nonlinear networks capable of learning a spatial pattern of arbitrary complexity. *Proceedings of the National Academy of Sciences, 59,* 368–372.

Grossberg, S. (1968b). Some physiological and biochemical consequences of psychological postulates. *Proceedings of the National Academy of Sciences, 60,* 758–765.

Grossberg, S. (1969a). On learning of spatiotemporal patterns by networks with ordered sensory and motor components, I: Excitatory components of the cerebellum. *Studies in Applied Mathematics, 48,* 105–132.

Grossberg, S. (1969b). On the production and release of chemical transmitters and related topics in cellular control. *Journal of Theoretical Biology, 22,* 325–364.

Grossberg, S. (1969c). On the serial learning of lists. *Mathematical Biosciences, 4,* 201–253.

Grossberg, S. (1971). On the dynamics of operant conditioning. *Journal of Theoretical Biology, 33,* 225–255.

Grossberg, S. (1972a). A neural theory of punishment and avoidance, I: Qualitative theory. *Mathematical Biosciences, 15,* 39–67.

Grossberg, S. (1972b). A neural theory of punishment and avoidance, II: Quantitative theory. *Mathematical Biosciences, 15*, 253–285.

Grossberg, S. (1973). Contour enhancement, short-term memory, & constancies in reverberating neural networks. *Studies in Applied Mathematics, 52*, 217–257.

Grossberg, S. (1974). Classical and instrumental learning by neural networks. In R. Rosen and F. Snell (Eds.), *Progress in theoretical biology* (pp. 51–141). New York: Academic Press. Reprinted in S. Grossberg, *Studies of mind and brain* (pp. 65–156). Boston: Reidel Press, 1982.

Grossberg, S. (1975). A neural model of attention, reinforcement, & discrimination learning. *International Review of Neurobiology, 18*, 263–327.

Grossberg, S. (1976a). Adaptive pattern classification and universal recoding, I: Parallel development and coding of neural feature detectors. *Biological Cybernetics, 23*, 121–134.

Grossberg, S. (1976b). Adaptive pattern classification and universal recoding, II: Feedback, expectation, olfaction, & illusions. *Biological Cybernetics, 23*, 187–202.

Grossberg, S. (1978). A theory of human memory: Self-organization and performance of sensory-motor codes, maps, & plans. In R. Rosen & F. Snell (Eds.), *Progress in theoretical biology*, Volume 5 (pp. 233–374). New York: Academic Press. Reprinted in S. Grossberg, *Studies of mind and brain* (pp. 498–639). Boston: Reidel Press, 1982.

Grossberg, S. (1980). How does a brain build a cognitive code? *Psychological Review, 87*, 1–51.

Grossberg, S. (1982a). Processing of expected and unexpected events during conditioning and attention: A psychophysiological theory. *Psychological Review, 89*, 529–572.

Grossberg, S. (1982b). *Studies of mind and brain.* Boston: Reidel Press.

Grossberg, S. (1984). Some psychological correlates of a developmental, cognitive and motivational theory. *Annals of the New York Academy of Sciences, 425*, 58–142.

Grossberg, S. (1986). The adaptive self-organization of serial order in behavior: Speech, language, & motor control. In E. C. Schwab & H. C. Nusbaum (Eds.), *Pattern recognition by humans and machines, Vol. 1: Speech perception* (pp. 187–294). New York: Academic Press.

Grossberg, S. (Ed.) (1987). *The adaptive brain*, Volumes I and II. Amsterdam: Elsevier/North-Holland.

Grossberg, S. (Ed.). (1988a). *Neural networks and natural intelligence.* Cambridge, MA: MIT Press.

Grossberg, S. (1988b). Nonlinear neural networks: Principles, mechanisms, & architectures. *Neural Networks, 1*, 17–61.

Grossberg, S., Boardman, I., & Cohen, M. A. (1997). Neural dynamics of variable-rate speech categorization. *Journal of Experimental Psychology: Human Perception and Performance, 23*, 481–503.

Grossberg, S., & Gutowski, W. E. (1987). Neural dynamics of decision making under risk: Affective balance and cognitive-emotional interactions. *Psychological Review, 94*, 300–318.

Grossberg, S., & Kuperstein, M. (1986). *Neural dynamics of adaptive sensory-motor control: Ballistic eye movements.* Amsterdam: Elsevier/North-Holland.

Grossberg, S., & Kuperstein, M. (1989). *Neural dynamics of adaptive sensory-motor control: Expanded edition.* Elmsford, NY: Pergamon.

Grossberg, S., & Levine, D. S. (1987). Neural dynamics of attentionally modulated Pavlovian conditioning: Blocking, inter-stimulus interval, & secondary reinforcement. *Applied Optics, 26*, 5015–5030.

Grossberg, S., & Merrill, J. W. L. (1992). A neural network model of adaptively timed reinforcement learning and hippocampal dynamics. *Cognitive Brain Research, 1*, 3–38.

Grossberg, S., & Merrill, J. W. L. (1996). The hippocampus and cerebellum in adaptively timed learning, recognition, & movement. *Journal of Cognitive Neuroscience, 8*, 257–277.

Grossberg, S., & Pepe, J. (1970). Schizophrenia: Possible dependence of associational span, bowing, & primacy versus recency on spiking threshold. *Behavioral Science, 15*, 359–362.

Grossberg, S., & Pepe, J. (1971). Spiking threshold and overarousal effects in serial learning. *Journal of Statistical Physics, 3*, 95–125.

Grossberg, S., Roberts, K., Aguilar, M., & Bullock, D. (1997). A neural model of multimodal adaptive saccadic eye movement control by superior colliculus. *Journal of Neuroscience, 38*, 91–99.

Grossberg, S., & Schmajuk, N. A. (1987). Neural dynamics of Pavlovian conditioning: Conditioned reinforcement, inhibition, & opponent processing. *Psychobiology, 15*, 195–240.

Grossberg, S., & Schmajuk, N. A. (1989). Neural dynamics of adaptive timing and temporal discrimination during associative learning. *Neural Networks, 2*, 79–102.

Grossberg, S., & Somers, D. (1991). Synchronized oscillations during cooperative feature linking in a cortical model of visual perception. *Neural Networks, 4*, 453–466.

Grossberg, S., & Stone, G. O. (1986a). Neural dynamics of word recognition and recall: Attentional priming, learning, and resonance. *Psychological Review, 93*, 46–74.

Grossberg, S., & Stone, G. O. (1986b). Neural dynamics of attention switching and temporal order information in short term memory. *Memory and Cognition, 14*, 451–468.

Ham, F., & Han, S. (1993). Quantitative study of ARS complex using fuzzy ARTMAP and MIT/BIH arrythmia database. In *Proceedings of the world congress on neural networks* (Vol I, pp. 207–211). Hillsdale, NJ: Lawrence Erlbaum Associates.

Harries, M. H., & Perrett, D. I. (1991). Visual processing of faces in temporal cortex: Physiological evidence for a modular organization and possible anatomical correlates. *Journal of Cognitive Neuroscience, 3*, 9–24.

Harvey, R. M. (1993). Nursing diagnostics by computers: An application of neural networks. *Nursing Diagnostics, 4*, 26–34.

Henneman, E. (1957). Relation between size of neurons and their susceptibility to discharge. *Science, 26*, 1345–1347.

Henneman, E. (1985). The size principle: A deterministic output emerges from a set of probabilistic connections. *Journal of Experimental Biology, 115*, 105–112.

Hoehler, F. K., & Thompson, R. F. (1980). Effects of the interstimulus (CS–UCS) interval on hippocampal unit activity during classical conditioning of the nictitating membrane response of the rabbit (*Oryctolagus cuniculus*). *Journal of Comparative and Physiological Psychology, 94*, 201–215.

Holder, M. D., & Roberts, S. (1985). Comparison of timing and classical conditioning. *Journal of Experimental Psychology: Animal Behavior Processes, 11*, 172–193.

Ito, M. (1984). *The cerebellum and neural control*. New York: Raven.

Jones, R., & Keck, M. J. (1978). Visual evoked response as a function of grating spatial frequency. *Investigative Ophthalmology and Visual Science, 17*, 652–659.

Kamin, L. J. (1969). Predictability, surprise, attention, & conditioning. In B. A. Campbell & R. M. Church (Eds.), *Punishment and aversive behavior.* New York: Appleton-Century-Crofts.

Kandel, E. R., & O'Dell, T. J. (1992). Are adult learning mechanisms also used for development? *Science, 258*, 243–245.

Kasperkiewicz, J., Racz, J., & Dubrawski, A. (1995). HPC strength prediction using artificial neural network. *Journal of Computing in Civil Engineering, 9*, 279–284.

Keehn, J. D. (1963). Effect of shock duration on Sidman avoidance response rates. *Psychology Reports, 13*, 852.

Kehoe, E. J., Gibbs, C. M., Garcia, E., & Gormenzano, I. (1979). Associative transfer and stimulus selection in classical conditioning of the rabbit's nictitating membrane response to serial compound CS's. *Journal of Experimental Psychology: Animal Behavior Processes, 5*, 1–57.

Kehoe, E. J., Horne, P. S., Macrae, M., & Horne, A. J. (1993). Real-time processing of serial stimuli in classical conditioning of the rabbit's nictitating membrane response. *Journal of Experimental Psychology: Animal Behavior Processes, 19*, 265–283.

Kehoe, E. J., Marshall-Goodell, B., & Gormenzano, I. (1987). Differential conditioning of the rabbit's nictitating membrane response to serial compound stimuli. *Journal of Experimental Psychology: Animal Behavior Processes, 13*, 17–30.

Kehoe, E. J., & Morrow, L. D. (1984). Temporal dynamics of the rabbit's nictitating membrane response in serial compound conditioned stimuli. *Journal of Experimental Psychology: Animal Behavior Processes, 10*, 205–220.

Keyvan, S., Durg, A., & Rabelo, L. (1993). Application of artificial neural networks for development of diagnostic monitoring system in nuclear plants. *Transactions of the American Nuclear Society, 1*, 515–522.

Knight, R. T. (1994). Attention regulation and human prefrontal cortex. In A. M. Thierry, J. Glowsinksi, P. S. Goldman-Rakic, & Y. Christen (Eds.), *Motor and cognitive functions of the prefrontal cortex* (pp. 160–173). New York: Springer-Verlag.

Kohonen, T. (1984). *Self-organization and associative memory.* New York: Springer-Verlag.

Kuno, M. (1995). *The synapse: Function, plasticity, & neurotrophism.* New York: Oxford University Press.

Leonard, B., & McNaughton, B. L. (1990). Spatial representation in the rat: Conceptual, behavioral, & neurophysiological perspectives. In R. P. Kesner & D. S. Olton, (Eds.), *Neurobiology of comparative cognition* (pp. 363–422). Hillsdale, NJ: Lawrence Erlbaum Associates.

Levy, W. B. (1985). Associative changes at the synapse: LTP in the hippocampus. In W. B. Levy, J. Anderson, & S. Lehmkuhle (Eds.), *Synaptic modification, neuron selectivity, & nervous system organization* (pp. 5–33). Hillsdale, NJ: Lawrence Erlbaum Associates.

Levy, W. B., & Desmond, N. L. (1985). The rules of elemental synaptic plasticity. In W. B. Levy, J. Anderson, & S. Lehmkuhle (Eds.), *Synaptic modification, neuron selectivity, & nervous system organization* (pp. 105–121). Hillsdale, NJ: Lawrence Erlbaum Associates.

Lisberger, S. G. (1988). The neural basis for motor learning in the vestibulo-ocular reflex in monkeys. *Trends in Neurosciences, 11*, 147–152.

Logothetis, N., Pauls, J., Buelthoff, H., & Poggio, T. (1994). View-dependent object recognition by monkeys. *Current Biology, 4*, 401–414.

Lynch, G., McGaugh, J. L., & Weinberger, N. M. (Eds.). (1984). *Neurobiology of learning and memory.* New York: Guilford.

Malsburg, C. von der (1973). Self-organization of orientation sensitive cells in the striate cortex. *Kybernetik, 14*, 85–100.

Maricq, A. V., Roberts, S., & Church, R. M. (1981). Methamphetamine and time estimation. *Journal of Experimental Psychology: Animal Behavior Processes, 7*, 18–30.

Marr, D. (1969). A theory of cerebellar cortex. *Journal of Physiology, 202*, 437–470.

Marr, D. (1971). Simple memory: A theory for archicortex. *Philosophical Transactions of the Royal Society of London, B-262*, 23–81.

McClelland, J. L., McNaughton, B. L., & O'Reilly, R. C. (1994). *Why are there complementary learning systems in the hippocampus and neocortex: Insights from the successes and failures of connectionist models of learning and memory.* (Tech. Rep. PDP. CNS.94.1). Pittsburgh, PA: Carnegie Mellon University.

Meck, W. H., & Church, R. M. (1987). Cholinergic modulation of the content of temporal memory. *Behavioral Neuroscience, 101*, 457–464.

Metha, B., Vij, L., & Rabelo, L. (1993). Prediction of secondary structures of proteins using fuzzy ARTMAP. In *Proceedings of the world congress on neural networks: Vol. I* (pp. 228–232). Hillsdale, NJ: Lawrence Erlbaum Associates.

Millenson, J. R., Kehoe, E. J., & Gormezano, I. (1977). Classical conditioning of the rabbit's nictitating membrane response under fixed and mixed CS–US intervals. *Learning and Motivation, 8*, 351–366.

Miller, E. K., Li, L., & Desimone, R. (1991). A neural mechanism for working and recognition memory in inferior temporal cortex. *Science, 254*, 1377–1379.

Milner, P. (1989). A cell assembly theory of hippocampal amnesia. *Neuropsychologia, 27*, 23–30.

Mishkin, M. (1978). Memory in monkeys severely impaired by combined but not separate removal of the amygdala and hippocampus. *Nature, 273*, 297–298.

Mishkin, M. (1982). A memory system in the monkey. *Philosophical Transactions of the Royal Society of London, B-298*, 85–95.

Mishkin, M. (1993). Cerebral memory circuits. In T. A. Poggio & D. A. Glaser (Eds.), *Exploring brain functions: Models in neuroscience* (pp. 113–125). New York: Wiley.

Mishkin, M., & Appenzeller, T. (1987). The anatomy of memory. *Scientific American, 256*, 80–89.

Mishkin, M., & Delacour, J. (1975). An analysis of short-term visual memory in the monkey. *Journal of Experimental Psychology: Animal Behavior Processes, 1*, 326–334.

Mishkin, M., Malamut, B., & Bachevalier, J. (1984). Memories and habits: Two neural systems. In J. L. McGaugh, G. Lynch, & N. Weinberger (Eds.), *The neurobiology of learning and memory* (pp. 287–296). New York: Guilford.

Moya, M. M., Koch, M. W., & Hostetler, L. D. (1993). One-class classifier networks for target recognition applications. In *Proceedings of the world congress on neural networks: Vol 3* (pp. 797–801). Hillsdale, NJ: Lawrence Erlbaum Associates.

Musselwhite, M. J., & Jeffreys, D. A. (1985). The influence of spatial frequency on the reaction times and evoked potentials recorded to grating pattern stimuli. *Vision Research, 25*, 1545–1555.

Nowak, A. J., & Berger, T. W. (1992). Functional three-dimensional distribution of entorhinal projections to dentate granule cells of the *in vivo* rabbit hippocampus. *Society for Neuroscience Abstracts, 18,* 321.

O'Keefe, J. (1990). A computational theory of the hippocampal cognitive map. In J. Storm-Mathisen, J. Zimmer, & O. P. Ottersen (Eds.), *Progress in brain research* (pp. 301–312). Amsterdam: Elsevier.

O'Keefe, J., & Nadel, L. (1978). *The hippocampus as a cognitive map.* New York: Oxford University Press.

Optican, L. M., & Robinson, D. A. (1980). Cerebellar-dependent adaptive control of primate saccadic system. *Journal of Neurophysiology, 44,* 108–176.

Otto, T., & Eichenbaum, H. (1992). Neuronal activity in the hippocampus during delayed non-match to sample performance in rats: Evidence for hippocampal processing in recognition memory. *Hippocampus, 2,* 323–334.

Parker, D. M., & Salzen, E. A. (1977a). Latency changes in the human visual evoked response to sinusoidal gratings. *Vision Research, 17,* 1201–1204.

Parker, D. M., & Salzen, E. A. (1977b). The spatial selectivity of early and late waves within the human visual evoked response. *Perception, 6,* 85–95.

Parker, D. M., Salzen, E. A., & Lishman, J. R. (1982a). Visual-evoked responses elicited by the onset and offset of sinusoidal gratings: Latency, waveform, & topographic characteristics. *Investigative Ophthalmology and Visual Sciences, 22,* 657–680.

Parker, D. M., Salzen, E. A., & Lishman, J. R. (1982b). The early waves of the visual evoked potential to sinusoidal gratings: Responses to quadrant stimulation as a function of spatial frequency. *Electroencephalography and Clinical Neurophysiology, 53,* 427–435.

Perrett, D. I., Mistlin, A. J., & Chitty, A. J. (1987). Visual cells responsive to faces. *Trends in Neurosciences, 10,* 358–364.

Perrett, S. P., Ruiz, B. P., & Mauk, M. D. (1993). Cerebellar cortex lesions disrupt learning-dependent timing of conditioned eyelid responses. *Journal of Neuroscience, 13,* 1708–1718.

Plant, G. T., Zimmern, R. L., & Durden, K. (1983). Transient visually evoked potentials to the pattern reversal and onset of sinusoidal gratings. *Electroencephalography and Clinical Neurophysiology, 56,* 147–158.

Posner, M. I., & Keele, S. W. (1968). On the genesis of abstract ideas. *Journal of Experimental Psychology, 77,* 353–363.

Posner, M. I., & Keele, S. W. (1970). Retention of abstract ideas. *Journal of Experimental Psychology, 83,* 304–308.

Pribram, K. H. (1986). The hippocampal system and recombinant processing. In R. L. Isaacson & K. H. Pribram (Eds.), *The hippocampus: Vol. 4* (pp. 329–370). New York: Plenum.

Rauschecker, J. P., & Singer, W. (1979). Changes in the circuitry of the kitten's visual cortex are gated by postsynaptic activity. *Nature, 280,* 58–60.

Rickert, E. J., Bennett, T. L., Lane, P. L., & French, J. (1978). Hippocampectomy and the attenuation of blocking. *Behavioral Biology, 22,* 147–160.

Roberts, S. (1981). Isolation of an internal clock. *Journal of Experimental Psychology: Animal Behavior Processes, 7,* 242–268.

Ryle, G. (1949). *The concept of mind.* Hutchinson Press.

Schmajuk, N. A., Spear, N. E., & Isaacson, R. L. (1983). Absence of overshadowing in rats with hippocampal lesions. *Physiological Psychology, 11,* 59–62.

Schwartz, E. L., Desimone, R., Albright, T., & Gross, C. G. (1983). Shape recognition and inferior temporal neurons. *Proceedings of the National Academy of Sciences, 80,* 5776–5778.

Scoville, W. B., & Milner, B. (1957). Loss of recent memory after bilateral hippocampal lesion. *Journal of Neurology, Neurosurgery, & Psychiatry, 20,* 11–21.

Seibert, M., & Waxman, A. M. (1991). Learning and recognizing 3–D objects from multiple views in a neural system. In H. Wechsler (Ed.), *Neural networks for perception: Vol. 1* (pp. 426–444). New York: Academic Press.

Seibert, M., & Waxman, A. M. (1992). Adaptive 3–D object recognition from multiple views. *IEEE Transactions on Pattern Analysis and Machine Intelligence, 14,* 107–124.

Sillito, A. M., Jones, H. E., Gerstein, G. L., & West, D. C. (1994). Feature-linked synchronization of thalamic relay cell firing induced by feedback from the visual cortex. *Nature, 369,* 479–482.

Singer, W. (1983). Neuronal activity as a shaping factor in the self-organization of neuron assemblies. In E. Basar, H. Flohr, H. Haken, & A. J. Mandell (Eds.), *Synergetics of the brain* (pp. 89–101). New York: Springer-Verlag.

Skrandies, W. (1984). Scalp potential fields evoked by grating stimuli: Effects of spatial frequency and orientation. *Electroencephalography and Clinical Neurophysiology, 58,* 325–332.

Smith, M. C. (1968). CS–US interval and US intensity in classical conditioning of the rabbit's nictitating membrane response. *Journal of Comparative and Physiological Psychology, 3,* 679–687.

Solomon, P. R. (1977). The role of hippocampus in blocking and conditioned inhibition of the rabbit's nictitating membrane response. *Journal of Comparative and Physiological Psychology, 91,* 407–417.

Spitzer, H., Desimone, R., & Moran, J. (1988). Increased attention enhances both behavioral and neuronal performance. *Science, 240,* 338–340.

Squire, L. R., & Butters, N. (Eds.). (1984). *Neuropsychology of memory.* New York: Guilford.

Squire, L. R., & Cohen, N. J. (1984). Human memory and amnesia. In G. Lynch, J. McGaugh, & N. M. Weinberger (Eds.), *Neurobiology of learning and memory* (pp. 3–64). New York: Guilford Press.

Squire, L. R., & Zola-Morgan, S. (1991). The medial temporal lobe memory system. *Science, 253,* 1380–1386.

Staddon, J. E. R. (1983). *Adaptive behavior and learning.* New York: Cambridge University Press.

Strouthes, A. (1965). Effect of CS-onset, UCS-termination delay, UCS duration, CS-onset interval and number of CS–UCS pairings on conditioned fear response. *Journal of Experimental Psychology, 69,* 287–291.

Suzuki, Y. (1995). Self-organizing QRS-wave recognition in ECG using neural networks. *IEEE Transactions on Neural Networks, 6,* 1469–1477.

Suzuki, Y., Abe, Y., & Ono, K. (1994). Self-organizing QRS wave recognition system in ECG using ART 2. In *Proceedings of the world congress on neural networks: Vol 4* (pp. 39–42). Hillsdale, NJ: Lawrence Erlbaum Associates.

Thompson, R. F. (1988). The neural basis of basic associative learning of discrete behavioral responses. *Trends in Neurosciences, 11,* 152–155.

Thompson, R. F., Barchas, J. D., Clark, G. A., Donegan, N., Kettner, R. E., Lavond, D. G., Madden, J., Mauk, M. D., & McCormick, D. A. (1984). Neuronal substrates of associative learning in the mammalian brain. In D. L. Aldon & J. Farley (Eds.),

Primary neural substrates of learning and behavioral change (pp. 71–99). New York: Cambridge University Press.

Thompson, R. F., Clark, G. A., Donegan, N. H., Lavond, G. A., Lincoln, D. G., Maddon, J., Mamounas, L. A., Mauk, M. D., & McCormick, D. A. (1987). Neuronal substrates of discrete, defensive conditioned reflexes, conditioned fear states, & their interactions in the rabbit. In I. Gormenzano, W. F. Prokasy, & R. F. Thompson (Eds.), *Classical conditioning* (3rd ed., pp. 371–399). Hillsdale, NJ: Lawrence Erlbaum Associates.

Vassilev, A., Manahilov, V., & Mitov, D. (1983). Spatial frequency and pattern onset-offset response. *Vision Research, 23,* 1417–1422.

Vassilev, A., & Strashimirov, D. (1979). On the latency of human visually evoked response to sinusoidal gratings. *Vision Research, 19,* 843–846.

Warrington, E. K., & Weiskrantz, L. (1974). The effect of prior learning on subsequent retention in amnesic patients. *Neuropsychology, 12,* 419–428.

Wienke, D., Xie, P., & Hopke, P. K. (1994). An adaptive resonance theory based artificial neural network (ART 2A) for rapid identification of airborne particle shapes from their scanning electron microscopy images. *Chemometrics and Intelligent Laboratory Systems, 25,* 367–387.

Wilkie, D. M. (1987). Stimulus intensity affects pigeons' timing behavior: Implications for an internal clock model. *Animal Learning and Behavior, 15,* 35–39.

Williamson, S. J., Kaufman, I., & Brenner, D. (1978). Latency of the neuromagnetic response of the human visual cortex. *Vision Research, 18,* 107–110.

Zola-Morgan, S. M., & Squire, L. R. (1990). The primate hippocampal formation: Evidence for a time-limited role in memory storage. *Science, 250,* 288–290.

3

Can the Whole Be Something Other Than the Sum of Its Parts?

E. James Kehoe
University of New South Wales

Huge strides have been made in our empirical and theoretical understanding of associative learning over the past 25 years. Much of the experimentation has used a simple procedure in which a compound of two stimulus elements, often a tone and a light, signals the reinforcer. By varying the conditions of training for the compound and its elements in relatively small ways, a startlingly diverse set of outcomes has appeared. This empirical diversity has fostered an equally large theoretical diversity. At one time or another, each phenomenon has had its own special theory. Although such special theories are useful heuristic devices, many of them seem redundant with one another.

The progress made in absorbing compound–stimulus phenomena into a unified theory relying on a few basic principles is reviewed. In particular, this chapter traces the evolution of two theoretical issues concerning the integration of multiple stimuli. First, there is a long-standing issue as to whether compound stimuli are processed in an atomistic, element-by-element fashion or a configural, holistic fashion. Second, among configural hypotheses, there are a variety of conceptions as to how the configural process emerges from the conjunction of the stimulus elements. Although theoretical development is not complete, both issues are being resolved by the use of connectionist models. Most importantly, these models rely on atomistic summation as the central principle in both the expression and acquisition of associative strength.

This chapter is divided into four sections. The first section briefly defines the empirical scope of the chapter. The second section details the key experimental paradigms. In brief, they are (a) stimulus compounding, in which training is conducted primarily with the individual elements, (b) compound conditioning, in which a compound of the elements receives extensive training, and (c) explicit differentiation between a compound and its elements. The third section describes the theories of stimulus integration, focusing particularly on the role of summation principles. The major classes of theory include (a) configural theory, (b) quasi-atomistic hypotheses, and (c) connectionist models. The fourth section focuses on a particular connectionist model that has had some success in explaining a wide range of phenomena obtained using compound stimuli (Kehoe, 1988).

EMPIRICAL SCOPE

The research and theory described here have revolved around classical conditioning of the rabbit's nictitating membrane (NM) response (Gormezano, 1966). The NM preparation has yielded robust findings across a wide range of manipulations using compound stimuli. Moreover, the quantitative models for compound conditioning have been especially calibrated to the results of the NM experiments.

These experiments have used a consistent set of parameters. Specifically, the conditioned stimuli (CSs) have been a 1000-Hz tone and a 20-Hz flashing light. Their durations and intensities have been manipulated across experiments. The unconditioned stimulus (US) has been an electrical current of 2- or 3-mA intensity and 50- or 100-ms duration delivered to a region of skin 15 mm lateral to the rabbit's right eye. Each training session has contained 40 to 90 trials delivered at intertrial intervals of 40 to 60 s.

EXPERIMENTAL PARADIGMS

Stimulus Compounding

Primary evidence for atomistic summation comes from studies that have used a stimulus compounding procedure. In that procedure, the two stimulus elements receive separate reinforced training and are periodically tested by presenting them simultaneously. The usual outcome is *summation*, meaning that the compound yields a higher level of responding than either element (e.g., Couvillon & Bitterman, 1982; Kehoe & Graham, 1988; Weiss, 1972, 1978). As may be apparent, this definition refers only to the ordinal relationships among response levels, not to whether the response levels of the elements add up strictly to that of the compound.

Where summation is defined in only an ordinal fashion, there are two alternative explanations. According to an atomistic hypothesis, the summation outcome reflects the summation of the associative strengths of the elements (Kehoe & Graham, 1988; Weiss, 1972). In contrast, configural hypotheses contend that compounding produces a perceptual interaction between the elements. This interaction is often seen as a fusion that supplants the encoding of the individual elements (Heinemann & Chase, 1975; Kehoe & Gormezano, 1980; Pearce, 1987). Consequently, responding to the compound stimulus is explained as the summation of generalized associative strengths from the encodings of the separate elements. Where there is substantial generalization, responding to the compound could exceed those of the elements, yielding ordinal summation.

Although ordinal summation can be explained by atomistic or configural hypotheses, stimulus compounding in the rabbit NM preparation strongly favors an atomistic hypothesis. Responding follows a precise quantitative relationship. That is to say, the percentage conditioned responses (CRs) to the test compound (Pc) is well predicted by the statistical sum of the percentage CRs to the separate elements (Pa, Pb). The specific formula is Pa + Pb -[PaPb].

Figure 3.1 shows an example of stimulus compounding and the application of the statistical summation rule in the rabbit NM preparation (Kehoe, Horne, Horne, & Macrae, 1994). On each day of training, the rabbits received 20 pairings of a tone with the US (T+), 20 pairings of a light with the US (L+), and 2 test trials with a compound of the tone and light presented simultaneously (TL). Figure 3.1 depicts the mean percent CRs across days of stimulus compounding on T+, L+, and TL trials. In addition, the figure shows the mean predicted levels of responding to the compound when the statistical summation rule (Pt + Pl - [PtPl]) was applied to the responding of individual subjects on T+ and L+ trials.

As can be seen in Fig. 3.1, responding to the compound was closely paralleled by the statistical sum of responding to the elements throughout training. To determine how well the predictions from statistical summation agreed with actual responding to the compound, a correlation coefficient was calculated using each subjects predicted and actual response averaged over all days of acquisition. These correlation coefficients confirmed the correspondence seen in the group means. Specifically, the subject-by-subject correlation was .83, $p < .01$ ($N = 16$), which agrees with previous results in which the correlations have ranged from .68 to .99 (Kehoe, 1982, 1986; Kehoe & Graham, 1988; Kehoe, Horne, Horne, & Macrae, 1994). There was, however, a constant error, about 5%, in the direction of an underprediction. This small oversummation in actual responding is also a frequent feature of stimulus compounding in the NM preparation.

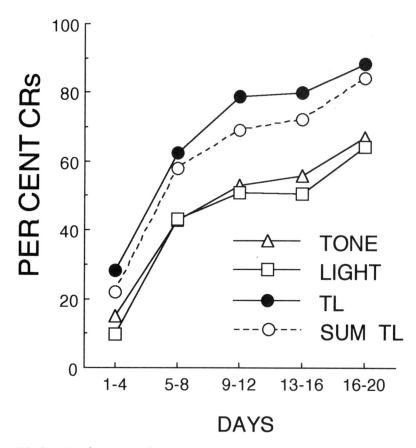

FIG. 3.1. Stimulus compounding. Mean percent CRs on tone (T), light (L), and tone +light compound (TL) trials. The figure also shows the summated responding to the individual stimuli (SUM TL) as combined according to the statistical rule, Pt + Pl - PtPl (Kehoe et al., 1994).

This precise additive relationship suggests that individual elements are represented in an atomistic fashion and that their associative strengths summate to determine responding. Moreover, the consistency of the rule across the entire response scale suggests that there is a relatively linear mapping between underlying associative strength and overt responding (cf. Anderson, 1982, pp. 4–5, 56–59). Conversely, the statistical summation rule contradicts the expectations from pure configural hypotheses. Unless there is complete generalization from the elements to their fused configural stimulus, the pure configural hypotheses would expect the statistical sum to overpredict the observed level of responding to the compound. In fact, the small systematic error is in the wrong direction, that is, an underprediction.

Compound Conditioning

In procedural terms, compound conditioning is the reverse of stimulus compounding. That is, in compound conditioning, training is conducted by pairing the compound with the US. Testing entails separate presentations of the elements. This reversal has produced outcomes that cannot be predicted by a statistical summation rule. In some cases, the levels of responding to the individual stimuli are so low that they literally do not add up to the level of responding to the compound. Such outcomes have seemed to require special principles, in particular, selective attention, associative competition, and configural fusion. Moreover, these outcomes are not haphazard. In the rabbit NM preparation, they occur systematically as a function of stimulus intensity and duration.

Unidirectional Overshadowing. Overshadowing is the oldest outcome of compound conditioning and was first identified by Pavlov (1927). As the name implies, CR acquisition to one stimulus is impaired by a more salient stimulus. The overshadowed stimulus itself does not need to be a weak stimulus. In the rabbit NM preparation, for example, CR acquisition to a salient stimulus can be hindered by training with another, even more salient stimulus (Kehoe, 1982). Figure 3.2 shows an example of such overshadowing. Specifically, a light (L) was held constant, while the intensity of a tone (T) was varied over values of 85, 89, and 93 dB sound pressure level (SPL). Three groups of rabbits were given compound conditioning in which the light plus one of the tones was paired with the US (LT+). These groups were designated as LT85, LT89, and LT93, respectively. In addition, a fourth group was trained only with the light (Group L).

Figure 3.2 shows the mean percentage CRs on test trials of the light that were presented during each training session. Inspection of the figure reveals that Group L showed rapid CR acquisition to the light and reached levels of virtually 100% CRs. Likewise, Group LT85 showed a high level of CR acquisition to the light. However, as the tone intensity was increased across Groups LT89 and LT93, CR acquisition to the light showed a progressive impairment. With regard to the tones themselves, there was no evidence of any reciprocal effect of the light. CR acquisition to the tone in all the compound groups reached levels of 90% CRs.

Reciprocal Overshadowing and Spontaneous Configuration. In the case of unidirectional overshadowing, CR acquisition to only one element is impaired. However, it is possible to impair and even eliminate CR acquisition to both elements. Such results have been described in two ways, namely as reciprocal overshadowing (Mackintosh, 1976) or spontaneous configuration (cf. Bellingham & Gillette, 1981; Razran, 1965). Regardless of its description,

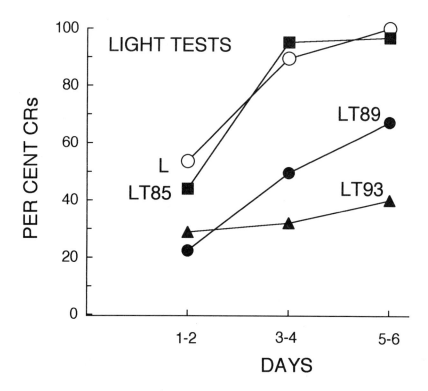

FIG. 3.2. Overshadowing. Mean percent CRs on light test trials in groups that were trained with light (L), light + 85-dB tone (LT85), light + 89-dB tone (LT89), and light + 93-dB tone (LT93) (Kehoe, 1982).

the rabbit NM preparation produces this outcome when the CS duration and CS–US interval are increased jointly to values greater than 400 ms (Kehoe, 1982, 1986; Kehoe & Schreurs, 1986).

Figure 3.3 shows an example of how unidirectional overshadowing can be converted into reciprocal overshadowing/spontaneous configuration (Kehoe, 1986). Three groups of subjects were trained with compounds in which the CS durations were 300, 800, and 1,300 ms. The CS–US interval equalled the CS duration. These groups were labelled as Groups C300, C800, and C1300, respectively. In addition, there were three control groups to determine the rate of CR acquisition for each CS duration. These control groups each received training with separate T+ and L+ trials. These groups were labelled as Groups S300, S800, and S1300 according to their CS durations.

As can be seen in Fig. 3.3, Group C300 showed some unidirectional overshadowing. That is, CR acquisition to the light in Group C300 was slower than the light in Group S300. Nevertheless, at the end of training, responding

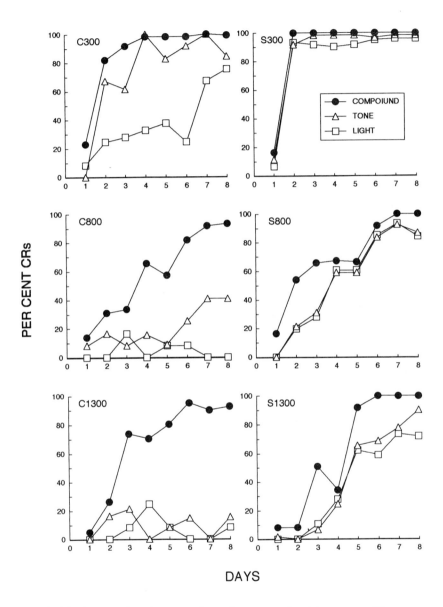

FIG. 3.3. Compound conditioning and stimulus compounding. Mean percent CRs on tone +light compound, tone, and light trials in six groups given training with either compounds (C) or separate tone and light trials (S). The CS duration and CS–US interval covaried across values of 300, 800, and 1,300 ms (Kehoe, 1986).

to the light had reached a level above 70% CRs and was still rising. CR acquisition to the tone was rapid in both Groups C300 and S300. For the 800-ms and 1300-ms CS durations, CR acquisition to both elements showed large impairments. In Group C800, responding on TL+ trials reached a level near 100% CRs, responding to the tone rose very slowly, and responding to the light was negligible. In Group C1300, the pattern was even clearer. Neither the tone nor the light evoked more than minimal responding, but responding to the compound remained intact.

In Groups C800 and C1300, the low level of responding to the separate stimuli depended solely on the reinforced training with the compound and not on any contribution by test presentations of the tone and light. Specifically, Kehoe and Schreurs (1986) conducted training with an 800-ms compound (TL+) in four groups of rabbits. Two groups received test presentations of the tone and light throughout training. In one of those groups, the stimuli were presented alone, and, in the other group, the stimuli were each paired with the US. In the remaining two groups, test trials were given only at the end of training. In all groups, the same pattern of responding appeared, namely, a high level of responding on TL+ trials (> 70% CRs), a moderate level of responding on tone trials (< 30% CRs), and negligible responding on light trials (< 10% CRs).

Stimulus Compounding and Compound Conditioning. The ability to eliminate responding to the individual elements in compound conditioning suggests that there is a large discontinuity between compound conditioning and stimulus compounding. In particular, some configural process may be engaged in compound conditioning that supersedes the summation process apparent in stimulus compounding. In operational terms, however, compound conditioning and stimulus compounding can be viewed as two extremes along a continuum of reinforced exposure to a compound (TL+) versus its elements (T+, L+) . Accordingly, there are intermediate possibilities, in which TL+ trials are intermixed with T+ and L+ trials.

To determine the degree of continuity or discontinuity between the two procedures, Kehoe (1986) intermixed TL+, T+, and L+ trials in various proportions. There were five groups, designated as C0, C10, C20, C30, and C40. Group C0 was a stimulus compounding group. It received 40 T+ and 40 L+ trials per session. Groups C10, C20, and C30 were the mixed groups. Specifically, in each session, Group C10 received 10 TL+, 30 T+, and 30 L+ trials; Group C20 received 20 TL+, 20 T+, and 20 L+ trials; Group C30 received 30 TL+, 10 T+, and 10 L+ trials. Finally, Group C40 was a compound conditioning group and received 40 TL+ trials per session. All groups also received three test trials each of TL, T, and L per session. All stimuli had a duration of 800 ms, and, on reinforced trials, the CS–US interval was 800 ms.

Figure 3.4 shows the overall mean level of responding on TL, T, and L trials as a function of the group variable. The figure also contains a fourth curve that represents the statistical sum of responding to T and L. Responding on TL trials was relatively constant across groups, hovering around 60% CRs. In contrast, responding on T and L trials declined as the proportion of TL+ trials increased. For Groups C0, C10, and C20, the summation formula slightly underpredicted the level of responding on TL trials. For Groups C30 and C40, however, the summation formula yielded progressively larger underpredictions.

As the proportion of TL+ trials in the mixture increased, statistical summation explained less of the responding to the compound. By a process of elimination, it appears that a process engaged on TL+ trials commanded an increasing share of the total responding. Nevertheless, there was no obvious discontinuity in the trends across mixtures. Consequently, the process engaged

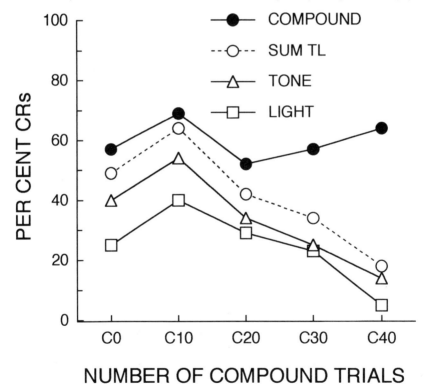

FIG. 3.4. Mixed compound and single element training. Mean percent CRs on tone+light compound, tone, and light trials as a function of the number of compound trials contained in the daily mixture of training trials. The figure also shows the statistical sum of responding to the individual elements (SUM TL) (Kehoe, 1986).

on TL+ trials may operate in concert with summation rather than superseding it.

Selective Attention and Associative Competition.

Overshadowing has been explained traditionally in atomistic terms by appealing to competition between the stimuli for access to either the subject's attentional resources or its associative apparatus. According to attentional accounts, the stronger stimulus captures the subject's fixed attentional resources to the detriment of the weaker stimulus (e.g., Sutherland & Mackintosh, 1971). Newer attentional hypotheses assume that the trade-offs between contending stimuli are more flexible (e.g., Mackintosh, 1975; Pearce & Hall, 1980; Schmajuk & Moore, 1988). The notion of competition between stimuli has also been captured by incorporating a summation rule into an associative learning rule. Specifically, Rescorla and Wagner (1972) contended that, on each reinforced presentation of a compound, changes in associative strength for each element depend on the summated associative strengths of all the elements in the compound. If one element is more intense than the others, it will gain associative strength quickly, leaving less for other stimuli on future trials.

These atomistic accounts face considerable difficulty in explaining the low levels of responding to elements at the longer stimulus durations. Attentional accounts of all varieties can explain both unidirectional and reciprocal overshadowing, but they would still expect that the attentional resources would be allocated to ensure some learning to be apparent. In similar terms, the Rescorla–Wagner model always expects some reciprocal overshadowing to occur. For example, two elements with equal salience would split the available associative strength between them. Thus, responding to each element would be less than that of the compound. Nevertheless, both elements would have enough associative strength to account for the total associative strength of the compound. Thus, neither of the atomistic approaches would predict the near absence of responding to the elements in the face of high levels of responding to their compound.

Atomistic models could be defended by arguing that the mapping between associative strength and responding is nonlinear. Specifically, responding to the compound may reflect summation by subthreshold associative strengths of the elements. However, this hypothesis encounters a problem in explaining stimulus compounding in the rabbit NM preparation. In stimulus compounding, there was no evidence for summation of subthreshold associative strengths. Had there been, there would have been a period early in training in which responding on compound trials rose suddenly while responding to the elements remained negligible. In fact, the summation was constant throughout training. That is, responding to the compound rose in a fixed relationship to the elements.

Explicit Differentiation Between a Compound and Its Elements

The spontaneous differentiation between a compound and its elements appears to be something other than reciprocal overshadowing, namely configural processing. For the rabbit NM preparation, configural processes appear to be engaged by reinforcing a compound that contains relatively long stimuli. Long stimuli may provide a greater opportunity for them to become fused into a perceptual unit. Nevertheless, there is always a lingering suspicion that there could be summation of subthreshold associative strengths. However, explicit differentiation between a compound and its elements can be obtained in a way that is contrary to summation of any associative strengths.

The most important of these explicit differentiations is negative patterning (Pavlov, 1927). In the training schedule for negative patterning, the individual elements are paired with the US (T+, L+) , but the compound is presented alone (TL-). Operationally, this procedure is closely related to stimulus compounding. In stimulus compounding, test presentations of the compound are relatively rare. In the negative patterning schedule, however, the compound is presented much more frequently.

Figure 3.5 shows the acquisition of negative patterning by a group of rabbits. Initially, each training session contained 30 T+ trials, 30 L+ trials, and 3 TL-trials. As can be seen in the figure, summation appeared when the tone and light were compounded on TL- trials. After responding was well established,

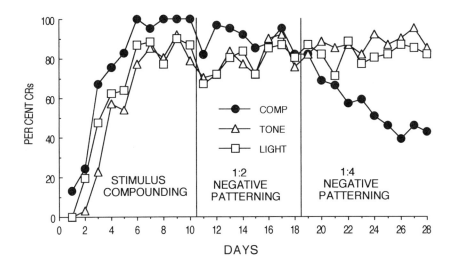

FIG. 3.5. Stimulus compounding and negative patterning. Mean percent CRs on tone+light compound, tone, and light trials in a group given stimulus compounding followed by negative patterning (Kehoe & Graham, 1988).

the number of TL- trials was increased to 30 per session, and the number of T+ and L+ trials was reduced to 15 each per session. Later, the mixture of trials per session was altered further to 40 TL-, 10 T+, and 10 L+ trials. Eventually, responding on TL-trials declined to a low level whereas responding on T+ and L+ trials remained at high levels. In this experiment and others, about six out of every eight animals showed the negative patterning outcome, that is, less responding on TL- trials than on T+ and L+ trials (Kehoe & Graham, 1988).

For those animals that did show the negative patterning outcome, there is no way that the low level of responding on TL- trials could have resulted from the summation of the obvious excitatory associative strengths possessed by the tone and light elements. In fact, to obtain negative patterning, the animal had to suppress the tendency toward summation and withhold responding to the compound. An explanation of negative patterning requires that the animals have a representation of the compound that is distinct from both of its elements. However, as discussed in the theoretical section of this chapter, there are a variety of schemes to explain how a configural representation of a compound emerges and interacts with those of the elements.

Empirical Summary

Three major findings have been obtained using compounds of tone and light. First, as seen in stimulus compounding, summation of responding is a fundamental, consistent outcome. In the rabbit NM preparation, this summation in performance follows a simple quantitative relationship. Second, as seen in compound conditioning, differentiation between a compound and its elements emerges spontaneously if the compound is given reinforced training and the stimulus duration is 800 ms or greater. Third, as seen in negative patterning, summation can be overcome and differentiation occurs if the compound is given frequent nonreinforced exposures relative to the number of reinforced presentations of the elements.

THEORIES

As mentioned previously, the variety of phenomena using compound stimuli has yielded an equally large theoretical diversity. Nevertheless, there have been two persistent themes that run through many of the theories. They concern the role that summation of associative strengths plays in (a) the expression of associative strength in responses to compounds and (b) the learning rules that govern acquisition of associative strength during compound stimuli. Neither of these developments is fully settled. In relation to performance, comparator models contend that ratios and/or differences among associative strengths govern responding (Miller & Matzel, 1989). Similarly, in relation to learning rules, attentional models have preferred to use differences in associative strength to determine which stimulus gains further strength (Mackintosh, 1975; Pearce & Hall, 1980). Nevertheless, associative summation has been sufficiently consistent that its further developments are worth considering.

Summation of associative strengths has been used almost universally in performance rules. Moreover, summation can be separated from atomism. As described later, configural theories as well as atomistic theories rely on a summation principle. In general, the associative summation rule can be expressed as: $R = f(\Sigma V_i x_i)$, where R is the overt measure of responding, f refers to a mapping function, V_i refers to the associative strength of the ith stimulus, and x_i is a binary variable (1, 0) that indicates whether or not the ith stimulus is currently present. Most theories have not formulated an explicit mapping function. They have been content to assume that there is a monotonic relationship between overt responding and the summated associative strengths.

Associative summation was first used in a learning rule by Rescorla and Wagner (1972) to account for overshadowing and related phenomena. The Rescorla–Wagner rule has thus been recognized as an example of a rule that is fundamental to self-organizing systems (Sutton & Barto, 1981). Other examples are known variously as the delta rule, the least-mean square rule, and the Widrow–Hoff rule (Gluck & Bower, 1988; Rumelhart, Hinton, & Williams, 1986). Regardless of its name, the equation for a change in the associative strength of the ith stimulus (ΔV_i) uses the formula:

$$\Delta V_i = \alpha_i \beta (\lambda - \Sigma V_i x_i) x_i \quad (1)$$

where:

α is a direct function of the salience of the ith stimulus ($0 < \alpha < 1$).

β is related to the intensity of the US ($0 < \beta < 1$). For trials on which the US has an intensity of zero, β is usually set to a positive value, for example, 0.33.

λ is the total associative strength that can be supported by the US on any given trial.

$\Sigma V_i x_i$ is the net sum of the associative strengths of active stimuli.

x_i is a binary variable (1, 0) for the presence or absence of the ith stimulus.

As may be apparent, associative summation is encapsulated in the expression $\Sigma V_i x_i$. Thus, associative summation is a central feature of the learning rule.

With these two theoretical themes in mind, this section reviews three classes of theory. First, there are theories that attempt to reconcile concepts of perceptual fusion with atomistic summation. This class of theory includes approaches from the perspectives of pure configuration, unique-stimulus, and configural synthesis. Second are atomistic theories that attempt to explain configural learning phenomena without resorting to perceptual fusion. They include a common components hypothesis and contextual encoding hypotheses. Third are connectionist models that use multiple layers of associative units based on atomistic summation rules. This last class of models can partially reconcile the other solutions.

Configural Theories

Pure Configuration. Pure configural hypotheses have been discussed previously. Here, they are described in greater detail to aid comparison with the other configural hypotheses. Pure configural hypotheses make two basic assumptions. They contend that perceptual fusion occurs whenever a compound is presented and that the configural stimulus supplants the representations of the individual elements. Thus, when a compound is presented, only the configural stimulus can receive associative strength. However, these hypotheses do not assume that the configural stimulus and the elements are completely isolated from one another. Rather, they assume that there will be mutual generalization between the configural stimulus and the elements. Furthermore, pure configural hypotheses do not abandon the assumption that associative strengths can summate. When a compound is presented, the generalized associative strengths from the elements are presumed to summate with any associative strength possessed by the compound. Conversely, responding to a single element is presumably governed by the sum of its own associative strength plus generalized strength from the configural stimulus (Bellingham, Gillette-Bellingham, & Kehoe, 1985; Heinemann & Chase, 1975; Pearce, 1987).

With these assumptions, a pure configural hypothesis can explain many key findings obtained with compound stimuli at an ordinal level. Overshadowing would occur when there is substantial generalization to the more salient element and little generalization to the less salient element. Spontaneous differentiation between the compound and its elements would occur when there is minimal generalization from the configural stimulus to either element. The acquisition of explicit differentiation as in negative patterning would occur when the configural stimulus gains enough inhibitory strength to cancel out the generalization of excitatory strength from the elements.

Pure configural hypotheses have considerable appeal. However, they seem to overstate the influence of configural processes. As described previously, pure configuration has difficulty explaining statistical summation in stimulus compounding. If there were no generalization from the elements to the configural stimulus, then there would be no responding on test trials for the compound. If there were substantial generalization from the elements to the configural stimulus, responding to the compound could exceed responding to either element, yielding ordinal summation. However, unless there was complete generalization, responding to the compound could never reach the level predicted by the statistical sum of response levels to the elements.

Unique Stimulus Hypothesis.   The unique stimulus hypothesis is an addition to the Rescorla–Wagner model. Like a pure configural hypothesis, the unique stimulus hypothesis assumes that perceptual fusion occurs whenever a compound is presented. However, unlike a pure configural hypothesis, the configural stimulus coexists with the elements. In terms of the Rescorla–Wagner model, the configural stimulus is assumed to compete for associative strength with the elements (Gluck, 1991; Gluck & Bower, 1988; Rescorla, 1973). The salience of the configural stimulus relative to those of the elements determines its success in gaining associative strength.

A unique stimulus hypothesis readily explains the quantitative summation seen in stimulus compounding. In training with the separate elements, only they would gain associative strength. On test trials with the compound stimulus, the unique stimulus would occur, but it would have no associative strength. Thus, the summation in stimulus compounding reflects only the associative strengths of the individual elements. Consequently, responding to the compound would be readily predictable from responding to the individual elements.

The unique stimulus hypothesis can also explain the results of compound conditioning and configural learning. In compound conditioning, the configural stimulus would compete with the elements. When the configural stimulus is highly salient, it would capture a large share of the available associative strength to the detriment of the individual elements. Hence, on test trials, separate elements would elicit little responding, thus producing spontaneous differentiation between the compound and the elements. In the rabbit NM preparation, it would appear that the configural stimulus becomes more salient as the duration of the elements increases (Kehoe & Schreurs, 1986).

For the explicit configural learning in negative patterning, the elements would gain excitatory associative strength on A+ and B+ trials. On AB- trials, the elements would lose some associative strength; more importantly, the configural stimulus would be driven into the inhibitory range. Eventually, the inhibitory value of the configural stimulus would balance the excitatory strengths of the elements, and responding would cease on AB-trials.

Configural Synthesis.   For both the pure configural hypothesis and the unique stimulus hypothesis, the configural process itself occurs automatically every time the compound is presented. In contrast, configural synthesis hypotheses contend that perceptual fusion occurs gradually as a result of experience with a compound. Representations of compound stimuli are constructed as they are needed and perhaps superseded when no longer needed.

Configural synthesis avoids the difficulties of assuming that an animal comes hardwired with representations of all possible compounds that it may encounter. Where there is only a handful of possible stimulus inputs, it is easy to assume that all the possible compounds are uniquely represented. However, hardwired models become implausible when extended to a large number of stimulus inputs and their explosively large number of potential compounds.

Although configural synthesis has some attractive features, the process of synthesis has been difficult to articulate in a rigorous fashion. In the conditioning literature, Razran (1939, 1965, 1971) consistently advocated configural synthesis but never went beyond vague appeals to *afferent-afferent learning.* Konorski (1967) was able to make more progress. He devoted a large portion of his book to configural synthesis, in the form of *gnostic units.* He contended that each sensory analyzer entails layers of neurons, with receptor fields on the input side and a layer of gnostic units as the outputs to other neural systems (Konorski, 1967). Each gnostic unit is a single neuron activated uniquely by a converging set of inputs from the previous layer. The gnostic units are not innate but are "a product of a particular type of *learning process*" (p. 86). This learning process occurs when a potential gnostic unit is under the joint influence of a set of specific inputs and "unspecific impulses delivered by the emotive system" (p. 87). However, Konorski recognized that he lacked the mathematical tools to articulate more clearly the formation of the gnostic units. As described later, connectionist modelling techniques have now provided the needed tools.

Quasi-Atomistic Hypotheses

As alternatives to the configural hypotheses, there are at least two quasi-atomistic approaches to the phenomena of compound stimuli. In brief, one approach relies on hypothetical common components presumably shared by the individual elements (Kehoe & Gormezano, 1980; Rescorla & Wagner, 1972). The second approach assumes that each element has different encodings depending on the presence or absence of other elements (Bellingham et al., 1985; Hull, 1943).

Common Components Hypothesis. A common components hypothesis tries to preserve the basic assumption of atomistic hypotheses that no new stimuli arise from the joint presentation of two elements. Inspired by the common element theories of discrimination and generalization (Bush & Mosteller, 1951; Guthrie, 1930; Thorndike & Woodworth, 1901), Rescorla and Wagner (1972) proposed that each element can be treated as an atomistic compound itself, containing distinctive and common components. Accordingly, stimulus A and stimulus B can be represented as ax and bx, respectively, and the AB compound as abx. This scheme can explain negative patterning. Specifically, the composite strengths of A+ and B+ ($[Va + Vx]$, $[Vb + Vx]$) would each approach the positive asymptotic strength associated with the US (λ). The composite strength of AB- ($Va + Vb + Vx$) would approach zero. To obtain this outcome, the strength of the distinctive components (Va and Vb) would have to take on a negative value ($-\lambda$), whereas the strength of the common component (Vx) would have to assume a high positive value (2λ). Thus, for the reinforced elements, $Va + Vx = Vb + Vx = (-\lambda) + (2\lambda) = \lambda$, and, for the nonreinforced compound, $Va + Vb + Vx = (-\lambda) + (-\lambda) + (2\lambda) = 0$.

Although the common components hypothesis provides an atomistic account of negative patterning, it paradoxically has difficulty explaining summation in stimulus compounding. In stimulus compounding, addition of the response levels on A and B trials should overpredict the response level to the compound. Specifically, addition of responding to A and to B should reflect the quantity ($[Va + Vx] + [Vb + Vx]$). Hence, the mathematical summation of separate response levels would double count the associative strength of the common component (Vx). However, actual responding to the compound would reflect the quantity $[Va + Vb + Vx]$, thus counting the common component only once. Even if the mapping of associative strength to overt responding is nonlinear, some overprediction would be expected, not a small underprediction as seen in the rabbit NM preparation.

A common components hypothesis has even greater difficulty in explaining spontaneous differentiation in compound conditioning. On reinforced compound trials, Va, Vb, and Vx would all gain associative strength in proportion to their relative saliences. There is no possible combination of saliences and associative strengths that would yield a low level of responding to both elements. If, say, the distinctive components a and b were the more salient components, then responding to A and B would both be high. If the common component x was more salient, then both A and B would still evoke high levels of responding.

Contextual Encoding. The earliest version of a contextual encoding hypothesis is Hull's (1943) hypothesis of neural afferent interaction (cf. Bellingham et al., 1985; Meehl, 1945). Contextual encoding hypotheses avoid perceptual fusion, and instead assume that concurrent stimuli modify each other's encoding. That is, different stimulus encodings are switched on depending on whether an element is presented by itself or in compound with another element. Thus, A can be represented as *a* on A trials and *a'* on AB trials. By the same token, B can be represented as *b* on B trials and *b'* on AB trials. Each of these alternative encodings can gain its own associative strength as dictated by the US.

Despite the atomistic flavor of contextual encoding, this kind of hypothesis is virtually identical to a pure configural hypothesis. That is, the compounding of two elements innately produces distinctive encodings that exclude alternative encodings. In stimulus compounding, for example, responding would depend on generalization between the alternative encodings of each element. Consequently, contextual encoding hypotheses have the same virtues and vices as pure configural hypotheses.

Connectionist Models

Although hypotheses about compound stimuli vary dramatically, they all contain two stages. The first stage typically concerns the encoding of the stimulus inputs, and the second concerns the associative connections of the encoded inputs to a behavioral output. The variations between hypotheses appear primarily in the encoding stage. Most hypotheses make the simplifying assumption that the encoding process is fixed in its functioning. Only configural synthesis hypotheses assume that the encoding stage itself is adaptive and entails the acquisition of new linkages.

These two-stage architectures can be portrayed as special cases of a larger class of models, namely, layered connectionist networks. The two sets of linkages can be viewed as different layers of a network. Like Konorski (1967), connectionist models suppose that there are layers of neuron-like associative units that intervene between sensory input and response output (Anderson & Rosenfeld, 1988; Barto, Anderson, & Sutton, 1982; McClelland, Rumelhart, & Hinton, 1986). Such networks are particularly attractive because they usually function according to atomistic summation. That is to say, each unit operates using atomistic summation to govern its output. Moreover, the learning rule for each unit is often a version of the Rescorla–Wagner rule, which also relies on atomistic summation. Hence, connectionist models contain no new principles, but do use the old principles in a novel fashion. Whereas Rescorla and Wagner (1972) used atomistic summation to explain learning in the organism as a whole, layered network models repeatedly apply the same rules in a distributed fashion to the local connections between units.

CONNECTIONIST MODEL FOR THE RABBIT
NM PREPARATION

Background

Connectionist models promise to provide an enriched, rigorous form of associationism. Moreover, they build on principles well understood in conventional theories of conditioning. However, connectionist models often use dozens of units. This large number of units can create daunting computational problems. More importantly, they can turn a network into a powerful but vacuous engine for curve-fitting (Massaro, 1988). In order to use connectionist modelling techniques and diminish their risks, a model was developed using a minimal number of units. The development of this model specifically for compound stimulus phenomena was triggered by three considerations:

Perceptual Processes. To explain configural phenomena, theorists have frequently created the needed stimuli simply by postulating fixed perceptual processes, for example, perceptual fusion, the unique stimulus, or neural afferent interaction. This theoretical strategy has been worthwhile as an interim device. Nevertheless, it puts configural encoding itself outside the scope of associative learning theory. In contrast, layered networks offered the possibility that configural encoding could be explained in a rigorous fashion by an associative theory as envisaged by Razran (1965, 1971) and Konorski (1967).

Performance Rules. Many conditioning models make no strong assumptions about the mapping of associative strength to overt responding. They assume only a monotonic relationship. This strategy permits the models to be applied across a large number of distinctive response systems. At the same time, this strategy deliberately limits the models to making ordinal predictions about response levels. Although this strategy has been successful, the results of research with compound stimuli in the rabbit NM preparation suggest the possibility of a more exact performance rule, at least for the percentage CR measure. Among connectionist techniques, the explicit statement of an output rule for each unit is an essential part of any model. As a consequence, probabilistic outputs from an entire network can be modelled without any additional assumptions.

Previous Success in Using a Connectionist Model. In addition to the general virtues of connectionist modelling, there has been success in applying them to some previously intractable phenomena, namely, learning to learn. In the rabbit NM preparation, learning to learn has been seen in transfer across sensory modalities. Specifically, initial CS–US pairings using a stimulus from one modality (e.g., tone) causes a large acceleration in CR acquisition when training is transferred to a new stimulus from another modality (e.g., light). Control conditions have revealed that this acceleration cannot be attributed to stimulus generalization. Moreover, the acceleration in CR acquisition to the new stimulus occurs even after extinction of the CR to the initial stimulus (Holt & Kehoe, 1985; Kehoe & Holt, 1984; Kehoe, Morrow, & Holt, 1984; Schreurs & Kehoe, 1987).

The results of transfer across modalities have been explained rigorously in a connectionist model of mediated generalization. In brief, there is assumed to be a sequence of associative linkages between stimulus input and response output. Some of these linkages come to be shared by otherwise distinctive stimuli (cf. Kimble, 1961). Figure 3.6 depicts the minimal network of associative linkages for explaining transfer across modalities. As can be seen, the network contains three sensory units, one each for stimulus A, stimulus B, and the US. The inputs for A and B project to an intermediate associative unit (X), which in turn projects to a second associative unit (R). This unit gives rise to the CR and UR. Both associative units receive an input from the US.

According to this scheme, observable CR acquisition in a naive subject to stimulus A requires the successive strengthening of both the element-specific link (A–X) and the shared link (X–R). In subsequent training with stimulus B, transfer appears, because CR acquisition to B requires only establishment of the specific B–X link, which then capitalizes on the established X–R link. This network can also explain the ability of cross-modal transfer to survive extinction of CRs to stimulus A. Specifically, the X–R link is partly protected from extinction when the A–X connection declines to a point at which it can no longer activate the X unit. If there is any residual associative strength in the X–R linkage, it will be available during subsequent pairings of stimulus B with the US.

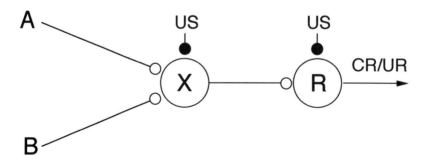

FIG. 3.6. Two-unit layered network. This network contains three sensory inputs (A, B, US) and two associative units (X, R) (Kehoe, 1988).

First Model

The first attempt to use a network model to explain compound stimulus phenomena was based on the network shown in Fig. 3.6. Specifically, the X unit operated according to the Rescorla–Wagner rule as shown in Equation 1. Thus, when both the A and B inputs were simultaneously active, they would compete with each other and thus split the available associative strength between them. When only the summated A–X and B–X connections were strong enough to trigger the X unit, the X unit acted like a *unique stimulus*. The key to making the X unit act like a unique stimulus was in its output rule. The output rule was based on summation of the associative strengths and a noisy threshold. Specifically, the output of X was determined as follows:

$$\text{Output } X = 1 \text{ if } (\Sigma V_i x_i) > T_{x(t)}, \text{ otherwise Output } X = 0 \quad (2)$$

where $\Sigma V_i x_i$ is the sum of the associative strengths of the current stimulus inputs.

$T_{x(t)}$ is the value of the threshold for the X unit on trial t. This value was randomly selected from a distribution that was approximately normal with a mean of 0.50, and a standard deviation of 0.167. However, the minimum value was 0. In many simulations, the mean threshold value was altered by adding or subtracting a constant.

The R unit also operated according to the same rules. However, because it received only one input, the X–R connection never had any competition. The value of the R unit's threshold on each trial was independent from that of the X unit. The output of the R unit over successive trials yielded a percentage CR measure.

As may be apparent, the threshold of the X unit determined whether or not it acted as a unique stimulus. If the A–X and B–X inputs each attained an associative strength of 0.50, then mean thresholds of 0.50 or greater prevented the separate inputs from reliably triggering the X unit. Only the summated strengths of the compounded inputs reliably triggered the X unit. In this way, the X unit became tuned to the AB compound. Conversely, if the mean threshold was less than 0.50, then either a weak A–X or B–X connection triggered the X unit with some degree of regularity.

The initial simulations focused on compound conditioning (AB+), stimulus compounding (A+, B+), and negative patterning (A+, B+, AB-). A series of simulations entailed systematic variations in the mean threshold value of the X unit. Figure 3.7 shows the level of simulated responding to A, B, and AB as a function of X's mean threshold value. The figure also shows the statistical sum of simulated responding to A and B. The levels of responding themselves are averages across 40 training trials.

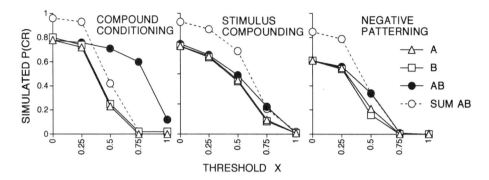

FIG. 3.7. Simulations based on two-unit network. Mean level of simulated responding to A, B, and AB as a function of the mean threshold value of the X unit for three compound stimulus phenomena. The figure also shows the statistical sum of simulated responding to A and B.

Compound Conditioning. The left-hand panel in Fig. 3.7 shows the simulations for compound conditioning. For mean thresholds of 0.00 and 0.25, responding to the compound and its elements were uniformly high, because either the A–X or B–X connection was strong enough to trigger the X unit. However, as anticipated, the mean thresholds of 0.50 and 0.75 produced spontaneous configuration. That is, responding to the AB compound remained high, whereas responding to the A and B elements was low. Finally, at the extreme threshold of 1.00, acquisition was very slow; even the A–X and B–X connections together did not acquire enough strength to trigger the X unit.

Stimulus Compounding. Although the simulations did yield spontaneous configuration, the results for stimulus compounding were less robust. As can be seen in the center panel of Fig. 3.7, summation was slight. Responding to the compound was only slightly higher than that of the elements for all threshold values. More importantly, for mean threshold values from 0.00 to 0.75, the level of simulated responding on AB trials was substantially below that of the statistical sum of responding on A and B trials. For the mean threshold value of 1.00, responding on AB trials was greater than the statistical sum. With extended training, this latter difference became more pronounced.

In light of the crossover in results between the threshold values of 0.75 and 1.00, further simulations of stimulus compounding revealed that a mean threshold of 0.95 could reproduce the close relationship between responding on AB trials and the statistical sum. This same value also yielded perfect spontaneous configuration in compound conditioning. That is, simulated responding on AB trials reached a level of 0.94, whereas responding on A and B trials remained at 0.00.

Negative Patterning.   The two-unit network could not reproduce negative patterning. Inspection of the right-hand panel shows that, for all threshold values, the simulated level of responding on AB- trials was equal to or slightly greater than responding on either A+ or B+ trials. This pattern of results closely paralleled those seen in the simulations of stimulus compounding. The overall level of simulated responding in the negative patterning simulation was lower than in stimulus compounding. This overall reduction occurred because AB trials produced a decrement in the strength of the A–X and B–X connections in proportion to the value of the β parameter (0.33). These decrements were offset on A+ and B+ trials, on which β was set to 1.00. Nevertheless, the AB- trials had a dampening effect on simulated responding.

Interim Conclusions.   The simulations revealed that a two-unit layered network could reproduce compound conditioning and stimulus compounding. In particular, an X unit with a high mean threshold appeared to be essential to both spontaneous configuration and summation outcomes. This high threshold ensured that, in compound conditioning, the X unit would act as a unique stimulus. That is, the X unit would be triggered only by the summated A–X and B–X connections, which the Rescorla–Wagner rule had capped at values around 0.50. In stimulus compounding, the high threshold of the X unit meant that summation would appear when the strengths of the A–X and B–X connections were growing during the middle portions of acquisition. However, as seen in the failure to obtain negative patterning, the X unit could not overcome the summation process and suppress responding to the AB compound.

Second Model

Subsequent models retained the X unit with its high threshold but added a second intermediate unit (Y). Figure 3.8 shows the architecture of the expanded network. The inputs for stimulus A and stimulus B each project to two associative units, labelled as X and Y. In turn, the X and Y units project to another associative unit labelled as R. The US input sends a fixed excitatory connection to the X, Y, and R units. All the units used the Rescorla–Wagner rule and an output rule based on summation plus a noisy threshold.

The addition of the Y unit accomplished two things. First, it allowed the network to use one of the intermediate units to act as a unique stimulus and the other intermediate unit to encode the separate stimuli. However, neither unit was constructed to respond to any particular stimulus or combination of stimuli. It was through the action of the learning and output rules that a unit became tuned to combinations of inputs as needed. Second, each of the intermediate units could acquire its associative connection to the R unit. For

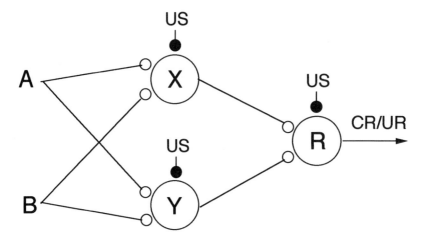

FIG. 3.8. Three-unit layered network. This network contains three sensory inputs (A, B, US) and three associative units (X, Y, R) (Kehoe, 1988).

example, the X–R connection can become excitatory while the Y–R connection becomes inhibitory.

Simulations of Basic Configural Outcomes. The added Y unit considerably increased the flexibility of the network. At the same time, however, the Y unit increased the risk that the network would act merely as a curve-fitting device. To hold the network in check and make it a realistic model for compound stimulus phenomena, Kehoe (1988) conducted exploratory simulations to identify a single set of parameter values that simultaneously would reproduce three distinct examples of configural learning. Those examples were compound conditioning (AB+), positive patterning (AB+, A-, B-), and negative patterning (AB-, A+, B+).

The left-hand panels of Fig. 3.9 show the course of differentiation between the compound and its components in those three schedules. The right-hand panels show simulated acquisition curves. As can be seen in Fig. 3.9, the simulations captured the rapid differentiation seen in compound conditioning and positive patterning. Moreover, the simulations also captured the slower acquisition of negative patterning, which was characterized by an initial rise in responding on AB- trials followed by a gradual decline.

The simulations were obtained when (a) the X unit had a higher mean threshold than the Y unit ($T_x = 0.70$, $T_y = 0.15$), and (b) the X unit had a higher learning rate than the Y unit ($\alpha_x = 0.050$, $\alpha_y = 0.001$). The R unit had an intermediate threshold [$T_r = 0.50$] and intermediate learning rate [$\alpha_r =$

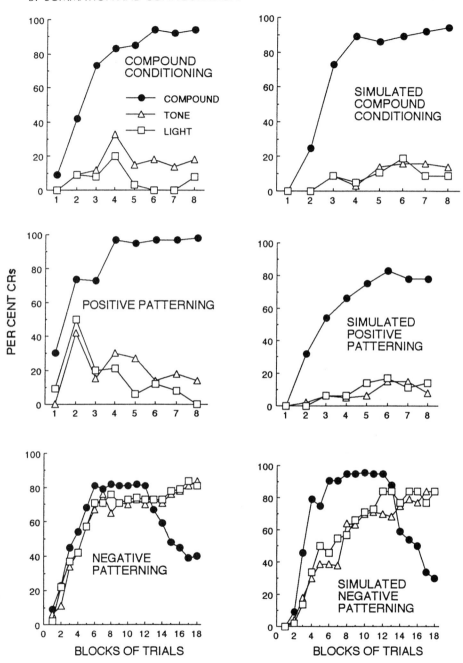

FIG. 3.9. Configural learning; actual and simulated responding. Percent CRs on compound, tone, and light trials in compound conditioning (TL+), positive patterning (TL+, T-, L-), and negative patterning (TL-, T+, L+) (Kehoe, 1988).

0.010]. On reinforced trials, the value of β was 1.00, and, on nonreinforced trials, the value of β was 0.33.

The high threshold of the X unit made it very sensitive to the effects of any competition between the A and B inputs. When A and B divided the total strength evenly ($V_{a-x} = V_{b-x} = 0.50$), X was triggered only by the AB compound. Conversely, the low threshold of the Y unit made it insensitive to the effects of competition between A and B. That is, the Y unit was reliably triggered by either A or B at all but the lowest connection strengths. The difference between the learning rates of the X and Y units gave the X unit an advantage in competing with the Y unit for access to the R unit. With its high learning rate, the X unit would start to fire earlier in training and thus ensure that the X–R connection began to strengthen before the Y–R connection became eligible for change.

To illustrate how the interior connections of the model operated, Fig. 3.10 shows the terminal associative strengths for the three configural schedules. In the compound conditioning schedule and positive patterning schedule, there was essentially the same pattern of connections. The A–X and B–X connections were at or near the limit of 0.50 permitted by competition between the A and B stimuli. The A–Y and B–Y connections reached much lower strengths as a consequence of the Y unit's slow learning rate. In the competition between the X and Y inputs to the R unit, the X–R connection gained most of the total strength ($V_{x-r} = 0.68$). The remaining strength gained by the Y–R connection ($V_{y-r} = 0.33$) was too low by itself to trigger the R unit with any reliability. Because the A–Y, B–Y, and Y–R connections were so weak, the network operated for compound conditioning in virtually the same way as the network that contained only the X and R units.

With regard to negative patterning, however, the importance of the Y unit becomes clear. Further inspection of Fig. 3.10 reveals that negative patterning produced a highly distinctive pattern of connections. Most notably, the X–R connection was strongly inhibitory ($V_{x-r} = -0.57$), and the Y–R connection was strongly excitatory ($V_{y-r} = 0.87$). Furthermore, the A–Y and B–Y connections were individually strong enough to trigger the Y unit, which, it should be remembered, had a low mean threshold ($T_y = 0.15$). Hence, the combined A and B stimuli triggered both the X and Y units. The opposite strengths of the X–R and Y–R connections largely cancelled each other and thus precluded CRs on AB– trials. On A+ and B+ trials, however, the separate stimuli by themselves were insufficient to trigger the X unit but did trigger the Y unit. Consequently, only the excitatory Y–R connection was activated, and it was more than enough to evoke CRs.

COMPOUND CONDITIONING

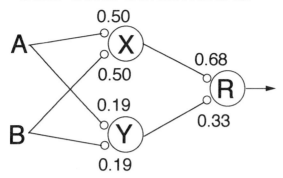

POSITIVE PATTERNING

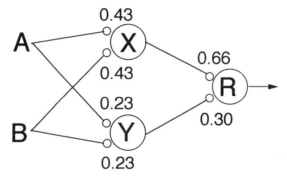

NEGATIVE PATTERNING

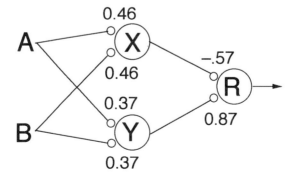

FIG. 3.10. Terminal associative strengths. Associative weights in the three-unit model at the end of the simulations for compound conditioning, positive patterning, and negative patterning (Kehoe, 1988).

Simulations of Summation and Configural Outcomes

Having identified parameter values that could explain configural learning, they were tested in a variety of training schedules, especially stimulus compounding. Specifically, Fig. 3.11 shows the results of the simulations for mixtures of AB+, A+, and B+ trials. Inspection of Fig. 3.11 reveals that the model was able to reproduce the key features of the actual findings, which are shown in Fig. 3.4. Specifically, the simulated responding to the compound remained relatively constant, whereas responding to the elements showed a progressive decline as the proportion of AB+ trials was increased across groups.

Figure 3.11 also shows the statistical sum of the simulated responding to the elements. In agreement with findings of stimulus compounding (C0), the statistical sum matched the level of responding to the compound. This

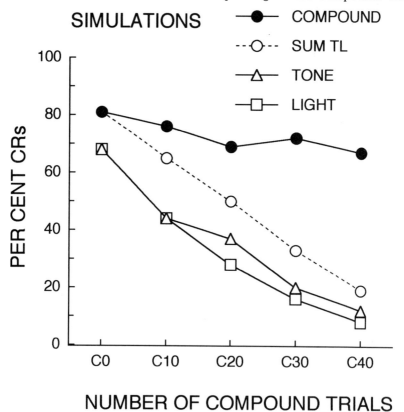

FIG. 3.11. Simulations of mixed compound and single element training. Percent CRs on compound, tone, and light trials as a function of the number of compound trials contained in the mixture of trials. The figure also shows the statistical sum of responding to the individual elements (SUM TL) (Kehoe, 1988).

outcome was very encouraging, because statistical summation in performance was not built into the network. Statistical summation emerged from the ordinary operation of the network as whole, that is, the combined effects of six different connection strengths and three different mean thresholds.

To illustrate further the operation of the model in both atomistic and configural settings, Fig. 3.12 shows the results of simulations for stimulus compounding followed by negative patterning. The actual findings are shown in Fig. 3.5.

For stimulus compounding, the simulated response levels in Fig. 3.12 compared favorably with the actual levels of responding. That is, there was summation that conformed to the statistical summation rule. An examination of the associative strengths for stimulus compounding reveals that the X and Y units operated in a purely atomistic fashion. Both the A and B inputs had strong connections with the X unit ($V_{a-x} = V_{b-x} = 0.85$). Thus, either input triggered the X unit with equal reliability. In turn, the X unit had a strong connection with the R unit ($V_{x-r} = 0.68$). Summation occurred because the summated A and B inputs to X increased the probability of exceeding the threshold on any given trial. For the Y unit, with its low learning rate, the A and B inputs had relatively weak subthreshold connections. However, the joint A and B inputs to the Y unit contributed to summation on AB trials.

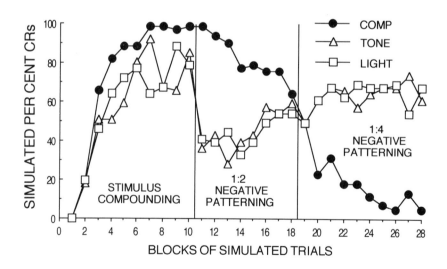

FIG. 3.12. Simulations of stimulus compounding and negative patterning. Simulated mean percent CRs on tone + light compound, tone, and light trials in a group given stimulus compounding followed by negative patterning. The weights of the associative linkages at the end of stimulus compounding and negative patterning (Kehoe, 1988) are shown.

As seen in the right-hand side of Fig. 3.12, the model could reproduce the gradual crossover from summation to the negative patterning outcome. The crossover occurred at the switch from a 1:2 to 1:4 schedule as it did in the actual results. During negative patterning, the connection strengths rearranged themselves in a dramatic fashion and attained the values needed for negative patterning. However, the simulation of negative patterning revealed some limitations to the original set of parameter values. The growth rates and mean threshold values did not need alteration, but it was necessary to alter the value of β on nonreinforced trials. In the original set, the value of β was 0.33 for all connections. To obtain negative patterning after stimulus compounding, it was necessary to set β to 0.60 for the X–R connection and 0.10 for all other connections.

Other than the difficulties encountered in the switch from stimulus compounding to negative patterning, the original parameter set proved robust. In addition to the phenomena already described, Kehoe's (1988) model has reproduced all the basic phenomena of compound stimuli as they appear in the rabbit NM preparation. They include overshadowing, blocking, conditioned inhibition, and a feature positive discrimination (Kehoe, 1988, 1989).

Simulation of CS Duration Effects

Kehoe's (1988) parameter estimates and subsequent simulations were directed at delay conditioning in which the duration of the stimuli prior to the US was 800 ms. However, as described previously, stimulus duration has a profound effect on the relative levels of responding to the compound and its elements. As was shown in Fig. 3.3, stimulus durations of 300 ms yielded high levels of responding to the components as well as the compound.

In order to reproduce the effects of stimulus duration, trace theories of conditioning offered guidance (Desmond, 1990; Gormezano, 1972; Gormezano & Kehoe, 1981; Grossberg & Schmajuk, 1989; Hull, 1943; Sutton & Barto, 1981, 1990). Among other things, they have been concerned with the effects of CS–US intervals on CR acquisition. In brief, they assumed that CS onset initiates one or more *stimulus traces*. Furthermore, the intensity of these traces is assumed to grow and then decay across time. The rate of CR acquisition is directly related to their intensity at the time of US presentation. In the estimates based on CR acquisition in the rabbit NM preparation, the trace intensity is just past its maximum at 300 ms after CS onset and declines to a moderate value by 800 ms after CS onset.

The trace theories suggest that increasing the value of the learning rate parameters (α) will model the effects of using a 300-ms stimulus duration relative to the 800-ms duration. Some additional considerations point particularly to the value of α in the Y unit. First, increasing the α values of all three units together increases the overall rate at which the network operates but does not change the basic pattern of results. Second, in a similar fashion, increasing the α value of the R unit also only alters the overall rate of CR acquisition. Third, increasing the α value for only the high-threshold X unit exaggerates configural outcomes, when in fact they disappear at the 300-ms stimulus duration. This exaggeration of configuration occurs because the A–X and B–X connections reach their asymptotic values sooner and thereby give the X–R connection a further head start in competing with the Y–R connection. Fourth, increasing the α value for the low-threshold Y unit increases the likelihood of atomistic outcomes. The acquisition of A–Y and B–Y connections would be accelerated and thus put the Y–R connection on a more competitive footing versus the X–R connection. Consequently, there would be greater opportunity for the separate A and B inputs to gain access to the R unit via the Y unit.

Figure 3.13 shows the results of simulations in which the value of α for the Y unit was manipulated. The left-hand panel shows the simulated level of responding on A, B, and AB trials in compound conditioning, and the right-hand panel shows the simulated levels in stimulus compounding. Both panels also show the statistical sum of responding on A and B trials.

As anticipated for compound conditioning, increases in the α value for the Y unit produced substantial increases in the simulated levels of responding to the elements. For the α value of 0.01, the summated levels of responding to A and B closely approximated responding to AB. In contrast, the results of

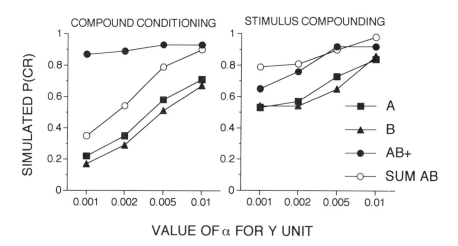

FIG. 3.13. Simulations for different learning rates for Y unit. Simulations of mean percent CRs on AB, A, and B trials in compound conditioning and stimulus compounding. The low values of Y unit are calibrated for CS durations of 800 ms, and the higher values are intended to duplicate the results of shorter CS durations.

simulated stimulus compounding were largely unaffected by increases in α value for the Y unit. Although there was a general increase in the level of responding, the relationship between A, B, and AB remained constant. For all α values, the statistical sum of responding to A and B closely approximated that of responding to AB.

In conclusion, raising the α value of the Y unit captures the results of stimulus duration manipulations in the rabbit NM preparation. In agreement with the actual results of compound conditioning, the higher α values suitable for shorter stimulus durations abolished the spontaneous differentiation seen at the lower α values that were suitable for longer stimulus durations. At the same time, the α value in stimulus compounding had little effect on summation just as manipulations of stimulus duration had little effect.

Although manipulation of the α value of the Y unit can reproduce the effects of stimulus duration, it may seem to be an ad hoc solution. Why would the learning rate of one unit be more sensitive to the intensity of stimulus traces than another unit? In fact, both units may be sensitive to stimulus traces but with different time constants. The stimulus traces for the X unit may have a slower rate of recruitment and decay compared to those of the Y unit. From the perspective of the network model, it is the relative values of α that are important. Thus, for shorter CS–US intervals, a reduction in the α value of the X unit would have much the same effect as an increase in the α value of the Y unit.

Kehoe's (1988) model is hardly the last word on compound stimuli (cf. Schmajuk & DiCarlo, 1992). Among other things, it was designed for a limited domain, namely, two discrete stimuli presented simultaneously. Nevertheless, within that domain, the model with a relatively fixed set of parameter values can explain a large number of important phenomena. Moreover, the model illustrates how the use of familiar associative principles in a distributed form can widen their scope of explanation.

CONCLUSIONS

In summary, three basic principles appear to govern the operation of the associative apparatus. First, the most secure principle is the summation of associative strengths as the rule governing output. Although theories differ as to which associative strengths are summated, this principle is pervasive, even in pure configural hypotheses (Bellingham et al., 1985; Pearce, 1987). Second, the summation of associative strengths also plays a key role in acquisition during compound stimulus training. Specifically, the Rescorla–Wagner rule has been applied both at the level of the whole organism and more recently as the rule for local associative linkages in connectionist models. Most importantly, it has been used to tune associative units to compound stimuli. With this rule in hand, configuration ceases to be a mysterious perceptual process and enters the domain of associative learning. This conclusion, however, may be limited to compounds of simultaneous elements and to compounds using stimuli from different sensory modalities. Within sensory modalities, there are numerous opportunities for interactions between stimulus inputs that do not entail associative learning (Honey & Hall, 1989; Kehoe et al., 1994).

A third basic principle concerns the architecture of the associative system. All the available theories assume that, in eliciting a CR, the stimulus inputs undergo a transformation that can be expressed in two stages. Many theories postulate an automatic transformation in their encodings that creates special stimuli. These transformations are thought to either fuse the elements, split them into common and distinctive components, or yield context-dependent representations. Other theories, including configural synthesis hypotheses and connectionist models, assume that the first-stage transformation itself entails associative learning. As may be apparent, this two-stage architecture dissolves the historic division between atomistic and holistic approaches. Every stimulus undergoes some degree of transformation. Of course, some transformations preserve the individuality of the input more than others. Nevertheless, atomistic and holistic outcomes can increasingly be viewed as consequences of shared mechanisms.

Although a comprehensive theory of compound stimulus effects is still some way off, the available theories have increased in their rigor, scope, and elegance. A bit of historical reflection reveals the extent of this increase. The state of affairs up to the late 1960s was summarized by Baker (1968). Using the limited available literature, he distinguished between compounds containing simultaneous elements, sequentially presented elements, and sequential but overlapping elements. For the simultaneous elements, he reviewed the literature on what we would now recognize as overshadowing, mutual overshadowing, and spontaneous differentiation. Stimulus compounding and patterning received only a passing mention. There was no overarching theoretical framework. However, there was discussion of Razran's (1965) notions of configural synthesis, which Baker elaborated in terms similar to the unique stimulus hypothesis. With regard to sequential elements, the review was almost entirely empirical, trying to put some order on the tangled interactions between elements.

In the same year as Baker's (1968) review, the foundation for the empirical and theoretical advances described in this chapter was laid down. The key findings were the first reports concerning blocking (Kamin, 1968) and the relative frequency of reinforcement for compounds and their elements (Wagner, Logan, Haberlandt, & Price, 1968). These findings were followed closely by Rescorla's (1969) review of conditioned inhibition, and the appearance of draft versions of the Rescorla–Wagner model. These developments primarily concerned the relationship between the elements that yielded instances of stimulus selection (Rudy & Wagner, 1975). However, the relationship between the elements and the compound was not ignored. Stimulus compounding and summation received systematic attention by Weiss (1972). Configural learning, particularly negative patterning, came under experimental examination for the first time since the 1940s, which led to the unique stimulus hypothesis (Kehoe & Gormezano, 1980; Rescorla, 1972, 1973; Whitlow & Wagner, 1972).

The 1980s saw the appearance of connectionist modelling techniques, which provided the means for solving the previously intractable problem of configuration. Just as the Rescorla–Wagner model provided an associative account of selective attention, connectionist modelling provided an associative account of configuration. The key writers in this area are Barto, Sutton, and their associates. Their work has consistently been directed at instrumental and classical conditioning, rather than issues of verbal learning and cognition. Moreover, they write with a clarity that is more accessible than many other writers on connectionist modelling (Barto et al., 1982; Sutton & Barto, 1981, 1990).

So, what of the future? Startling new discoveries are not the sort of thing that can be foreseen. It is easier, however, to see what areas of research still offer worthwhile possibilities. In experimentation using compound stimuli, the basic parameters have been mapped out. However, transfer from one training schedule to another, say, from compound conditioning to negative patterning, should provide a useful basis for refining models of compound stimuli.

In addition to these incremental advances, a major issue in both research and theory concerns the mechanism underlying conditional discriminations obtained using sequentially presented stimuli. In a schematic form, one sequence of stimuli (A → X+) is always paired with the US, and the other sequence is always presented alone (B → X–). The X stimulus is usually a conventional discrete stimulus, but the A and B stimuli may be either discrete stimuli or more prolonged, *tonic* stimuli. In all these procedures, the main focus concerns whether the subject learns to respond differentially to the shared X stimulus on the two types of trial.

In the rabbit NM preparation, research with serial stimuli has focused largely on discrete stimuli (e.g., Kehoe, Marshall-Goodell, & Gormezano, 1987). However, conditional control by tonic stimuli has been obtained recently in the rabbit NM preparation (Macrae & Kehoe, 1995) and in the outer eyelid preparation (Brandon & Wagner, 1991). For example, in one experiment, a 45-s white noise and a 45-s 1,000-Hz tone were used as the tonic stimuli (A, B). They were separated by an intertrial period without the noise or tone. The discrete stimulus (X) was a 100-ms light flash, which occurred about 25 s after the onset of the tonic stimulus. The light flash was paired with the US at a 400-ms CS–US interval during one tonic stimulus (A $\rightarrow$ X+), and the light flash was presented alone during the alternative tonic stimulus (B $\rightarrow$ X-) (Macrae & Kehoe, 1995).

Conditional discriminations appeared rapidly in the rabbit NM preparation. Figure 3.14 shows the mean percentage CRs on A $\rightarrow$ X+ and B $\rightarrow$ X- trials. As can be seen, there was rapid CR acquisition during X on A $\rightarrow$ X+ trials and low levels of responding during X on B $\rightarrow$ X- trials. This conditional discrimination appeared early in training and steadily grew larger. The tonic stimuli themselves did not evoke CRs. Following the onset of the tonic stimuli (A, B), the frequency of an NM response was always less than 10%. Moreover, further measurements in the period immediately before the X stimulus also revealed response levels less than 10%. The few responses that were seen occurred at the baseline level of spontaneous responses.

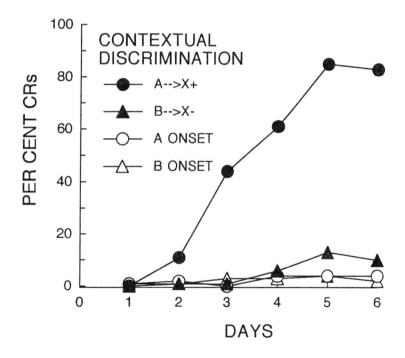

FIG. 3.14. Contextual discrimination. Mean percent CRs during the discrete stimulus (X) on reinforced trials (A $\rightarrow$ X+) and nonreinforced trials (B $\rightarrow$ X-). The percent CRs immediately following the onset of the alternative contextual stimuli (A ONSET, B ONSET) (Macrae & Kehoe, 1995).

Conditional discriminations using serial stimuli have suggested another approach to the integration of elements. In this approach, the stimuli become related to each other in a hierarchical fashion. The first stimulus in a sequence does not become directly involved in the elicitation of CRs but rather acquires the ability to modulate responding to the second stimulus. This indirect control capacity of a stimulus has been described variously as its gating, preparatory, occasion-setting, or facilitating function (e.g., Balsam & Tomie, 1985; Holland, 1983; Konorski & Lawicka, 1959; Rescorla, 1985).

As may be apparent, hierarchical hypotheses represent the latest special theory arising from research with compound stimuli. However, these hypotheses fall into a now familiar pattern. In some respects, hierarchical hypotheses are like the configural synthesis hypotheses. Specifically, hierarchical hypotheses make no special assumptions about encoding and assume two sets of associative linkages, one of which is only indirectly expressed in behavior. Accordingly, the superordinate and subordinate linkages can be viewed as an example of a two-stage architecture. From this perspective, hierarchical hypotheses should be amenable to connectionist modelling techniques. A network as simple as Kehoe's (1988) model probably will not suffice, but hierarchical control structures can be used in connectionist models (e.g., Klopf, Weaver, & Morgan, 1993).

ACKNOWLEDGMENTS

Preparation of this chapter was supported by Australian Research Council Grant AC9231222. The author thanks A. Horne and M. Macrae for their assistance in preparing the manuscript.

REFERENCES

Anderson, J. A., & Rosenfeld, E. (Eds.). (1988). *Neurocomputing: Foundations of research.* Cambridge, MA: MIT Press.

Anderson, N. H. (1982). *Methods of information integration theory.* New York: Academic Press.

Baker, T. W. (1968). Properties of compound-conditioned stimuli and their components. *Psychological Bulletin, 70,* 611–625.

Balsam, P. D., & Tomie, A. (Eds.). (1985). *Context and learning.* Hillsdale, NJ: Lawrence Erlbaum Associates.

Barto, A. G., Anderson, C. W., & Sutton, R. S. (1982). Synthesis of nonlinear control surfaces by a layered associative search network. *Biological Cybernetics, 43,* 175–185.

Bellingham, W. P., & Gillette, K. (1981). Spontaneous configuring to a tone-light compound using appetitive training. *Learning and Motivation, 12,* 420–434.

Bellingham, W. P., Gillette-Bellingham, K., & Kehoe, E. J. (1985). Summation and configuration in patterning schedules with the rat and rabbit. *Animal Learning and Behavior, 13*, 152–164.

Brandon, S. E., & Wagner, A. R. (1991). Modulation of a discrete Pavlovian conditioned reflex by a putative emotive Pavlovian conditioned stimulus. *Journal of Experimental Psychology: Animal Behavior Processes, 17*, 299–311.

Bush, R. R., & Mosteller, F. A. (1951). A mathematical model for simple learning. *Psychological Review, 58*, 313–323.

Couvillon, P. A., & Bitterman, M. E. (1982). Compound conditioning in honeybees. *Journal of Comparative and Physiological Psychology, 96*, 192–199.

Desmond, J. E. (1990). Temporal adaptive responses in neural models: The stimulus trace. In M. Gabriel & J. W. Moore (Eds.), *Learning and computational neuroscience* (pp. 421–456). Cambridge, MA: MIT Press.

Gluck, M. A. (1991). Stimulus generalization and representation in adaptive network models of category learning. *Psychological Science, 2*, 50–55.

Gluck, M. A., & Bower, G. H. (1988). Evaluating an adaptive network model of human learning. *Journal of Memory and Language, 27*, 166–195.

Gormezano, I. (1966). Classical conditioning. In J. B. Sidowski (Ed.), *Experimental methods and instrumentation in psychology* (pp. 385–420). New York: McGraw-Hill.

Gormezano, I. (1972). Investigations of defense and reward conditioning in the rabbit. In A. H. Black & W. F. Prokasy (Eds.), *Classical conditioning II: Current research and theory* (pp. 151–181). New York: Appleton-Century-Crofts.

Gormezano, I., & Kehoe, E. J. (1981). Classical conditioning and the law of contiguity. In P. M. Harzem & M. D. Zeiler (Eds.), *Advances in analysis of behavior Vol. 2: Predictability, correlation, and contiguity* (pp. 1–45). New York: Wiley.

Grossberg, S., & Schmajuk, N. A. (1989). Neural dynamics of adaptive timing and temporal discrimination during associative learning. *Neural Networks, 2*, 79–102.

Guthrie, E. R. (1930). Conditioning as a principle of learning. *Psychological Review, 37*, 412–428.

Heinemann, E. G., & Chase, S. (1975). Stimulus generalization. In W. K. Estes (Ed.), *Handbook of learning and cognitive processes* (pp. 305–349). Hillsdale, NJ: Lawrence Erlbaum Associates.

Holland, P. (1992). Occasion setting in Pavlovian conditioning. In D. L. Medin (Ed.), *The psychology of learning and motivation* (pp. 69–125). San Diego, CA: Academic Press.

Holland, P. C. (1983). Occasion-setting in Pavlovian feature positive discriminations. In M. L. Commons, R. J. Herrnstein, & A. R. Wagner (Eds.), *Quantitative analyses of behavior: Discrimination processes* (pp. 182–206). Cambridge, MA: Ballinger.

Holt, P. E., & Kehoe, E. J. (1985). Cross-modal transfer as a function of similarities between training tasks in classical conditioning of the rabbit. *Animal Learning & Behavior, 13*, 51–59.

Honey, R. C., & Hall, G. (1989). Attenuation of latent inhibition after compound pre-exposure: Associative and perceptual explanations. *The Quarterly Journal of Experimental Psychology, 41B*, 355–368.

Hull, C. L. (1943). *Principles of behavior.* New York: Appleton-Century-Crofts.

Kamin, L. J. (1968). "Attention-like" processes in classical conditioning. In M. R. Jones (Ed.), *Miami symposium on the prediction of behavior: Aversive stimulation* (pp. 9–31). Miami, FL: University of Miami Press.

Kehoe, E. J. (1982). Overshadowing and summation in compound stimulus conditioning of the rabbit's nictitating membrane response. *Journal of Experimental Psychology: Animal Behavior Processes, 8,* 313–328.

Kehoe, E. J. (1986). Summation and configuration in conditioning of the rabbit's nictitating membrane response to compound stimuli. *Journal of Experimental Psychology: Animal Behavior Processes, 12,* 186–195.

Kehoe, E. J. (1988). A layered network model of associative learning: Learning-to-learn and configuration. *Psychological Review, 95,* 411–433.

Kehoe, E. J. (1989). Connectionist models of conditioning: A tutorial. *Journal of the Experimental Analysis of Behavior, 52,* 427–440.

Kehoe, E. J., & Gormezano, I. (1980). Configuration and combination laws in conditioning with compound stimuli. *Psychological Bulletin, 87,* 351–378.

Kehoe, E. J., & Graham, P. (1988). Summation and configuration in negative patterning of the rabbit's conditioned nictitating membrane response. *Journal of Experimental Psychology: Animal Behavior Processes, 14,* 320–333.

Kehoe, E. J., & Holt, P. E. (1984). Transfer across CS–US intervals and sensory modalities in classical conditioning in the rabbit. *Animal Learning & Behavior, 12,* 122–128.

Kehoe, E. J., Horne, A. J., Horne, P. S., & Macrae, M. (1994). Summation and configuration between and within sensory modalities in classical conditioning of the rabbit. *Animal Learning & Behavior, 22,* 19–26.

Kehoe, E. J., Marshall-Goodell, B., & Gormezano, I. (1987). Differential conditioning of the rabbit's nictitating membrane response to serial compound stimuli. Journal of Experimental Psychology: *Animal Behavior Processes, 13,* 17–30.

Kehoe, E. J., Morrow, L. D., & Holt, P. E. (1984). General transfer across sensory modalities survives reductions in the original conditioned reflex in the rabbit. *Animal Learning & Behavior, 12,* 129–136.

Kehoe, E. J., & Schreurs, B. G. (1986). Compound conditioning of the rabbit's nictitating membrane response: Test trial manipulations. *Bulletin of the Psychonomic Society, 24,* 79–81.

Kimble, G. A. (1961). *Hilgard and Marquis' conditioning and learning.* New York: Appleton-Century-Crofts.

Klopf, A. H., Weaver, S. E., & Morgan, J. S. (1993). A hierarchical network of control systems that learn: Modelling nervous system function during classical and instrumental conditioning. *Adaptive Behavior, 1,* 263–319.

Konorski, J. (1967). *Integrative activity of the brain: An interdisciplinary approach.* Chicago: University of Chicago Press.

Konorski, J., & Lawicka, W. (1959). Physiological mechanisms of delayed reactions: 1. The analysis and classification of delayed reactions. *Acta Biologiae Experimentalis, 19,* 175–197.

Mackintosh, N. J. (1975). A theory of attention: Variation in the associability of stimuli with reinforcement. *Psychological Review, 82,* 276–298.

Mackintosh, N. J. (1976). Overshadowing and stimulus intensity. *Animal Learning & Behavior, 4,* 186–192.

Macrae, M., & Kehoe, E. J. (1995). Transfer between conditional and discrete discriminations in conditioning of the rabbit nictitating membrane response. *Learning and Motivation, 26,* 380–402.

Massaro, D. W. (1988). Some criticisms of connectionist models of human performance. *Journal of Memory and Language, 27,* 213–234.

McClelland, J. L., Rumelhart, D. E., & Hinton, G. E. (1986). The appeal of parallel distributed processing. In D. E. Rumelhart & J. L. McClelland (Eds.), *Parallel distributed processing: Explorations in the microstructures of cognition* (pp. 3–44). Cambridge, MA: MIT Press.

Meehl, P. E. (1945). An examination of the treatment of stimulus patterning in Professor Hull's "Principles of Behavior." *Psychological Review, 52*, 324–332.

Miller, R. R., & Matzel, L. D. (1989). Contingency and relative associative strength. In S. B. Klein & R. R. Mowrer (Eds.), *Contemporary learning theories: Pavlovian conditioning and the status of traditional learning theory* (pp. 61–84). Hillsdale, NJ: Lawrence Erlbaum Associates.

Pavlov, I. P. (1927). *Conditioned reflexes* (G. V. Anrep, Trans.). London: Oxford University Press.

Pearce, J. M. (1987). A model for stimulus generalization in Pavlovian conditioning. *Psychological Review, 94*, 61–73.

Pearce, J. M., & Hall, G. (1980). A model for Pavlovian learning: Variations in the effectiveness of conditioned but not of unconditioned stimuli. *Psychological Review, 87*, 532–552.

Razran, G. (1939). Studies in configural conditioning: 1. Historical and preliminary experimentation. *Journal of General Psychology, 21*, 307–330.

Razran, G. (1965). Empirical codifications and specific theoretical implications of compound-stimulus conditioning: Perception. In W. F. Prokasy (Ed.), *Classical conditioning* (pp. 226–248). New York: Appleton-Century-Crofts.

Razran, G. (1971). *Mind in evolution.* New York: Houghton Mifflin.

Rescorla, R. A. (1969). Pavlovian conditioned inhibition. *Psychological Bulletin, 72*, 77–94.

Rescorla, R. A. (1972). "Configural" conditioning in discrete-trial barpressing. *Journal of Comparative and Physiological Psychology, 79*, 307–317.

Rescorla, R. A. (1973). Evidence for "unique stimulus" account of configural conditioning. *Journal of Comparative and Physiological Psychology, 85*, 331–338.

Rescorla, R. A. (1985). Inhibition and facilitation. In R. R. Miller & N. E. Spear (Eds.), *Information processing in animals: conditioned inhibition* (pp. 299–326). Hillsdale, NJ: Lawrence Erlbaum Associates.

Rescorla, R. A., & Wagner, A. R. (1972). A theory of Pavlovian conditioning: Variations in the effectiveness of reinforcement and nonreinforcement. In A. H. Black & W. F. Prokasy (Eds.), *Classical conditioning II* (pp. 64–99). New York: Appleton-Century-Crofts.

Rudy, J. W., & Wagner, A. R. (1975). Stimulus selection in associative learning. In W. K. Estes (Ed.), *Handbook of learning and cognitive processes* (pp. 269–303). Hillsdale, NJ: Lawrence Erlbaum Associates.

Rumelhart, D. E., Hinton, G. E., & Williams, R. J. (1986). Learning internal representations by error propagation. In D. E. Rumelhart & J. L. McClelland (Eds.), *Parallel distributed processing: Explorations in the microstructures of cognition* (pp. 318–362). Cambridge, MA: MIT Press.

Schmajuk, N. A., & DiCarlo, J. J. (1992). Stimulus configuration, classical conditioning and hippocampal formation. *Psychological Review, 99*, 268–305.

Schmajuk, N. A., & Moore, J. W. (1988). The hippocampus and the classically conditioned nictitating membrane response: A real-time attentional–associative model. *Psychobiology, 16*, 20–35.

Schreurs, B. G., & Kehoe, E. J. (1987). Cross-modal transfer as a function of initial training level in classical conditioning with the rabbit. *Animal Learning & Behavior, 15*, 47–54.

Sutherland, N. S., & Mackintosh, N. J. (1971). *Mechanisms of animal discrimination learning.* New York: Academic Press.

Sutton, R. S., & Barto, A. G. (1981). Toward a modern theory of adaptive networks: Expectation and prediction. *Psychological Review, 88,* 135–171.

Sutton, R. S., & Barto, A. G. (1990). Time-derivative models of Pavlovian reinforcement. In M. Gabriel & J. W. Moore (Eds.), *Learning and computational neuroscience* (pp. 497–537). Cambridge, MA: MIT Press.

Thorndike, E. L., & Woodworth, R. S. (1901). The influence of improvement in one mental function upon the efficiency of other functions. *Psychological Review, 8,* 247–261.

Wagner, A. R., Logan, F. A., Haberlandt, K., & Price, T. (1968). Stimulus selection in animal discrimination learning. *Journal of Experimental Psychology, 76,* 171–180.

Weiss, S. J. (1972). Stimulus compounding in free operant and classical conditioning: A review and analysis. *Psychological Bulletin, 78,* 189–208.

Weiss, S. J. (1978). Discriminated response and incentive processes in operant conditioning: A two-factor model of stimulus control. *Journal of the Experimental Analysis of Behavior, 3,* 361–381.

Whitlow, J. W., & Wagner, A. R. (1972). Negative patterning in classical conditioning: Summation of response tendencies to isolable and configural components. *Psychonomic Science, 27,* 299–301.

4

The First Principle
of Reinforcement

Peter R. Killeen
Arizona State University

THE THREE PRINCIPLES

Three basic processes are central to the field of learning and performance: motivation, response competition, and association. A recent theory of reinforcement—a *mechanics of reinforcement* (Killeen, 1994, 1995)—epitomizes these processes in three principles:

1. Activation. The first principle states a simple relation: An incentive activates a seconds of responding. If incentives are delivered at a rate of R per second, then it follows that aR seconds of responding are activated per second:

$$A = aR, \quad (1)$$

where A is the arousal level, and a is a constant called the *specific activation*. A is not the same as the measured *rate* of responding, because there may be inadequate time to emit all of the responses that are elicited by the incitement. It is the issue of how to map arousal levels onto dependent variables that is the first concern of this chapter.

2. Temporal Constraint. The execution of other responses, including responses of the same type, may block the emission of a response—either through competition for emission within a limited time, or because it ends the trial, or because the experimenter records only the percentage of trials with at least one response. The second principle formally recognizes such temporal

constraints on the dependent variable. If it requires δ seconds to emit a response, then the maximum response rate is clearly $1/\delta$. Less obvious is how rates change as they approach their asymptote. Killeen (1994) employed *blocked-counter* models as a way of accommodating them. Other models are possible. A second goal of this chapter is to restate that model, and to examine its relation to response probability.

3. Coupling. According to the mechanics, reinforcement occurs when an incentive occupies the same time frame as a response. The more responses that are in memory at the time an incentive is acquired, the more responses of that class will be reinforced. Those responses of special interest to the experimenter are called *target responses*. If memory is filled with target responses, the incentive will have a large impact on that class; if it is filled with nontarget responses, it will strengthen those to the detriment of the target class. To the extent that the design of the chamber or of the contingencies of reinforcement encourage nontarget responses, the motivation provided by the incentive will be wasted on them. To be maximally effective, incentives should be delivered when the subject's memory is filled with the target response. Experimental contingencies between responses and incentives determine the degree to which this is achieved. The efficacy of reinforcement thus depends on the degree of correlation between our definition of the act to be reinforced, as reflected in the contingencies, and the animal's memory of its behavior at the instant the incentive is delivered. Killeen (1994) provided rules for calculating that correlation, called a *coupling coefficient*, for various schedules of reinforcement. It is this recognition of the importance of coupling, determined by the conjunction of responses and incentives in memory, that constitutes the third principle. The third goal of this chapter is to clarify and reinforce the distinction between *arousal* due to incitement and *association* due to coupling.

THE INTERACTION OF THE PRINCIPLES

Incentives incite behavior diffusely, according to the first principle. *Reinforcement* occurs when that excitement is coupled to target responses according to the third principle. Manifestation of these associations as behavior is *constrained* in time according to the second principle.

The distinctive action of the first and third processes—excitement and association—must be distinctively recognized in our theories of behavior. Incentives that are not coupled to a particular response will still arouse an animal, but are unlikely to reinforce the instrumental response of interest to the experimenter—that is, are unlikely to reinforce the *target* response. Instead we may see the elicitation of substantial adjunctive, superstitious, or frustrative responses; in fact, it is the emission of such responses that we interpret as a hallmark of arousal.

Only emitted responses are reinforcable: Incipient responses not made because of response competition under temporal constraints are not available in memory to be reinforced (cf. Staddon & Zhang, 1989). The repeated delivery of incentives engenders a state of *arousal*; depending on the degree of coupling to a response, it may also *reinforce* associations. The cumulating arousal engendered by a sequence of incentives is manifest in the increasing response rate called *warm-up*. Omission of incentives *quenches* that arousal, manifesting what we may call, by symmetry, *cool-down*. Repeated coupling of a stimulus or response with an incentive strengthens the association between the two, a process called *conditioning* or *learning*. Skinner (1938) showed that association between a simple response and reinforcement may be complete after only one or two pairings. Responding may continue to increase after association is complete because the cumulation of arousal may continue over dozens of incitements. A sequence of incentives will *arouse* an organism and may also *condition* a response; withholding reinforcers will eventually *calm* an organism and may also *decondition* the response. Rescorla (1993a) showed that some degree of association may persist through cool-down; such residual associations may become manifest as spontaneous recovery when subjects are reintroduced to the chamber, or reexcited through the conditioning of other responses.

Traditionally these separate motivational and associational processes have been lumped under the singular terms conditioning and extinction, but their separate treatment as activation and association is essential. These distinctions are summarized in Table 4.1.

This chapter develops the provenance and implications of the first principle. It does not address coupling in any detail, as that does not interact with the considerations of central concern here; whenever data are modelled, the coupling coefficient is assumed constant at a value of $\zeta = 1$.

TABLE 4.1
Motivational and Associational Processes

CAUSE	PROCESS	EFFECT	OLD NAME FOR EFFECT
Deliver Incentives	Arousal	Warm-up	Conditioning, Pseudoconditioning
Correlate in Memory	Coupling	Association	Conditioning, Reinforcement, Learning
Withhold Incentives	Quenching	Cool-down	Extinction
Debase Correlation	Decoupling	Disassociation	Extinction

DERIVING THE FIRST PRINCIPLE FROM BASIC AXIOMS

In earlier work on a theory of incentive motivation (e.g., Killeen, 1979), it was shown that incentives energize animals, with each incentive adding to the arousal level, which then decays exponentially with time. At some time t after the delivery of an incentive,

$$A_t = A_0 e^{-\alpha t}, \quad (2)$$

where A_0 is the level of arousal at $t = 0$. Figure 4.1 shows the general activity of pigeons after a single feeding. The coordinates are semilogarithmic, within which Equation 2 plots as a straight line. Equation 2 is one solution of an

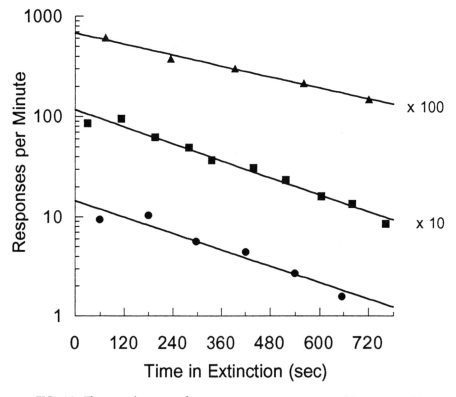

FIG. 4.1. The general activity of pigeons in extinction as measured by six moveable floorboards. The pigeons were fed once near the beginning of each session. The straight lines in these semilogarithmic coordinates indicate that the course of extinction conformed to Equation 2. The time constants of the extinction curves were all around 6 minutes, and the area under the curves was about 64 responses per incentive. The data from 3 different experiments are shown, offset by factors of 10. The figure is redrawn from the data of Killeen, Hanson, and Osborne (1978).

elementary differential equation called *Newton's law of cooling*: The rate of change in the temperature difference between an object and its context is proportional to that difference. If we call the absolute temperature difference A, then the differential equation is:

$$\frac{dA}{dt} = -\alpha A, \quad (3)$$

and one of its solutions is Equation 2.

The total arousal derived from a single feeding is the integral of Equation 2 from 0 to infinity, which is simply $A_{Total} = A_0/\alpha$.

If additional incentives occur before the arousal from the first has been exhausted, activation will accumulate, as shown in Fig. 4.2. Killeen, Hanson and Osborne (1978) showed that the average level of arousal in the interval

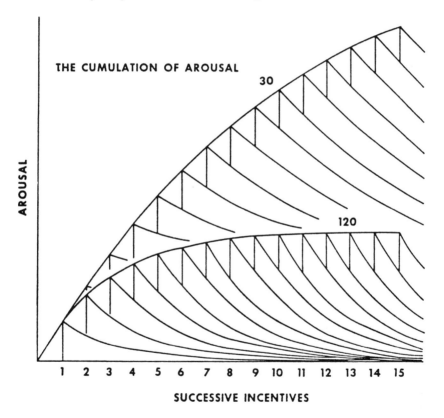

FIG. 4.2. The hypothetical accumulation of arousal after successive feedings at two different interreinforcement intervals. The component extinction curves are given by Equation 2 and their composite envelope by Equation 4. The figure is reprinted from Killeen, Hanson, & Osborne (1978) with permission of the APA.

between incentives (T)—will change with the number of incentives (n) as:

$$A_n = \frac{A_0}{\alpha T}(1 - e^{-\alpha Tn}), \qquad \alpha T > 0. \quad (4)$$

The changes in response rates of real pigeons when shifted from one feeding per day to one feeding every T seconds are shown in Fig. 4.3.

This model assumes that there is no carry-over of arousal from previous sessions, nor excitement conditioned to the experimental context. Sometimes this is justified, as seen in the first exposures in Fig. 4.3 and Figs. 4.5 and 4.6 following.

The parenthetical phrase in Equation 4 goes to 1.0 asymptotically with the number of trials, leaving $A_\infty = \alpha/T$. If we write $a = A_0/\alpha$ and rate of

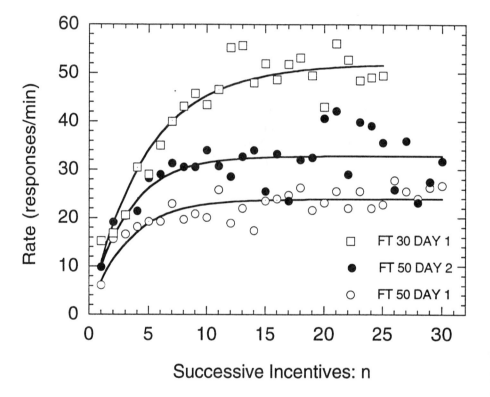

FIG. 4.3. The increase in activity measured when some of the pigeons, whose data are shown in Fig. 4.1 were switched to multiple feedings at 30- or 50-sec intervals. The functions through the unfilled symbols are the cumulative-exponential increases in arousal drawn by Equation 4, inserted into a correction for ceilings on response rate (Equation 7), and for the second day (filled symbols) for conditioning of the context that carried-over $c = 23\%$ of the previous session's arousal. The data are from Killeen, Hanson, & Osborne (1978).

reinforcement as $R = 1/T$, then the asymptotic arousal level is $A_\infty = aR$. This is the origin of Equation 1. Notice from the preceeding that the total arousal from a single incentive is also A_0/α. Therefore a is not the only constant of proportionality that translates rate of reinforcement into arousal level, it is also the energy derived from a single incentive.

This development shows why arousal is (asymptotically) proportional to the rate of reinforcement (Equation 1); that is the level that arousal settles to when the accumulation of activation from new incentives and its decumulation over time comes to equilibrium. Hanson showed that Equation 4 holds for situations in which: (a) the time between incentives is random, and (b) the incentives are of random magnitude (Killeen, Hanson, & Osborne, 1978).[1] This is an important generalization, because it extends the theory to variable schedules of incitement and variable amounts of reward, and to situations in which the impact of an incentive varies because of shifting levels of deprivation.

If response rate is proportional to arousal level, we should be able to predict the asymptotic levels of arousal given the data shown in Fig. 4.1. Those curves were described by Equation 2, which on the average had an intercept of $A_0 = 11$ responses per minute and a rate of decay of $\alpha = 1/(6 \text{ min})$; the total arousal conferred by a single incentive is therefore: $A_0/\alpha = a = 66$ responses. Several years before collecting the data shown in Figs. 4.1 and 4.3, the ebb and flow of general activity during the intervals between reinforcers had been measured by Killeen (1975). The same chamber, incentives, and subjects were employed in both studies. One of the key parameters of the resulting model was the asymptotic rate of responding projected by factoring out the competition from nontarget responses. This parameter was found to be proportional to the rate of reinforcement, $A = aR$. Figure 4.1 provided the specific data needed for an independent estimate of that constant of proportionality, a. Figure 4.4 shows the empirical values of asymptotic responding reported in 1975, along with the values predicted by Equation 1 with $a = 66$ responses/reinf, as derived from Fig. 4.1. For a true prediction across paradigms involving no free parameters, the accuracy is gratifying. This development of models for the cumulation of arousal was originally called *Incentive Theory*.

Relation to Skinner's reflex reserve. This hydraulic model of motivation is similar to Skinner's (1938) construct of the *reflex reserve*, both invoking the notion of the total energy available for the response. But Skinner understood this capacity of the reflex reserve as: (a) an attribute of the response/reflex,

---

[1]To avoid some confusion, note that in that paper a was used to signify the *time constant* of the process; here it is used as the *rate constant*, which is the reciprocal of the time constant. We continue to mention the time constant, now $1/a$, whose value takes the convenient units of seconds (rather than per-seconds), and represents the time at which an exponential process has completed $1/e = 63\%$ of its course.

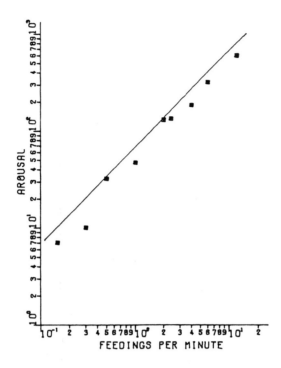

FIG. 4.4. Levels of arousal under schedules of periodic feeding, from Killeen (1975). The straight line is the level predicted by the parameters derived from Fig. 4.1 (viz., $A = 66R$).

(b) an associative ceiling, and (c) independent of rate of incitement. Because Skinner took the total capacity to specify the maximum associational strength of a response, he had to postulate other mechanisms in order to explain the large number of responses supported by intermittent schedules of reinforcement. These were the chaining of responses—the sensory consequences of a response act as conditioned reinforcers for previous responses—and the unitization of groups of responses (Skinner, 1938, p. 300). These *ad hoc* hypotheses are inadequate; on the same page on which he introduced the response unit hypothesis, Skinner noted its shortcomings. The chaining hypothesis is both empirically insufficient and logically uncertain—from whence do the putative conditioned reinforcers derive their strength? If truly a chain, why should not the early components, as signs of nonreinforcement, *punish* responding? Why should only target responses that precede a (reinforcing) lever-press be reinforced by it, and not any gratuitous response that precedes it?

For Killeen, A is: (a) a motivational factor, (b) a cumulative function of the number of incitements, whose rate (R) appears as a multiplier in Equations 1 and 4, and (c) a positive function of the magnitude and quality of incitements and level of drive. High rates of reinforcement thus naturally empower many more responses than a single reinforcer can muster. The associative factors, not emphasised here, are represented by the coupling coefficient which multiplies the motivational factor A.

For Skinner, decreases in strength were caused by the emission of responses, wheras for Killeen they decayed over time (Equation 2)—although exponential decay over time is predicted by both models. Because responses that are prevented from occurring do not usually undergo extinction, Skinner's etiology for Equation 2 may be the correct one.

APPLICATIONS OF THE MODEL TO WARM-UP

Warm-up effects are especially noteworthy in studies of avoidance, where the delivery of disincentives such as shock generates fear or anxiety, which cumulates over trials and dissipates between them. Whereas the subjects in appetitive conditioning experiments usually start the session quite hungry or thirsty, subjects in aversion experiments must acquire the drive to respond, through conditioning of fear to time and place (Hineline, 1972; Killeen, 1979). For some strains of rats this conditioning is not very effective, and there is an "almost complete lack of transfer from the end of one session to the beginning of the next" (Nakamura & Anderson, 1962, p. 747). The data of Hoffman and associates (Hoffman, Fleshler, & Chorny, 1961) are especially noteworthy because after 54 sessions of avoidance conditioning the investigators exposed the animals to additional sessions of 40 trials during which the lever was inaccessible, and then reintroduced the lever. Figure 4.5 shows the resulting data (circles), averaged for two such sessions, along with the performance of rats that were permitted to avoid shock during the first 40 trials (disks). The comparability of groups shows that the warm-up has nothing to do with relearning, but instead involves the reinstigation of fear motivation—or perhaps the habituation of competing species-typical defense reactions (Hineline, 1978a). This figure displays most graphically why the distinction between arousal and association is essential; after the first session of such experiments, most of the conditioning is remotivation, not relearning.

This assertion of minimal contextual conditioning is not novel: "The cumulation of shock over a session would gradually increase the motivation to respond but this motivation would dissipate between sessions" (Nakamura & Anderson, 1962, p. 747). "The decrement in performance at the start of each session represented a motivational phenomenon; apparently as shocks occur, their motivational after effects persist and summate to create an emotional state which somehow facilitates avoidance. When, however, the S was re-

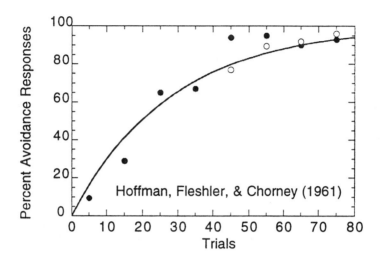

FIG. 4.5. The percentage of trials with a response in an aversive conditioning experiment conducted by Hoffman et al. (1961). The subjects contributing the data denoted by unfilled circles had no opportunity to respond during the first 40 trials, yet thereafter responded at rates comparable to the control subjects, indicating that the initial rise of the curve was due to warm-up, not associative conditioning. The curve through the data comes from Equation 4, converted to a probability by inserting it into Equation 6. The time constant was 15 min.

moved from the apparatus, this motivational state dissipated" (Hoffman et al., 1961, p. 512).

It seems strange that after 54 sessions, organisms will not come to fear the chamber, and in fact there is evidence that the conditioning session and its cues constitute an aversive event (Hineline, 1972). But the time-course of arousal drawn in Fig. 4.5 is measured using an operant manipulative response that the rat is relatively unprepared to make in such aversive situations (Bolles, 1970; Fanselow & Lester, 1988); freezing and flight are much more easily aroused by fear and coupled to cues of safety than is lever-pressing. Similar curves drawn for species-specific defense reactions (SSDRs) would reach asymptote within several trials, and relatively small amounts of carryover of arousal would engender them at high rates at the beginning of subsequent sessions. SSDRs are intrinsically more tightly coupled to fear than are the manipulative responses involved in lever-pressing. Even when the operant response is learned, it is often accomplished by the redirection of aggressive responses, as witnessed by the chewed and shattered levers found in many shock-avoidance chambers. By requiring greater motivation to be manifested, unprepared responses give us a picture of the cumulation of arousal that is not available from SSDRs, which reach their ceiling almost immediately.

Historically, issues of response *preparedness* were treated as issues of differential readiness to form an associative bond. The issue here is one of performance, not just association. After 54 sessions, the animals will have learned all they ever will about the association of lever-pressing and surcease from shock. It takes more fear motivation to engender a lever-press than a jumping-out of the box response; just as it may take more food motivation to engender jumping than lever-pressing. Arousal is treated as a general construct here, as models for its accumulation and decumulation may be general; however there are important differences between different types of arousal, and the most important may be the differential ease with which they support different action patterns. This constitutes a difference in coupling, not a central issue here.

Data on acquisition of the eyeblink response using classical conditioning were published by Kimble and associates (Kimble, Mann, & Dufort, 1955; see Fig. 4.6). They are strikingly similar to those shown in Fig. 4.5, because a period of trials in which there was no CS (and thus no CR possible) seemed just as effective as trials in which a CS was present. Such continued strengthening of the CR by UCS-only trials had been called *pseudoconditioning*. Kimble (1961) suggested that "*pseudoconditioning may be a part of all conditioning; in which*

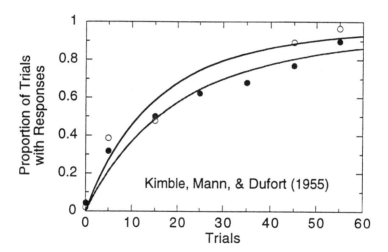

FIG. 4.6. Proportion of trials with a conditioned eyeblink response, from data collected by Kimble et al. (1955). The subjects contributing the data denoted by unfilled circles had no opportunity to avoid the puff during trials 21 to 40, yet thereafter responded at rates comparable to the control subjects, indicating that the initial rise of the curve was due to warm-up, not associative conditioning. The curve through the data comes from Equation 4, converted to a probability by inserting into Equation 6. The time constant was 11 min. The model required that the aversive value, a, of a puff to an open eye be three times the value of one to a closed eye (which was 0.036 s^{-1}).

a noxious stimulus is employed" (p. 62). A more appropriate name for it is *remotivation,* or *warm-up.* Kimble acknowledged that his data provided:

> support for the existence of two, independently manipulable, processes in eyelid conditioning. The first is an *associative* factor which depends on the presentation of the CS and UCS together. Apparently it develops rapidly. Certainly, after 20 conditioning trials, whether the CS is presented or not is irrelevant so long as the UCS is presented. The behavioral changes which are usually plotted as learning curves for eyelid conditioning probably reflect mainly the development of a second, *performance,* factor [which is] a motivational one. (Kimble, Mann, & Dufort, 1955, p. 415)

Figure 4.6 shows the "learning" curves for the eyeblink response for a group of subjects that received CS–UCS pairing on every trial (disks) and a group for which the CS was omitted on trials 21 to 40 for the experimental group (circles). If anything, the experimental group performed better. Kimble and associates (1955) allowed that "the most likely explanation for this appears to be that Ss in the experimental groups receive the air puff on the open eye more often than Ss in the control group" (p. 415). Despite this recognition, their additional comparisons of instrumental and classical conditioning of the eyeblink were interpreted as showing the superiority of classical conditioning for that reflex; but the data are equally well interpreted as being due to the greater number of presentations of the UCS that the classical conditioning paradigm guaranteed. That is, the critical distinction between operant and respondent effectiveness was not a fundamental one of differential association. but rather an incidental one of differential motivation due to differential rates of incitement.

These considerations involve arousal level, but what is the relation of that to dependent variables such as probability or rate of responding? It is time to develop a model that will permit us to convert arousal levels predicted by Equation 1 into more traditional dependent variables.

From Arousal to Responding

This theory of the arousing effects of single incentives and of the cumulation of arousal when additional incentives follow closely in time provides a grounding of Equation 1. But we need to be more explicit about how this arousal is converted into behavior. We treat two cases, the relation of arousal to probability of a response, and then to response rate.

The early work on Incentive Theory did not take into account the duration of a response. Response duration plays two roles: (a) It establishes a ceiling that response rates cannot exceed; this is what we mean by temporal restraint. (b) It requires more arousal to motivate extended or ponderous response, so that Equation 1 becomes $A = aR/\delta$. This may not always be the case (Bizo &

Killeen, 1997), but data are not available for a more precise formulation. In this model A represents responses/second, a response-seconds/reinf, R reinf/second, and δ seconds/response.

Response Probability. Consider a very brief epoch of time, Δ, during which no more than one response can occur. Then the probability of making a response during that epoch is proportional to the length of the epoch and the level of arousal: $p_d = \Delta A$. This implies an exponential distribution of times between responses, which is in rough conformity with the distributions of interresponse times typically found, and also with the extensive data set and model of Palya (1992). How do we compute the probability of making a response over a longer time, such as during the 4-second trials used by Hoffman and associates? Once the duration of the epoch becomes long enough so that there is some chance of more than one response occurring, we must calculate the probability of a response at each instant, preceded by its nonoccurrence during the prior instants, and sum that instantaneous probability over all instants in the trial. This results in a geometric progression which, in the limit as Δ becomes very small, converges to the continuous exponential:

$$\frac{1}{A}e^{-dA}. \quad (5)$$

The probability of a response during a trial of duration d is the integral of Equation 5, a cumulative exponential distribution on trial duration d:

$$p_d = 1 - e^{-dA}. \quad (6)$$

The curves through the data in Figs. 4.5 and 4.6 are derived by inserting Equation 4 into Equation 6. As trial duration increases, the probability of a response should also increase. Changing the value of d in Equation 6 suitably generates curves parallel to the one shown, above it for $d > 4$, and below it for $d < 4$. If the organism is under good control by the trial stimulus, then the value of T in Equation 4 is also equal to d. After many trials, the asymptotic probability of a response is $p_d = 1 - e^{-daR}$.

Response Rate. Data such as those shown in Figs. 4.5 and 4.6 derived from the occurrence or nonoccurrence of a single response during a trial. As that dependent variable contributes its maximum value (1.0) as soon as a single response occurs, there is no constraint due to competition from other target responses within a limited time span. How do we predict free-operant responding in situations where animals may be spending virtually all of their time emitting the target response? There is simply not enough time for an aroused animal to make responses of finite duration at a rate strictly proportional to its arousal level, as incipient responses cannot occur if other responses are already in progress. A *Type I blocked-counter model* has been developed to accommodate the problem of *ceilings* on rate (Killeen, 1979, 1981, 1994). It assumes that any incipient responses which cannot be emitted because another response is in progress are simply lost to the counter. A related *Type II* model (Killeen, 1994, Equation B8) is appropriate if the incipient responses are not only blocked, but also extend the down-time as though they had actually been emitted. Schwarz (1989) provided an extended analysis of Type II counters, along with a good introduction to the issues and bibliography. For Type I counters where the average time required to emit a response is d, the measured response rate, B, is related to the instantaneous probability of of a response, A, by:

$$B = \frac{A}{1+\delta A} \quad (7)$$

Appropriate expressions for arousal level may be inserted into this equation to predict response rates. If we insert Equation 4 into Equation 7, the curves shown through the unfilled circles in Fig. 4.3 result. (These data are sufficiently below their ceiling that a comparable fit is obtained with Equation 4 alone). A very simple derivation of Equation 7 is to assume that responses are initiated at a rate of A, but because it then requires δ seconds to emit the response, the average interresponse time is $\delta + 1/A$. The average response rate is the reciprocal of the average interresponse time, which is $A/(1 + \delta A)$.

In the case of asymptotic arousal levels, we substitute $A = aR/\delta$ into Equation 7 to find:

$$B = \frac{R}{\delta(R+\frac{1}{a})}, \qquad\qquad \delta, a > 0, \quad (8)$$

which is Herrnstein's (1979) hyperbola. This model was also derived by Staddon (1977), from different premises.

Because we have not taken into account how the coupling of incentive motivation to behavior is affected by the particular contingencies of reinforcement, Equations 7 and 8 are not quite complete. However, the exact predictions for ratio and interval schedules inherit the basic hyperbolic shape of Equation 8, with the main difference being a new scaling factor, the coupling coefficient. Because this coefficient is usually less than 1.0, this keeps rates below their theoretical maximum of $1/\delta$ (Killeen, 1994).

Savings.   Arousal existing at the start of a session (call it s) that is stimulated by handling or transportation to the experimental chamber will elevate the responding at the beginning of a session and wash out exponentially (according to $se^{-\alpha d}$) as time elapses. Adding a savings of s to Equation 4, the response strength d seconds into a session becomes:

$$A_d = A(1 - e^{-\alpha d}) + sAe^{-\alpha d}, \quad (9)$$

where, as before, $A = aR/\delta$. In order to write this more concisely, introduce the quenching parameter $q = 1 - s$, which gives the proportion of the arousal lost between sessions, and we can rewrite:

$$A_d = A(1 - qe^{-\alpha d}), \qquad\qquad 0 \leq q \leq 1. \quad (10)$$

If $q = 0$ there is no loss, and each session starts at its asymptotic level A. If $q = 1$, each session starts from scratch. We apply these results to the data of Powell (1972), who studied warm-up and cool-down in various rodents. In one condition, Powell conducted sessions of 1, 5, 15, 30, 45, or 60 minutes, and thereafter measured avoidance responses of laboratory rats in "extinction" when no shocks were delivered. The term *extinction* is hedged to remind us that absence of shocks *cools-down* the organism, but it does not necessarily decondition the response, as is often implied by the term extinction. Each rat experienced repeated extinction after these sessions, of varying duration in different orders. The initial levels of responding in extinction are well-predicted by Equation 10 with a time-constant of 12 minutes (see Fig. 4.7), and quenching of $q = 67\%$. For these rats, $\delta = 0.25$ sec, a typical value for the minimum lever-press interresponse time.

It is also possible that some fraction of the level of arousal throughout a session will be conditioned to the experimental and temporal context. The relevant model is a simple geometric progression, but will not be detailed here. It results in the multiplication of Equation 10 by the sum of the progression, which is $(1 - c^N)/(1 - c)$, where c is the proportion of context conditioning that is retained from one session to the next, N is the number of the session, and $0 \leq c < 1$. For the data from the second session shown in Fig. 4.3, $c = 0.37$.

EXTINCTION AS COOL-DOWN

What happens when the shock- or pellet-dispenser, is unplugged? The basic theory (Equation 2) requires that arousal decay exponentially by the factor $e^{-\alpha t}$, where t is the time in extinction. Multiplying Equation 10 by this factor yields the predicted changes in arousal level through periods of warm-up and cool-down:

$$A_{d,t} = A(1 - qe^{-\alpha d})e^{-\alpha t}. \quad (11)$$

Inserting this into Equation 7, we may predict the cool-down of Powell's (1972) rats, *using the same parameters* derived in fitting the data in Fig. 4.7: a takes the value 63 response-seconds per incitement, the time-constant 12

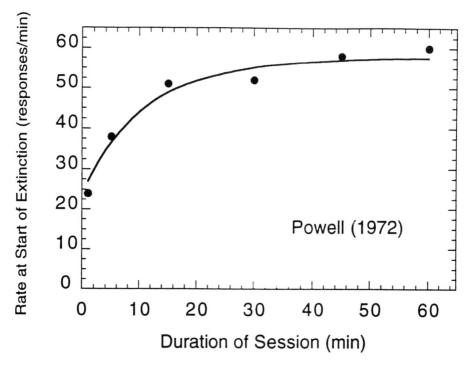

FIG. 4.7. Response rates of albino rats in a shock-avoidance experiment, shown as a function of the amount of time into the session. The data are from Powell (1972), and the curve from Equation 10 inserted into Equation 7.

minutes, δ = 0.25 seconds, and q = 0.67. The results are shown in Fig. 4.8. The predictions are within the standard error bars of the data (not shown).

Just as warm-up may be manipulated by varying the session length, cooldown may be manipulated by varying the duration of the intertrial interval (ITI) or the events occurring during same. In a careful series of experiments, Miller (1982) demonstrated how noncontingent presentation of reinforcers (sucrose and footshock) during the ITI enhanced performance. Hineline (1978a) showed that time-outs inserted in avoidance sessions decreased response rates. The same Equation 11 may be used to fit the data. In this case, the decay of arousal during the interruption follows the course $q = e^{-\alpha t}$, where t is the duration of the interruption. When t = 0, there is no interruption and the session continues with no loss of arousal. As t increases, more of the momentum is lost; in the limit of a very long interruption, the savings falls toward zero and must be rebuilt as though a new session were starting, as was the case for the rats of Hoffman and associates. The functions describing the change in response and shock rates found by Hineline are well-described by

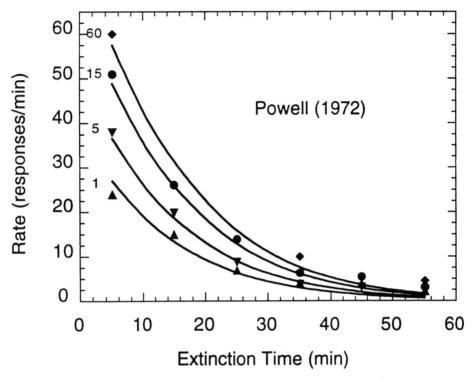

FIG. 4.8. Response rates of albino rats in a shock-avoidance experiment, shown as a function of the amount of time into extinction, after warm-up sessions of various durations. The data are from Powell (1972), and the curves from Equation 11 inserted into Equation 7, using the same parameters as those determined from Fig. 4.7.

Equation 11 with a time constant of 15 minutes. In another study, Hineline (1978b) found uniform half-lives for warm-up averaging about 18 minutes for shock intervals (SS = RS) ranging from 5 to 60 seconds.

We may write the explicit equation for extinction from asymptotic responding by inserting Equation 11 into Equation 7. For asymptotic rates of responding, the parenthetical term representing warm-up has gone to 1, and we find:

$$B = \frac{aR/\delta}{aR+e^{\alpha t}}, \qquad\qquad \delta > 0. \quad (12)$$

At the start of extinction $t = 0$ and this reduces to the asymptotic rate given by Equation 8. As time passes the denominator grows exponentially, and rates fall toward zero. If the rate of reinforcement during the conditioning phase (R) was low, this decrease is approximately exponential, like the curves shown in Fig. 4.1. But when aR is large, response rates bend under their ceiling, and in extinction they unfold from it as an ogival function of time. Equation 12 can be recognized as a decreasing logistic equation if we rearrange it into its classic logistic form by setting its mean $\mu = \alpha\ln(aR)$, and its time constant $k = 1/\alpha$. Then:

$$B = \frac{1}{\delta(1+e^{(t-\mu)/k})}, \qquad\qquad k,\delta > 0. \quad (13)$$

When $t = \mu$, the rate has fallen to half its maximal value ($1/\delta$). In the case of higher rates of reinforcement before the start of extinction, the half-maximal point falls farther to the right ($\mu = \alpha\ln(aR)$), making the ogival form more apparent (see Figs. 4.10 and 4.11).

Extinction from Positive Reinforcement

Most of the many studies of extinction to be found in the literature report only the total number of responses in extinction (Nevin, 1988). Because this conflates the roles of level and rate of decay, they are unuseful in assessing the generality of the *Partial Reinforcement Extinction Effect* (PREE), which states that extinction takes longer after schedules of partial reinforcement. This is important because the development thus far has assumed constant rate of decay. Nevin's reanalyses of much of the operant data suggest that the rate of decay through the course of extinction is invariant, as hypothesized: After correcting for differences in response rate at the beginning of extinction, rates fall off as parallel functions of time. Figure 4.9 displays one set of data collected by Nevin and associates; in both studies responses maintained by different rates or amounts of reinforcement were extinguished simultaneously, as components of multiple schedules. The rate of extinction (α) is the same across conditions. But it is not always so simple. In extinction from different Fixed-Interval schedules, Hanson and Killeen (unpublished data) found decay rates to vary as a function of training conditions, so that the rate-constants of the extinction curves were proportional to the rate of reinforcement during training. Figure 4.10 shows their data, fit with Equation 12 with $a = 250$, and with α equal to 10% of the FI value under which the animals were trained. Notice that the shapes of the curves change from mildly convex at the highest baseline rate of reinforcement to concave at the lowest rate of reinforcement, as predicted by the logistic model. The x-axes are scaled in proportion to the interreinforcement interval during training; the visual similarity of the extinction slopes in these normalized coordinates therefore reflects their proportionality to the rates of reinforcement received during training. This exemplifies the PREE.

Figure 4.11 shows a summary of extinction rates found in selected studies. These data were not captured by Nevin's survey of the literature because they did not include a condition of continuous reinforcement and multiple sessions of extinction, one of his criteria for inclusion in his review.

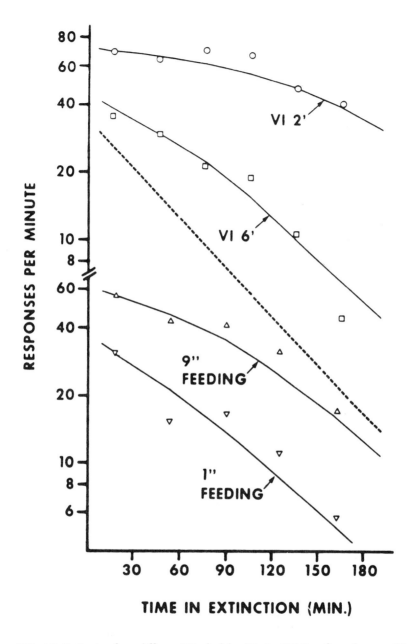

FIG. 4.9. Extinction from different VI schedules (Nevin, 1974) or from the same VI schedule with different amounts of reinforcement (Shettleworth & Nevin, 1965). The theoretical curves are from Equation 12, with a time constant of 60 min and d of 0.8 for all curves. Note that training and extinction occurred in the context of multiple schedules. The figure is from Killeen (1979).

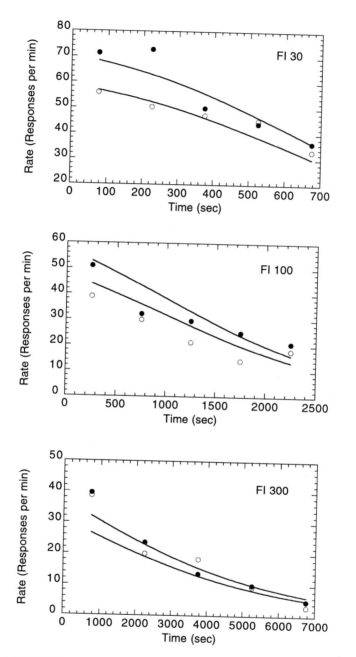

FIG. 4.10. Extinction curves averaged over four pigeons after Fixed Interval feeding of 30, 100, and 300 sec. Data are from the first (filled circles) and second (open circles) sessions of extinction. Equation 12 governed the curves through the data.

146

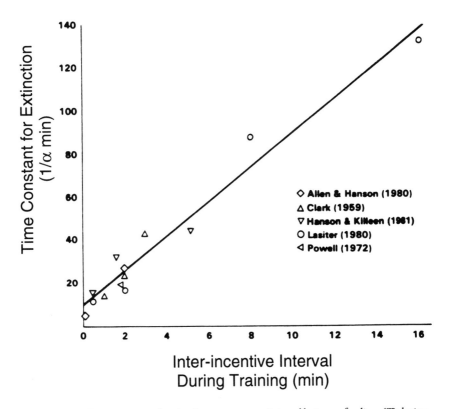

FIG. 4.11. Time constant of extinction curves versus interval between feedings (*T*) during conditioning. The linear function through the data is consistent with Equation 18. Figure from Killeen (1981).

Life would be simpler if responding always decreased at the same rate, but such is not the case. Nevin (1988) also noted that the PREE *is* regularly found in instrumental (runway) studies: "Between-groups research has demonstrated repeatedly that running persists at high speeds for many more trials after training on a partial reinforcement schedule than after CRF" (p. 53). Figure 4.12 gives one example, from a study by Jensen and Cotton (1950) in which rats were trained to run alleys under either 100% or 50% reinforcement. The differences in rates of extinction are obvious—speed decreses twice as fast under the CRF condition as under the 50% condition. The partially-reinforced subjects manifest higher running rates in conditioning as well as a slower decrease in extinction (PREE). The PREE, the conditions necessary for its manifestation, and the considerations necessary for their explanation provides important constraints in the development of all theories of conditioning and performance.

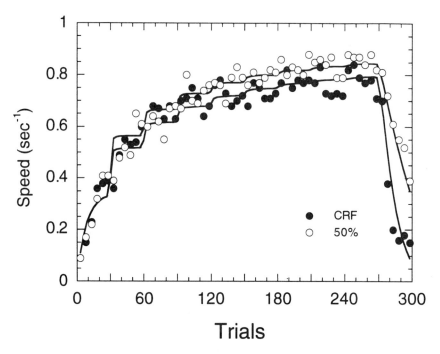

FIG. 4.12. Running speeds for groups of rats trained on a 100% reinforcement (CRF) on days 1–9 and put into extinction on the 10th day; or on 100% for Day 1, 50% reinforcement for Days 2–9, and put into extinction on the 10th day. Thirty trials were run per day. The data are from Jensen & Cotton (1950). The curves through the data are generated by Equation 11 inserted into Equation 7, applied within and across trials. The time constant took a value of 5 min for the 100% animals and 10 min for the 50% animals.

Reconciling Incentive Theory with the PREE

The original model for the accumulation of arousal—Incentive Theory—assumed an exponential rate of decay of arousal with a constant rate parameter; yet the PREE provides evidence that decay rate increases with the rate or probability of incitement during conditioning. This constitutes a fundamental problem for the theory.

The exponential decay derives from the basic intuition that the rate of change in arousal with respect to time will be proportional to the current arousal level—Newton's law of cooling again: $dA/dt = -\alpha A$ and is the origin of Equation 2. Figures 4.10, 4.11, and 4.12 show that the rate is slower after larger values of T, and in particular that approximately $1/\alpha = kT$. This is as it should be for optimal responsiveness to the environment (McNamara & Houston, 1987). What would be a mechanism for such an adjustment, and how can that be reconciled with the logic of Incentive Theory? Let us examine various hypotheses.

1. If the rate of decay decreased *continuously* as a function of time since the last reinforcer, we could write the above differential as $\frac{dA}{dt} = \frac{-\alpha A}{t+c}$. The solution of this equation is a power function, which predicts that arousal would unfold as a linear function in log-log coordinates:

$$A_t = \frac{A_0}{(t/c+1)^\alpha} \quad (14)$$

where A_0 is arousal level at the beginning of extinction. If rates are at all close to their ceiling, this would need to be embedded in Equation 7 to predict response rates. Equation 14 has been used for the discounting of delayed rewards by Rachlin (1989), and by Green, Fry, and Myerson (1994). This is a viable model of the extinction process, but is not explored further here.

2. In 1984 a model was developed for the adjustment of the rate constant a that accounted for much of the data on rate of conditioning in the auto-shaping paradigm (Killeen, 1984). It was proposed that the time constant $1/\alpha$ is itself an exponentially-weighted moving average of the time between incentives.[2] But later (1991) it was realized that a special updating of the time-constant is not necessary. Consider what happens if the estimate of T is updated upon the delivery of the incentive. The estimate will be updated more frequently at higher rates of reinforcement than at lower ones. This guarantees the proportionality of α to the rate of reinforcement during the acquisition process. It provides for the quicker acquisition under high rates of reinforcement than low, which was needed to model the auto-shaping data. Let us see how that works.

The equations developed for the cumulation and decumulation of arousal are continuous versions of an averaging process called an *Exponentially-Weighted Moving Average*, or EWMA (Killeen, 1981), a central model of the more general Incentive Theory (Killeen, 1979). According to this treatment, the animal's estimate of the interval between reinforcers, τ, is updated as a linear average of the old estimate and the newest sample:

$$\tau_{new} = \alpha T + (1-\alpha)\tau_{old}. \quad (15)$$

Arousal level is proportional to the updated estimate of the rate of reinforcement $(1/\tau_{new})$.

If the animal's estimate of the time between incentives is updated on the occasion of each incitement—if Equation 15 is iterated with each reinforce-

---

[2]It is the time between incentives, and a fortiori the time constants, that must be averaged; if either the rates of reinforcement or the rate constants are updated, the time constant will asymptotically approach proportionality with the harmonic mean of the inter-food intervals. This would imply faster extinction under variable interval schedules, which is counterfactual.

ment—the *effective* rate constant will assume proportionality with the rate of reinforcement (Killeen, 1991). To show this, insert the new estimates of τ derived from Equation 15 into the right-hand side, and iterate the equation again and again, once for each of the j incentives in a fixed interval of time. Some algebra shows that:

$$\tau_{new} = \alpha T(1+(1-\alpha)+(1-\alpha)^2+\ldots+(1-\alpha)^{j-1})+(1-\alpha)^j \tau_{old}.$$

Summing the series over the j reinforcers gives:

$$\tau_{new} = (1-(1-\alpha)^{j-1})T+(1-\alpha)^j \tau_{old}.$$

For small values of α, we may approximate the coefficient of T by replacing $(1-\alpha)^{j-1}$ with the first two terms of its series expansion, resulting in:

$$\tau_{new} = \alpha'T+(1-\alpha')\tau_{old}, \quad (15')$$

with the effective $\alpha' = j \log[1/(1-\alpha)]$. This is Equation 15 with a currency parameter α' proportional to j. α is the intrinsic rate constant, measurable in experiments such as those whose results are shown in Fig. 4.1.

Thus, iteration of the update equation with each reinforcement yields a new EWMA, one whose effective currency parameter α' is proportional to the number of reinforcers per unit time. This treatment provides a mechanism for the quicker acquisition under high rates of reinforcement than low. It provides a mechanism for the apparent changes in the rate constant that were merely posited in the previous section.

Under this interpretation the underlying rate constant α is not adjusted: It is constant both during acquisition and extinction. But what about Figs. 4.10 to 4.12? There are no deliveries of reinforcement in extinction, and so updating should not occur. Under this model the proportionality of α' with the rate of reinforcement was an artifact of the lock-step with the deliveries of incentives; when they are gone, the update is no longer driven by reinforcement, and decay of arousal should then unfold according to the underlying value α; this analysis thus fails to predict the PREE.

3. What if the updating occurs with each tick of an internal clock, whose time constant is proportional to the estimated interval between reinforcers? Now reinforcement rate drives clock rate, but, because the clock is endogenous, it continues to pace the extinction process. τ then becomes the period of the clock, and $\tau_{new} = \alpha' (kT) + (1 - \alpha') \tau_{old}$. This analysis saves the PREE, as the clock will continue to tick in extinction, and thus show a greater rate of decrease after high rates of reinforcement than after low. The extinction curves are no longer logistic: Rather than describe straight lines in semilogarithmic coordinates after the curvature due to ceilings on rate has played out, they

would be concave-up, a series of segments of decreasing slope, each ended by a tick of the clock, with each tick taking longer to occur.

4. What about the data in Fig. 4.1? If there ever were empirical exponential decay processes, those functions exemplify them, showing no departure from a straight line in those semilog coordinates.

Perhaps the speed of the clock is a *linear* function of the rate of reinforcement with a nonzero intercept; after all, there is always some minimal, nonzero level of activation in living organisms. That would allow the straight lines shown in Fig. 4.1, because the speed of the clock will be at its minimal value in those experiments and extinction would play out at its constant minimal rate. This idea is developed next.

5. Rather than maintain the clock speed as a mediating variable, it is somewhat more parsimonious to develop the model in terms of the effect of reinforcement rate on the organism's arousal level. In addition, some experiments suggest analogous effects with other variables that affect arousal level: Gracely and Church (1976), for instance, found that performance adjustments to intense shocks occurred more quickly than adjustments to less intense shocks.

If we replace the rate constant $-\alpha$ in Equation 3 with a linear function of arousal level $-(\beta A + \alpha)$, we get:

$$\tfrac{dA}{dt} = -(\beta A + \alpha)A. \quad (16)$$

For $\beta = 0$, Equation 16 reduces to Equation 3.
For $\alpha = 0$,

$$\tfrac{dA}{dt} = -\beta A^2,$$

whose solution is a hyperbola:

$$A_t = \tfrac{A_0}{A_0 \beta t + 1}. \quad (17)$$

Equation 17 has been proposed as a model of the extinction process by Mazur (1984). When A_t is converted to response strength (divided by δ) and inserted in Equation 7, the model for extinction becomes:

$$B_t = \tfrac{A_0}{\delta[1 + A_0(\beta t + 1)]}, \qquad \delta > 0. \quad (18)$$

Equation 18 provides a fine model of the extinction process. Figure 4.13 shows the average data from Fig. 4.10 replotted with the curves generated by Equation 18, with values of 0.01 min for δ and 1/(600 sec) for β for all curves, and with different initial arousal levels A_0 for each of the curves. The x-axis is logarithmic to permit all curves to fit in the same panel. Unlike the model drawn in Fig. 4.10, the rate constant β is invariant—as would be hoped, given

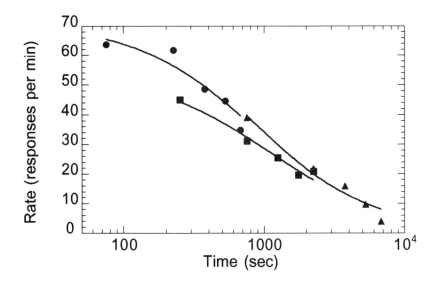

FIG. 4.13. Extinction curves averaged over four pigeons after Fixed Interval feeding of 30, 100, and 300 sec. Data are averaged over the first two sessions of extinction, which were shown in Fig. 4.10. Equation 18 governed the curves through the data.

that that was the motivation for developing the model. The hyperbola shows its tail at the rightmost part of the graph, where it hangs higher than the data. This is because Equation 17 has the rate of decay go to zero as arousal level goes to zero. This is not the case for Equation 16, where a minimum rate of decay is maintained by α, which lets the curves progress to zero more quickly at the lowest levels of arousal.

The complete model—Equation 16—with both α and β greater than zero—may be separated using the method of partial fractions and solved to yield:

$$A_t = \frac{\alpha A_0}{(\beta A_0 + \alpha)e^{\alpha t} - \beta A_0} \qquad (19)$$

Although an unfamiliar equation, it has the required properties: For low levels of initial arousal (A_0) it is an approximately exponential decay. As β goes to zero, it also reduces to an exponential decay and promises to retain parameter invariance across extinction from different intermittencies of reinforcement during training.

Equation 19 may be inserted into Equation 7 to predict response rates; however, the simpler Equation 18 provided a sufficiently good fit to these few data that there is not enough power here to yet justify the more general model.

6. What then of Nevin's (1988) data that suggest an invariant rate of extinction for free-operant experiments? Is it not futile to devise a model for the PREE, if that only occurs in runway experiments, themselves nowadays in the last stages of extinction? Equation 19 has the flexibility to deal with both constant and varying rates of extinction. Because all of the runway studies are with rats, it may turn out that rats simply have a larger value for β than pigeons, and that would be that. Conversely, most (but not all) of Nevin's analyses concerned changes in response rate over successive sessions of extinction, not over time within a session. If the experimental context loses the fraction 1 - c of its conditioned arousal value from one session to the next, then we would expect a geometric decrease in responding from one session to the next equal to c^n, independent of training conditions. This decrease would plot as a straight line on semilogarithmic coordinates, as Nevin found his data to array themselves. The common rates of extinction shown in Fig. 4.9 for within-session data may have been due to the conditioning of arousal to the chamber and its simultaneous decay for both schedules, which were extinguished contemporaneously.

Where do these models leave us for the acquisition of arousal—for warm-up, one of the key concerns of this chapter? The warm-up process will be slightly different—accelerating more slowly when arousal levels are low at the beginning of conditioning, and faster as conditioning proceeds, and will reach asymptote sooner under high rates of reinforcement. The analytic expression for the curves appears to be an unfriendly continued fraction. The shape of the curves will in any case be a less sensitive discriminator than parameter invariances across different schedule transitions. These analyses—within versus between session decay functions, rat versus pigeon data, and correlation of the pacemaker posited here with those assumed in studies of timing—now have a theoretical structure to motivate them. Using that map to find our way through the maze of relevant data is an adventure for another day.

CONCLUSIONS

This chapter constitutes a reworking of Incentive Theory, developed to take into account ceilings on the dependent variable, due either to the measurement of behavior as a probability, or to its measurement as a rate in competition with the emission of other responses. A straightforward model of savings was outlined and applied to data. The incorporation of acquisition and extinction into Killeen's (1994) mechanics of behavior was not straightforward. In large part this is because psychologists have not formulated effective low-level treatment of the independent variables they use. After many years of research, the simplest laws relating latency, probability, and rate of responding have not been developed, let alone integrated with runway speed or acceleration. Many of the studies bearing on partial reinforcement effects were painstakingly thorough, but the data were reported as total number of responses in extinction, a virtually useless DV. Those that reported running rates in alleys often did not separately report rates from the different segments of the alley, thus confounding nonmonotonically related performances (Killeen & Amsel, 1987).

The most difficult challenge was reconciling the partial reinforcement extinction effect with the basic theory of accumulation of arousal. A picture of this struggle was provided, along with its outcome: The recognition that the rate constant may not in fact be constant, but may covary with the arousal level of the subject. The mathematical analysis is still inchoate, but has moved closer to consistency with the data as a whole. Equation 18 provided a good fit to some extinction data while maintaining admirable parameter invariance; now Equation 18, the alternative Equation 14, and the more general Equation 19 need to be tested against a wide range of data.

Perhaps the most important qualitative message was the necessity of distinguishing between motivational and associative factors. This is a distinction the field has often made, only then to forget, and confuse warm-up effects with conditioning, and cool-down effects with extinction. In the current mechanics arousal is a state variable, a function of deprivation, rate of reinforcement, and contextual conditioning. Arousal is focused on responses (and stimuli) by arranging contingencies so that they co-occur with incentives in memory. This associative conditioning may occur at a faster rate than the equilibration of arousal levels; and arousal levels may deplete and replete independently of those associations. Separate treatment of these effects will speed us to more effective models of action.

ACKNOWLEDGMENTS

This research was supported by grants ISBN-9408022, BNS 9021562, and KO5 MH01293. I thank Armando Machado who checked the manuscript's mathematics, logic, and prose, and improved it on all counts.

REFERENCES

Bizo, L. A., & Killeen, P. R. (1997). Models of ratio schedule performance. *Journal of Experimental Psychology: Animal Behavior Processes, 23,* 351–367.
Bolles, R. C. (1970). Species-specific defense reactions and avoidance learning. *Psychological Review, 71,* 32–48.

Fanselow, M. S., & Lester, L. S. (1988). A functional behavioristic approach to aversively motivated behavior: Predatory imminence as a determinant of the topography of defensive behavior. In R. C. Bolles & M. D. Beecher (Eds.), *Evolution and learning* (pp. 185–212). Hillsdale, NJ: Lawrence Erlbaum Associates.

Gracely, R. H., & Church, R. M. (1976). Adjustment of speed in repeated shifts of a negative reinforcer. *Bulletin of the Psychonomic Society, 1,* 455–457.

Green, L., Fry, A. F., & Myerson, J. (1994). Discounting of delayed rewards: A life-span comparison. *Psychological Science, 5,* 33–36.

Herrnstein, R. J. (1979). Derivatives of matching. *Psychological Review, 86,* 486–495.

Hineline, P. N. (1972). Avoidance sessions as aversive events. *Science, 176,* 430–432.

Hineline, P. N. (1978a). Warmup in avoidance as a function of time since prior training. *Journal of the Experimental Analysis of Behavior, 29,* 87–103.

Hineline, P. N. (1978b). Warmup in free-operant avoidance as a function of the response–shock=shock–shock interval. *Journal of the Experimental Analysis of Behavior, 30,* 281–291.

Hoffman, H. S., Fleshler, M., & Chorny, H. (1961). Discriminated bar-press avoidance. *Journal of the Experimental Analysis of Behavior, 4,* 309–316.

Jensen, G. D., & Cotton, J. W. (1950). Successive acquisitions and extinctions as related to percentage of reinforcement. *Journal of Experimental Psychology, 63,* 41–49.

Killeen, P. R. (1975). On the temporal control of behavior. *Psychological Review, 82,* 89–115.

Killeen, P. R. (1979). Arousal: Its genesis, modulation, and extinction. In M. D. Zeiler & P. Harzem (Eds.), *Advances in analysis of behavior: Vol. 1. Reinforcement and the organisation of behavior* (pp. 31–78). Chichester, England: Wiley.

Killeen, P. R. (1981). Averaging theory. In C. M. Bradshaw, E. Szabadi, & C. F. Lowe (Eds.), *Quantification of steady-state operant behavior* (pp. 21–34). Amsterdam: Elsevier.

Killeen, P. R. (1984). Incentive theory III: Adaptive Clocks. In J. Gibbon & L. Allen (Eds.), *Timing and time perception* (Vol. 423, pp. 515–527). New York: New York Academy of Sciences.

Killeen, P. R. (1991). Behavior's time. In G. H. Bower (Ed.), *The Psychology of Learning and Motivation* (Vol. 27, pp. 295–334). New York: Academic Press.

Killeen, P. R. (1994). Mathematical principles of reinforcement. *Behavioral and Brain Sciences, 11,* 105–172.

Killeen, P. R. (1995). Economics, ecologics and mechanics: The dynamics of responding under conditions of varying motivation. *Journal of the Experimental Analysis of Behavior, 64,* 405–431.

Killeen, P. R., & Amsel, A. (1987). The kinematics of locomotion toward a goal. *Journal of Experimental Psychology: Animal Behavior Processes, 13,* 92–101.

Killeen, P. R., Hanson, S. J., & Osborne, S. R. (1978). Arousal: Its genesis and manifestation as response rate. *Psychological Review, 85,* 571–581.

Kimble, G. A. (1961). *Hilgard and Marquis' conditioning and learning* (2nd ed.). New York: Appleton-Century-Crofts.

Kimble, G. A., Mann, L. I., & Dufort, R. H. (1955). Classical and instrumental eyelid conditioning. *Journal of Experimental Psychology, 49,* 407–417.

Mazur, J. E. (1984). Tests of an equivalence rule for fixed and variable delays. *Journal of Experimental Psychology: Animal Behavior Processes, 10,* 426–436.

McNamara, J. M., & Houston, A. I. (1987). Memory and the efficient use of information. *Journal of Theoretical Biology, 125,* 385–395.

Miller, R. R. (1982). Effects of intertrial reinstatement of training stimuli on complex maze learning in rats: Evidence that "acquisition" curves reflect more than acquisition. *Journal of Experimental Psychology: Animal Behavior Processes, 8,* 86–109.

Nakamura, C. Y., & Anderson, N. H. (1962). Avoidance behavior differences within and between strains of rats. *Journal of Comparative and Physiological Psychology, 55,* 740–747.

Nevin, J. A. (1974). Response strength in multiple schedules. *Journal of the Experimental Analysis of Behavior, 21,* 389–408.

Nevin, J. A. (1988). Behavioral momentum and the partial reinforcement effect. *Psychological Bulletin, 103,* 44–56.

Palya, W. L. (1992). Dynamics in the fine structure of schedule-controlled behavior. *Journal of the Experimental Analysis of Behavior, 51,* 267–287.

Powell, R. W. (1972). Analysis of warm-up effects during avoidance in wild and domesticated rodents. *Journal of Comparative and Physiological Psychology, 18,* 311–316.

Rachlin, H. (1989). *Judgement, decision, and choice.* New York: Freeman.

Rescorla, R. A. (1993). Preservation of response-outcome associations through extinction. *Animal Learning & Behavior, 21,* 238–245.

Schwarz, W. (1989). A general counter model for refractory phenomena. *Journal of Mathematical Psychology, 33,* 473–495.

Shettleworth, S., & Nevin, J. A. (1965). Relative rate of response and relative magnitude of reinforcement in multiple schedules. *Journal of the Experimental Analysis of Behavior, 8,* 199–202.

Skinner, B. F. (1938). *The behavior of organisms.* New York: Appleton-Century-Crofts.

Staddon, J. E. R. (1977). On Hermstein's equation and related forms. *Journal of the Experimental Analysis of Behavior, 28,* 163–170.

Staddon, J. E. R., & Zhang, Y. (1989). Response selection in operant learning. *Behavioural Processes, 20,* 189–197.

5

Using Biology to Solve a Problem in Automated Machine Learning

John R. Koza
Stanford University

BACKGROUND

The goal of automatic programming is to create, in an automated way, a computer program that enables a computer to solve a problem. This goal, attributed to Arthur Samuel (1959), can be stated as follows:

How can computers learn to solve problems without being explicitly programmed?

Genetic programming is a domain-independent method for evolving a computer program for solving, or approximately solving, a problem (Koza, 1989). Genetic programming is an extension of the biologically motivated *genetic algorithm* that was first described in John Holland's pioneering *Adaptation in Natural and Artificial Systems* (1975). Goldberg (1989) and Mitchell (1996) surveyed the field of genetic algorithms.

Genetic programming starts with a primordial ooze of randomly generated computer programs composed of the available programmatic ingredients and then applies the principles of animal husbandry to breed a new (and often improved) population of programs. The breeding is done in a domain-independent way using the Darwinian principle of survival of the fittest and an analog of the naturally occurring genetic operation of crossover (sexual recombination). The crossover operation is designed to create syntactically valid

157

offspring computer programs from any selected pair of parental computer programs (given closure amongst the set of ingredients). Genetic programming combines the expressive high-level symbolic representations of computer programs with the near-optimal efficiency of learning of Holland's genetic algorithm.

Genetic Programming: On the Programming of Computers by Means of Natural Selection (Koza, 1992) provided evidence that genetic programming can solve, or approximately solve, a variety of problems from a variety of fields, including many benchmark problems from machine learning, artificial intelligence, control, robotics, optimization, game playing, symbolic regression, system identification, and concept learning. More advances in genetic programming were described in Kinnear (1994), Angeline and Kinnear (1996), and the proceedings of the annual genetic programming conferences (Koza, Goldberg, Fogel, & Riolo, 1996).

The sequence of the work-performing steps of the programs being evolved by genetic programming is not specified in advance by the user. Instead, the sequence of steps is evolved by genetic programming as a result of the competitive and selective pressures of the evolutionary process and the recombinative role of crossover. However, *Genetic Programming* had the limitation that the vast majority of its evolved programs are single-part (i.e., one result-producing main part, but no subroutines).

No approach to automated programming is likely to be successful on nontrivial problems unless it provides some hierarchical mechanism to exploit, *by reuse* and *parameterization*, the regularities, symmetries, homogeneities, similarities, patterns, and modularities inherent in problem environments. Subroutines do this in ordinary computer programs. Accordingly, *Genetic Programming II: Automatic Discovery of Reusable Programs* (Koza, 1994a) describes how genetic programming can be extended to evolve multipart programs consisting of a main program and one or more reusable, parameterized, hierarchically-called subprograms.

An *automatically defined function* (*ADF*) is a function (i.e., subroutine, subprogram, DEFUN, procedure, module) that is dynamically evolved during a run of genetic programming and that may be called by a calling program (or subprogram) that is concurrently being evolved. When automatically defined functions are being used, a program in the population consists of a hierarchy of one (or more) *reusable* function-defining branches (i.e., automatically defined functions) along with a main result-producing branch. Typically, the automatically defined functions possess one or more dummy arguments (formal parameters) and are reused with different instantiations of these arguments. As a run progresses, genetic programming evolves different subprograms in the function-defining branches, different main programs in the result-producing branch, different instantiations of the dummy arguments (formal parameters) of the automatically defined functions in the function-defining branches, and different hierarchical references between the automatically defined functions.

When automatically defined functions are being used in genetic programming, the initial random generation of the population is created so that every individual program in the population has a constrained syntactic structure consisting of a particular architectural arrangement of branches. When crossover is to be performed, a *type* is assigned to each potential crossover point in the parental computer programs either on a branch-wide basis (called *branch typing* or *like-branch typing*) or on the basis of the actual content of the subtree below each potential crossover point (called *point typing*). Crossover is then performed in a structure-preserving way (given closure) to preserve the syntactic validity of all offspring (Koza, 1994a).

Genetic programming with automatically defined functions has been shown to be capable of solving numerous problems. More importantly, the evidence so far indicates that, for many problems, genetic programming requires less computational effort (i.e., fewer fitness evaluations to yield a solution with a satisfactorily high probability) with automatically defined functions than without them (provided the difficulty of the problem is above a certain relatively low break-even point). Also, genetic programming usually yields solutions with smaller average overall size with automatically defined functions than without them (again provided that the problem is not too simple). That is, both learning efficiency and parsimony appear to be properties of genetic programming with automatically defined functions.

Moreover, there is also evidence that genetic programming with automatically defined functions is scalable. For a limited number of problems for which a progression of scaled-up versions was studied, the computational effort increases as a function of problem size at a *slower rate* with automatically defined functions than without them. In addition, the average size of solutions similarly increases as a function of problem size at a *slower rate* with automatically defined functions than without them. This observed scalability results from the profitable reuse of hierarchically callable, parameterized subprograms within the overall program.

There are five major preparatory steps required before genetic programming can be applied to a particular problem, namely determining:

1. The set of terminals (i.e., the actual variables of the problem, zero-argument functions, constants) for each branch,
2. The set of functions (e.g., primitive functions) for each branch,
3. The fitness measure (or other arrangement for implicitly measuring fitness),
4. The parameters to control the run, and
5. The termination criterion and the result designation method for the run.

When automatically defined functions are added to genetic programming, it is also necessary to determine the architecture of the yet-to-be-evolved programs. The specification of the architecture consists of:

1. The number of function-defining branches in the overall program,
2. The number of arguments (if any) possessed by each function-defining branch, and
3. If there is more than one function-defining branch, the nature of the hierarchical references (if any) allowed between the function-defining branches.

Sometimes these architectural choices flow directly from the nature of the problem. Sometimes heuristic methods are helpful. However, in general, there is no way of knowing a priori the optimal (or minimum) number of automatically defined functions that will prove to be useful for a given problem, or the optimal (or sufficient) number of arguments for each automatically defined function, or the optimal (or sufficient) arrangement of hierarchical references among the automatically defined functions.

If the goal is to develop a single, unified, domain-independent approach to automatic programming that requires that the user prespecify as little direct information as possible about the problem, the question arises as to whether these architectural choices can be automated. Indeed, the requirement that the user predetermine the size and shape of the ultimate solution to a problem has been a bane of automated machine learning from the earliest times (Samuel, 1959).

In other words, the question is whether genetic programming can be enabled to discover the architecture of a multipart program during a run. Presumably, as the evolutionary process proceeds from generation to generation, various architectures would be dynamically created during the run of genetic programming. The new architectures would then be tested as to how well they solve the problem at hand. Certain individuals with certain architectures will prove to be more fit than others at grappling with the problem. The more fit architectures might tend to prosper, whereas the less fit architectures might tend to wither away. Eventually a computer program with an appropriate architecture might emerge from this process of *evolution of architecture*.

This chapter describes the naturally occurring processes of gene duplication and gene deletion and automatically defined functions. Six new architecture-altering operations are described, and the implications of the new architecture-altering operations are addressed. The problem of symbolic regression of the Boolean even-5-parity function is solved as the architecture is simultaneously evolved by showing an example of an actual run. A comparison is given of the computational effort required for five different ways of solving the parity problem using genetic programming, including the new way described in this chapter involving the evolution of architecture.

GENE DUPLICATION AND DELETION IN NATURE

Nature points the way to perform the *evolution of architecture* that would enable genetic programming to discover dynamically the program architecture for solving a problem. A change in the architecture of a multipart computer program during a run of genetic programming corresponds to a change in genome structure in the natural world. Therefore, it seems appropriate to consider the different ways that a genomic structure may change in nature.

In nature, sexual recombination (crossover) exchanges alleles (gene values) at particular locations (loci) along the chromosome (a molecule of DNA). The DNA then controls the manufacture of various proteins that determine the structure, function, and behavior of the living organism (Stryer, 1988). The resulting organism then spends its life grappling with its environment. Some organisms in a given population do better than others in that they survive to the age of reproduction, produce offspring, and thereby pass on all or part of their genetic makeup to the next generation of the population. Over a period of time and many generations, the population as a whole evolves so as to give increasing representation to traits (and, more importantly, co-adapted combinations of traits) that contribute to survival of the organism to the age of reproduction and that facilitate large numbers of offspring. This process, which Charles Darwin (1859) called *natural selection,* tends to evolve near-optimal (perhaps even optimal) co-adapted sets of alleles in the chromosomes of the organism (given its environment).

Holland's pioneering *Adaptation in Natural and Artificial Systems* (1975) described how an analog of the naturally occurring evolutionary process can be applied to solving artificial problems using what is now called the *genetic algorithm.* Before applying the genetic algorithm to the problem, the user designs an artificial chromosome of a certain size and shape and then defines a mapping (encoding) between the points in the search space of the problem and the artificial chromosome. For example, in applying the genetic algorithm to a multidimensional optimization problem (where the goal is to find the global optimum of an unknown multidimensional function), the artificial chromosome is often a linear character string (modeled directly after the linear string of information found in DNA). A specific location (a gene) along this artificial chromosome is associated with each of the variables of the problem. Character(s) appearing at a particular location along the chromosome denote the value of a particular variable (i.e., the gene value or allele).

Each individual in the population has a fitness value (which, for a multidimensional optimization problem, is the value of an unknown function). The genetic algorithm then manipulates a population of such artificial chromosomes (usually starting from a randomly created initial population of strings) using the operations of reproduction, crossover, and mutation. Individuals are probabilistically selected to participate in these genetic operations based on their fitness. That is, individuals with better fitness are more likely to be selected in the genetic operations. The goal of the genetic algorithm in a multidimensional optimization problem is to find an artificial chromosome that, when decoded and mapped back into the search space of the problem, corresponds to a globally optimum (or near-optimum) point in the search space of the problem. The probabilistic aspect of the genetic algorithm is important. The best individuals are not guaranteed to be selected. Even poor individuals in the population are sometimes selected.

Both the natural and artificial evolutionary processes just described indicate how a globally optimum combination of alleles (gene values) within a fixed-size chromosome can be discovered by means of evolution. However, in both the natural and artificial processes, the crossover operation merely exchanges alleles (gene values) at particular locations along an already-existing chromosomal structure of fixed size and shape. This description does not address the question of how genome lengths change during the course of evolution. Neither does the description address the question of how totally new structures, new functions, new behaviors, and new species arise.

In nature, there is not only short-term optimization of alleles in their fixed locations within a fixed-size chromosome, but long-term emergence of new proteins (which, in turn, create new structures, functions, and behaviors, and thereby sometimes create new and more complex organisms). The emergence of new proteins alters the architecture of the chromosome. Indeed, genome lengths in nature have generally increased with the emergence of new and more complex organisms (Dyson and Sherratt, 1985; Brooks Low, 1988). In genetic algorithms, a change in the architecture and length of a chromosome corresponds to a dynamic alteration, during a run of the algorithm, of the user-created mapping (the encoding and decoding) between points from the search space of the problem and instances of the artificial chromosome. In genetic programming, a change in the architecture of the evolving program corresponds to a change in the number of automatically defined functions, the number of arguments possessed by each automatically defined function in an overall program, and in the pattern of hierarchical references among the automatically defined functions.

In considering how to solve the problem of how to evolve the architecture of the multipart computer program during a run of genetic programming, it was determined that an analogous mechanism must be operating in nature.

Therefore, it seemed appropriate to examine the mechanism by which genome structure is altered in nature over the course of millions of years of evolution.

Gene duplications are rare and unpredictable events in the evolution of genomic sequences. In gene duplication, there is a duplication of a lengthy portion of the linear string of nucleotide bases of the DNA in the living cell. After a sequence of bases that code for a particular protein is duplicated in the DNA, there are two identical ways of manufacturing the same protein. Thus, there is no immediate change in the proteins that are manufactured as a result of a gene duplication even though the genomic structure has changed.

Over time, however, some other genetic operation, such as mutation or crossover, may change one or the other of the two identical genes. Over short periods of time, the changes accumulating in the changed gene may be of no practical effect or value. As long as one of the two genes remains unchanged, the original protein manufactured from the unchanged gene continues to be manufactured and the structure and behavior of the organism involved may continue as before. The changed gene is simply carried along in the DNA from generation to generation.

Natural selection exerts a powerful force in favor of maintaining a gene that encodes for the manufacture of a protein that is important for the survival and successful performance of the organism. However, after a gene duplication has occurred, there is no disadvantage associated with the loss of the *second* way of manufacturing the original protein. Consequently, natural selection usually exerts little or no pressure to maintain a second way of manufacturing a particular protein. Over a period of time, the second gene may accumulate additional changes and diverge more and more from the original gene. Eventually the changed gene may lead to the manufacture of a distinctly new and different protein that actually does affect the structure and behavior of the living thing in some advantageous or disadvantageous way. When a changed gene leads to the manufacture of a viable and advantageous new protein, natural selection again works to preserve that new gene.

Ohno's *Evolution by Gene Duplication* (1970) corrected the mistaken notion that natural selection is a mechanism for promoting change. Instead, Ohno emphasized the essentially conservative role of natural selection in the evolutionary process:

> The true character of natural selection . . . is not so much an advocator or mediator of heritable changes, but rather it is an extremely efficient policeman which conserves the vital base sequence of each gene contained in the genome. As long as one vital function is assigned to a single gene locus within the genome, natural selection effectively forbids the perpetuation of mutation affecting the *active* sites of a molecule. (p. 59)

Ohno further pointed out that ordinary point mutation and crossover are insufficient to explain major evolutionary changes:

While allelic changes at already existing gene loci suffice for racial differentiation within species as well as for adaptive radiation from an immediate ancestor, they cannot account for large changes in evolution, because large changes are made possible by the acquisition of new gene loci with previously non-existent functions. (p. 59)

Ohno continued,

Only by the accumulation of *forbidden* mutations at the *active* sites can the gene locus change its basic character and become a new gene locus. An escape from the ruthless pressure of natural selection is provided by the mechanism of gene duplication. By duplication, a redundant copy of a locus is created. Natural selection often ignores such a redundant copy, and, while being ignored, it accumulates formerly forbidden mutations and is reborn as a new gene locus with a hitherto non-existent function. (p. 59)

Ohno concluded, "Thus, gene duplication emerges as the major force of evolution" (p.59). Ohno's provocative thesis is supported by the discovery of pairs of proteins with similar sequences of DNA and similar sequences of amino acids, but different functions.

Examples include trypsin and chymotrypsin; the protein of microtubules and actin of the skeletal muscle; myoglobin and the monomeric hemoglobin of hagfish and lamprey; myoglobin used for storing oxygen in muscle cells and the subunits of hemoglobin in red blood cells of vertebrates; and the light and heavy immunoglobin chains (Brooks Low, 1988; Dyson & Sherratt, 1985; Go, 1991; Hood & Hunkapiller, 1991; Maeda & Smithies, 1986; Nei, 1987; Patthy, 1991). For the *Escherichia coli* bacteria, a relatively simple organism, it is known that more than 30% of its proteins are the result of gene duplications (Lazcano & Miller, 1994; Riley, 1993). These proteins include its DNA polymerases, dehydrogenases, ferredoxins, glutamine synthetases, carbamoyl-phosphate synthetases, F-type ATPases, and DNA topoisomerases.

The midge, *Chironomus tentans*, provides an additional example of gene duplication (Galli & Wislander, 1993, 1994). In particular, we focus our attention on the particular contiguous sequence containing 3,959 nucleotide bases of the DNA of this midge that is archived under accession number X70063 in the European Molecular Biology Laboratory (EMBL) database and the Gen Bank database. The 732 nucleotide bases located at positions 918–1,649 of the 3,959 bases of the DNA sequence involved become expressed as a protein containing 244 (i.e., one third of 732) amino acid residues. The 759 nucleotide bases at positions 2,513–3,271 become expressed as a protein containing 253 residues. The 732-base subsequence is called the C. *tentans* Sp38–40.A gene and the 759-base subsequence is called C. *tentans* Sp38–40.B. The bases of DNA before position 918, the bases between positions 1,650 and 2,612, and the bases after position 3,371 of this sequence of length 3,959 do not become expressed as any protein.

Both the A and the B proteins are secreted from the midge's salivary gland to form two similar, but different, kinds of water-insoluble fibers. The two kinds of fibers are, in turn, spun into one of two similar, but different, kinds of tubes. One tube is for larval protection and feeding while the other tube is for pupation (the stage in the development of an insect in which it lies in repose and from which it eventually emerges in the winged form).

Table 5.1 shows the bases of DNA in positions 900 through 3,399 of the 3,959 nucleotide bases of X70063. In the DNA sequence, A represents the nucleotide base adenine, C represents cytosine, G represents guanine, and T represents thymine. Each group of three consecutive bases (a codon) of DNA becomes expressed as one of the 20 amino acid residues of the protein. The letters A, T, and G appearing at positions 918, 919, and 920, respectively in this reading frame, of the DNA sequence are translated into the amino acid residue methionine (denoted by the single letter M using the 20-letter coding for amino acid residues in proteins). Thus, methionine is the first amino acid residue (i.e., N-terminal) of protein A. Positions 921, 922, and 923 of the DNA contain the bases A, G, and A, respectively, and these three bases, in this reading frame, are translated into arginine (an amino acid residue denoted by the letter R). Thus, arginine is the second amino acid residue of protein A and the protein sequence begins with the residues M and R. The DNA up to position 1,649 encodes the first protein. Positions 1,647, 1648, and 1,649 code for the amino acid resident lysine (denoted by the letter K). Thus, lysine is the last (244th) residue (i.e., C-terminal) of protein A.

Table 5.2 shows the 244 amino acid residues of the C. *tentans* Sp38–40.A protein.

Table 5.3 shows the 253 amino acid residues of the C. *tentans* Sp38–40.B protein.

The two proteins are similar, but different. For example, the first 14 amino acid residues are identical. Residue 15 of the A protein is phenylalanine (F), while the residue 15 of the B protein is leucine (L), a chemically similar amino acid. Residues 16–50 are identical. Residue 51 of the A protein is glutamic acid (E), while residue 51 of the B protein is Aspartic acid (D). Both D and E are similar in that both are electrically negatively charged residues at normal pH values. However, for some positions, such as 76, the amino acid residues (T and A) are not chemically or electrically similar.

If we now read from the end of each protein, we see that the last few residues of each protein are identical. Since the proteins are of different length, identification of the similarity between the two protein sequences requires aligning the two proteins in some way. Protein alignment algorithms, such as the Smith–Waterman algorithm (Smith & Waterman, 1981), provided a way to align two proteins and to measure the degree of similarity or dissimilarity between two proteins. The Smith–Waterman algorithm is a progressive align-

TABLE 5.1.

Portion of a DNA sequence containing the two expressed proteins.

TGAAGTAATA	TTAAGCTATG M	AGAATTAAGT R I K F	TCCTAGTAGT L V V	ATTAGCAGTT L A V	950
AATCTGCTTGT I C L F	TTGCACATTA A H Y	TGCCTCAGCT A S A	AGTGGTATGG S G M G	GGGGTGATAA G D K	1000
AAAACCCAAA K P K	GATGCCCCAA D A P K	AACCCAAAGA P K D	TGCCCCAAA A P K	CCCAAAGAAG P K E V	1050
TGAAGCCTGT K P V	CAAAGCTGAG K A E	TCATCAGAGT S S E Y	ATGAGATAGA E I E	AGTCATTAAA V I K	1100
CACCAGAAAG H Q K E	AAAAGACCGA K T E	GAAGAAGGAG K K E	AAGGAGAAGA K E K K	AGACTCACGT T H V	1150
TGAAACCAAG E T K	AAAGAAGTTA K E V K	AAAAGAAGGA K K E	GAAGAAGCAA K K Q	ATCCCTTGTT I P C S	1200
CTGAAAAACT E K L	CAGGATGAA K D E	AAACTTGATT K L D C	GTGAGACCAA E T K	GGGCGTCCCT G V P	1250
GCAGGCTACA A G Y K	AAGCAATCTT A I F	CAAATTCACA K F T	GAAAACGAGG E N E E	AGTGCCGATTG C D W	1300
GACGTGCGAT T C D	TATGAAGCAC Y E A L	TTCCACCACC P P P	TCCAGGAGCA P G A	AAGAAAGACG K K D D	1350
ACAAGAAAGA K K E	AAAGAAGACA K K T	GTTAAAGTCG V K V V	TTAAGCCACC P P	AAAGGAGAAA K E K	1400
CCACCAAAGA P P K K	AGCTTAGAAA L R K	GGAATGCTCT E C S	GGCGAAAAAG G E K V	TGATCAAATT I K F	1450
CCAAAACTGT Q N C	CTCGTTAAGA L V K I	TTAGAGGACT R G L	TATTGCCTTT I A F	GGTGATAAGA G D K T	1500
CAAAGAACTT K N F	TGATAAGAAG D K K	TTCGCAAAGC F A K L	TTGTCCAAGG V Q G	AAAGCAGAAG K Q K	1550
AAGGGCGCAA K G A K	AAAAAGCTAA K A K	AGGCGGTAAG G G K	AAGGCAGCAC K A A P	CAAAACCAGG K P G	1600

TABLE 5.1 (cont'd)

ACCAAAAACCA	GGGCCAAAAC	AAGCTGATAA	ACCAAAAGAT	GCAAAAAAT	1650
P K P	G P K Q	A D K	P K D	A K K	
AAACTGACAT	AGTAAGAATA	ATAAAATAAA	CATTATTTGA	GCAACATCAC	1700
AACACAAGAA	AAAAATCATA	TCAACATAAT	TAAGACCTAA	AAATTCTCGC	1750
TATTCACTTT	TTTTCAAATG	AATATCCAAA	ACAACATCAT	TAGGGATCT	1800
TACACAATTT	TATCCCAAAT	TAGTTTTAAG	TCTATTTTTT	AGTTTTAAGT	1850
AAAACATTAG	TTAGAGAAAT	TTCAAATGCG	AAAAAAGAC	AACAATATTT	1900
TTAACTCCAA	CTAATTGTCT	AGATCTAATC	ACCACTGAAA	TTCAAAAAAA	1950
AAAAAACAAA	TATCTGAGAT	GAAAATTTTG	TAAGATACGA	AAAACTTACA	2000
TTTTCAATAA	AACTTAAATA	TTTTCTTTAT	AAGAAAGTAA	TTTAAAGAAA	2050
TGAACAACAA	GTAGACTAAG	GGCTTAAAAA	TACTAAGGAA	CATCCATTCA	2100
CTGAACCAAT	AACATCCAAT	AAATATAAGC	GTGTATTTAA	GAATATTTTT	2150
TGCAAAATTT	GACTTGTTTT	ATTCTAAACT	TTTGAATTGT	CAAGGAAACT	2200
GATGATTATT	GAATATTTTA	CAGCATTTTT	CGACAAAATC	CGAAAAACTG	2250
GTTTTGTTTA	ATATATACTA	CAGCTCAGTA	TCTATGCACA	GTATTTTAAA	2300
TAACAGACCA	GACCATAAAA	CCTACACATC	ACCAAGATAC	CATGAAAACC	2350
TTCATGTGAC	TGACAAAAGC	TGGAAACACT	TGTGTCACGT	TATAAAGAG	2400
TCGTTGAAAT	AAAACTTCTA	GAAAGGTTAT	CATGAAAGAG	TCATATGAAG	2450
ATCTCAAACG	AGGCTCAGTC	AGTTCAGTTT	AGCTTGGACT	CAGTTATCTG	2500
TAATATTTAG	CTATGAGAAT	TAAGTTCCTA	GTAGTATTAG		2550
		K F L	V V L A	V I C	
CTTGCTTGCA	CATTATGCCT	CAGCTAGTGG	TATGGGGGGT	GATAAAAAC	2600
L L A	H Y A S	A S G	M G G	D K K P	
CCAAAGATGC	CCCAAAACCC	AAAGATGCCC	CAAACCCAA	AGAAGTGAAG	2650
K D A	P K P	K D A P	K P K	E V K	
CCTGTCAAAG	CTGACTCATC	AGAGTATGAG	ATAGAAGTCA	TTAAACACCA	2700
P V K A	D S S	E Y E	I E V I	K H Q	

167

```
GAAAGAAAAG ACCGAGAAGA AGGAGAAGGA GAAGAAAGCT CACGTCGAAA   2750
 K  E  K    T  E  K  K   E  K  E    K  K  A    H  V  E  I

TCAAGAAAAA GATTAAAAAT AAGGAGAAGA AGTTTGTCCC ATGTTCTGAA   2800
 K  K  K    I  K  N    K  E  K  K   F  V  P    C  S  E

ATTCTCAAGG ATGAAAAACT TGAATGTGAG AAAAATGCTA CTCCAGGCTA   2850
 I  L  K  D   E  K  L    E  C  E    K  N  A  T   P  G  Y

TAAAGCACTC TTCGAATTCA AAGAAAGCGA AAGTTTTTGC GAATGGGAGT   2900
 K  A  L    F  E  F  K   E  S  E    S  F  C    E  W  E  C

GCGATTATGA AGCAATTCCA GGAGCAAAGA AAGACGAAAA AAAGGAGAAG   2950
   D  Y  E   A  I  P    G  A  K    D  E  K    K  E  K

AAGGTAGTTA AAGTCATTAA GCCACCAAAG GAAAAACCAC CAAAGAAGCC   3000
 K  V  V  K   V  I  K    P  P  K    E  K  P  P   K  K  P

TAGAAAGGAA TGCTCTGGCG AAAAAGTGAT CAAATTCCAA AACTGTCTCG   3050
 R  K  E    C  S  G  E   K  V  I    K  F  Q    N  C  L  V

TTAAGATTAG AGGACTTATT GCCTTTGGTG ATAAGACAAA GAACTTTGAT   3100
 K  I  R    G  L  I    A  F  G  D   K  T  K    N  F  D

AAGAAGTTTG CAAAGCTTGT CCAAGGAAAG CAAAGAGG   GCGCAAAAA    3150
 K  K  F  A   K  L  V    Q  G  K    Q  K  K  G  A  K  K

AGCTAAAGGC GGTAAGAAGG CAGAACCAAA ACCAGGACCA AAACCAGCAC   3200
 A  K  G    G  K  K  A   E  P  K    P  G  P    K  P  A  P

CAAAACCAGG ACCAAAACCA GCACCAAAAC CAGTACCAAA ACCAGCTGAT   3250
 K  P  G    P  K  P    A  P  K  P   V  P  K    P  A  D

AAACCAAAAG ATGCAAAAAA ATAAACTGAC ATAGTGAGAA TAATAAAATA   3300
 K  P  K  D   A  K  K
```

TABLE 5.2.
Protein Sequence of "A" Protein.

MRIKFLVVLA	VICLFAHYAS	ASGMGGDKKP	KDAPKPKDAP	KPKEVKPVKA	50
ESSEYEIEVI	KHQKEKTEKK	EKEKKTHVET	KKEVKKKEKK	QIPCSEKLKD	100
EKLDCETKGV	PAGYKAIFKF	TENEECDWTC	DYEALPPPPG	AKKDDKKEKK	150
TVKVVKPPKE	KPPKKLRKEC	SGEKVIKFQN	CLVKIRGLIA	FGDKTKNFDK	200
KFAKLVQGKQ	KKGAKKAKGG	KKAAPKPGPK	PGPKQADKPK	DAKK	244

ment method employing dynamic programming based on a scoring algorithm. Because the proteins being aligned are typically of different lengths, gaps may be introduced (and then lengthened) in an attempt to best align the residues making up the proteins. A penalty is assessed to open a gap (5 here) and another penalty is assessed to lengthen a gap (25 here). An additional penalty is assessed when one residue disagrees with another. This penalty is smaller for substitutions involving evolutionarily close amino acid residues. The PAM-250 (*Percentage of Accepted point Mutations*) matrix is used to reflect the likelihood of one amino acid residue being mutated into another. The overall scoring algorithm performs a tradeoff employing dynamic programming between the penalties assessed by the PAM-250 matrix, the gap-opening penalty, and the gap-lengthening penalty. The Smith–Waterman algorithm has been implemented in GeneWorks™, a software package available from Intelligenetics, Inc. of Mountain View, California.

Table 5.4 shows the alignment of the C. *tentans* Sp38–40.A protein and the C. *tentans* Sp38–40.B protein. Identical residues are boxed. The alignment shows that there is 81% identity between the two protein sequences. As can be seen, the first disagreement between the two aligned sequences occurs at position 15 and the second occurs at residue 51. The first gap is introduced at

TABLE 5.3.
Protein Sequence of "B" Protein

MRIKFLVVLA	VICLLAHYAS	ASGMGGDKKP	KDAPKPKDAP	KPKEVKPVKA	50
DSSEYEIEVI	KHQKEKTEKK	EKEKKAHVEI	KKKIKNKEKK	FVPCSEILKD	100
EKLECEKNAT	PGYKALFEFK	ESESFCEWEC	DYEAIPGAKK	DEKKEKKVVK	150
VIKPPKEKPP	KKPRKECSGE	KVIKFQNCLV	KIRGLIAFGD	KTKNFDKKFA	200
KLVQGKQKKG	AKKAKGGKKA	EPKPGPKPAP	KPGPKPAPKP	VPKPADKPKD	250
AKK					253

residue 112 where the A protein has an alanine (A) residue. A gap of length 3 is introduced at positions 147, 148, and 149 where the A protein has three proline (P) residues. Note that this alignment recognizes the identity between the last five residues of the two proteins. This alignment has a total cost of 265.

Galli and Wislander (1993) pointed out that these two similar proteins arise as a consequence of a gene duplication. Immediately after the gene duplication occurred at some time in the distant past, there were two identical copies of the duplicated sequence of DNA. Over a period of millions of years since the initial gene duplication, additional mutations accumulated so that the two proteins are now only 81% identical (after alignment). More importantly, the two proteins now perform different (but similar) functions in the midge.

More complex organisms have a general tendency to have more expressed proteins, more different kinds of structures, more complex structures, perform more different functions, and have longer genomes (Dyson & Sherratt, 1985). The rise of new functions as a consequence of gene duplication is consistent with the observed longer genomes of more complex organisms.

Gene deletion also occurs in nature. In gene deletion, there is a deletion of a portion of the linear string of nucleotide bases that would otherwise be translated and manufactured into work-performing proteins in the living cell. After a gene deletion occurs, some particular protein that was formerly manufactured will no longer be manufactured and there may be some change in the structure or behavior of the biological entity. The absence of the protein

TABLE 5.4.
Protein Alignment of the A and B Proteins

```
First.protein    MRIKFLVVLA VICLFAHYAS ASGMGGDKKP KDAPKPKDAP KPKEVKPVKA      5
Second.protein   MRIKFLVVLA VICLLAHYAS ASGMGGDKKP KDAPKPKDAP KPKEVKPVKA      5


First.protein    ESSEYEIEVI KHQKEKTEKK EKEKKIIHVET KKEVKKKEKK QIPCSEKLKD     10
Second.protein   ISSEYEIEVI KHQKEKTEKK EKEKKAHVEI KKKIKNKEKK FVPCSEILKD     10


First.protein    EKLDCETKGV PAGYKALFKF IIENEE-CLWT CDYEALPPHP GAKKDLKKEK     14
Second.protein   EKLECEKNAT PHGYKALFHF KESESFCEWE CDYEAI---P GAKKDEKKEK     14


First.protein    KIIVKVMKPPK EKPPKKLRKE CSGEKVIKFQ NCLVKIRGLI AFGDKTKNFD     19
Second.protein   KMVKVIIKPPK EKPPKKFRKE CSGEKVIKFQ NCLVKIRGLI AFGDKTKNFD     19


First.protein    KKFAKLVQGK QKKGAKKAKG GKKAAPKPGP KPGPK----Q ADKP------    23
Second.protein   KKFAKLVQGK QKKGAKKAKG GKKAEPKPGP KPAPKPGPKP AHKPVPKPAD    24


First.protein    --KDAKK                                                   24
Second.protein   KHKDAKK                                                   25
```

may then affect the structure and behavior of the living thing in some advantageous or disadvantageous way. If the deletion is advantageous, natural selection will tend to perpetuate the change, but if the deletion is disadvantageous, natural selection will tend to lead to the extinction of the change.

AUTOMATICALLY DEFINED FUNCTION
IN GENETIC PROGRAMMING

Automatically defined functions (ADFs) are the analog of subroutines in the genetic programming process. When automatically defined functions are being used, each program in the population contains one or more function-defining branches (each defining an automatically defined function) and one main result-producing branch. The automatically defined functions can perform arithmetic, conditional, and other types of operations, define constants, define subsets, and so forth. In addition, for certain problems, there may be other problem-specific types of branches (such as iteration-performing branches and iteration-terminating branches).

Figure 5.1 shows an overall program consisting of one two-argument automatically defined function (called ADF0 here) and one result-producing branch (RPB). The argument map describes the architecture of a multipart program in terms of the number of its function-defining branches and the

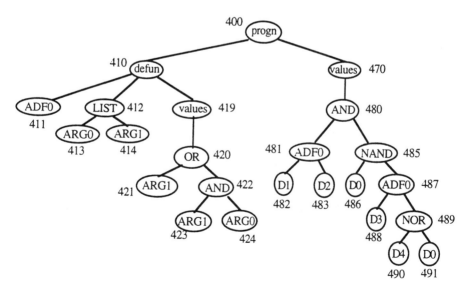

FIG. 5.1. Program with an argument map of {2} consisting of one 2-argument function-defining branch (ADF0) and one result-producing branch.

number of arguments that they each possess. The *argument map* of the set of automatically defined functions belonging to an overall program is the list containing the number of arguments possessed by each automatically defined function in the program. The argument map for the overall program in Fig. 5.2 is {2} because there is one function-defining branch that takes two arguments.

The program in Fig. 5.2 contains architecture-defining points of the following types:

1. The PROGN (labeled 400) appearing as the top-most point of the overall program,
2. A DEFUN (labeled 410) as the top-most point of the function-defining branch,
3. A name (i.e., ADF0 labeled 411) appearing as the first argument below the DEFUN,
4. The function LIST (labeled 412) appearing as the second argument of the DEFUN,
5. Dummy arguments (such as ARG0 and ARG1 labeled as 413 and 414, respectively) appearing below LIST,
6. The VALUES (labeled 419) of the function-defining branch appearing as the third argument of the DEFUN, and
7. The VALUES (labeled 470) of the result-producing branch appearing as the final argument of PROGN.

If the program in Fig. 5.2 were to have more than one automatically defined

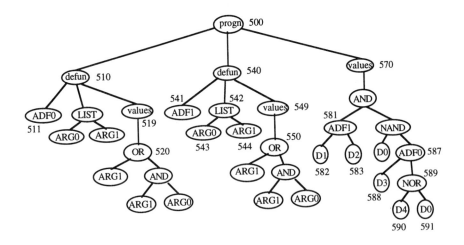

FIG. 5.2. Program with an argument map of {2, 2} consisting of two 2-argument function-defining branches (ADF0 and ADF1) and one result-producing branch.

function, there would be additional occurrences of items 2, 3, 4, 5, and 6 for each additional function-defining branch.

These architecture-defining points were called invariant points in *Genetic Programming II* because they were not subject to alteration by crossover or mutation. However, this terminology becomes obsolete with the introduction of the architecture-altering operations described herein.

The program in Fig. 5.2 also contains work-performing points. These work-performing points are the bodies of the result-producing branch and the function-defining branch(es).

The work-performing points of Fig. 5.2 include:

1. The five points labeled 420, 421, 422, 423, and 424 that are found below the VALUES (labeled 419) in the function-defining branch, and
2. The 11 points starting with the AND (labeled 480) that are found below the VALUES (labeled 470).

These work-performing points were called noninvariant points in *Genetic Programming II* because these points represented the sequence of steps of the to-be-evolved computer program and because they were almost always different from branch to branch within a program and from program to program within the population. Again, this terminology becomes obsolete with the introduction of the architecture-altering operations described herein.

The result-producing branch may invoke all, some, or none of the automatically defined functions that are present within the overall program. The result-producing branch does not contain dummy arguments (formal parameters). The result-producing branch typically contains the actual variables of the problem (e.g., D0, D1, D2).

The value returned by the overall program consists of the value returned by the result-producing branch. The automatically defined functions of a particular overall program are usually named sequentially as ADF0, ADF1, and so forth. The automatically defined functions typically each possess a certain number of dummy arguments (formal parameters). Here, ADF0 possesses two dummy arguments, ARG0 and ARG1. Typically, the actual variables do not appear in the function-defining branches.

If the overall program has more than one automatically defined function, there may (or may not) be hierarchical references between function-defining branches. For example, the function-defining branch of an overall program may be allowed to refer (nonrecursively) to all other previously-defined (i.e., lower numbered) function-defining branches.

References within a particular program to an automatically defined function are to the automatically defined function belonging to that particular program.

Actions (with side effects) may be performed within the function-defining branches, the result-producing branches, or both.

When automatically defined functions are being used, the initial random generation of the population must be created so that each individual overall program in the population has the intended constrained syntactic structure. The constrained syntactic structure in Fig. 5.1 calls for one result-producing branch and one function-defining branch. The function-defining branch for ADF0 is a random composition of functions from the function set, F_{adf}, and terminals from the terminal set, T_{adf}. Here the function set, F_{adf}, consists of the two-argument Boolean functions AND, OR, NAND, and NOR. The terminal set, T_{rpb}, of the function-defining branch consists of the two dummy arguments (formal parameters), ARG0 and ARG1. The result-producing branch is a random composition of functions from the function set, F_{rpb}, and terminals from the terminal set, T_{rpb}. In Fig. 5.2, the function set, F_{rpb}, of the result-producing branch consists of the two-argument Boolean functions AND, OR, NAND, and NOR as well as the now-defined automatically defined function, ADF0. The terminal set, T_{rpb}, of the result-producing branch consists of the five actual variables of the problem (i.e., D0, D1, D2).

Execution of genetic programming consists of the following steps. The six operations appearing as items (2)(c)(iii) through (2)(c)(ix) are the new architecture-altering operations described in detail in a later section.

The steps for executing genetic programming are as follows:

1. Generate an initial random population (Generation 0) of computer programs.
2. Iteratively perform the following substeps until the termination criterion of the run has been satisfied:
   a. Execute each program in the population and assign it (explicitly or implicitly) a fitness value according to how well it solves the problem.
   b. Select program(s) from the population to participate in the genetic operations in (c).
   c. Create new program(s) for the population by applying the following genetic operations.
      i.   *Reproduction*: Copy an existing program to the new population.
      ii.   *Crossover*: Create new offspring program(s) for the new population by recombining randomly chosen parts of two existing programs.
      iii. *Mutation*: Create one new offspring program for the new population by randomly mutating a randomly chosen part of one existing program.
      iv.   *Branch duplication*:
      v.   *Argument duplication*:

vi. *Branch deletion*:
vii. *Argument deletion*:
viii. *Branch creation*:
ix. *Argument creation*:

3. After satisfaction of the termination criterion (which usually includes a maximum number of generations to be run as well as a problem-specific success predicate), the single best computer program in the population produced during the run (the best-so-far individual) is designated as the result of the run. This result may (or may not) be a solution (or approximate solution) to the problem.

SIX NEW ARCHITECTURE-ALTERING GENETIC OPERATIONS

The six new architecture-altering genetic operations provide a new way of determining the architecture of a multipart program. When these operations are performed during a run of genetic programming, the architecture of the participating individuals changes during the run. Meanwhile, the Darwinian selection and the reproduction operation continues to favor the more fit individuals in the population to be modified by the usual operations of crossover and mutation.

Branch Duplication

The operation of *branch duplication* duplicates one of the branches of a program in the following way:

1. Select a program from the population to participate in this operation.
2. Pick one of the function-defining branches of the selected program as the branch-to-be-duplicated.
3. Add a uniquely-named new function-defining branch to the selected program, thus increasing, by one, the number of function-defining branches in the selected program. The new function-defining branch has the same argument list and the same body as the branch-to-be-duplicated.
4. For each occurrence of an invocation of the branch-to-be-duplicated anywhere in the selected program (e.g., the result-producing branch or any other branch that invokes the branch-to-be-duplicated), randomly choose either to leave that invocation unchanged or to replace that invocation with an invocation of the newly created function-defining branch.

The step of selecting a program for this operation (and all the other new operations described herein) is performed probabilistically on the basis of fitness, so that a program that is more fit has a greater probability of being selected to participate in the operation than a less fit program.

Figure 5.2 shows the program resulting after applying the operation of branch duplication to the program in Fig. 5.1 (consisting of one two-argument automatically defined function and one result-producing main branch). Specifically, the function-defining branch 410 of Fig. 5.1 defining ADF0 (also shown as 510 of Fig. 5.2) is duplicated and a new function-defining branch (defining ADF1) appears in Fig. 5.2.

There are two occurrences of invocations of the branch-to-be-duplicated, ADF0, in the result-producing branch of the selected program, namely ADF0 at 481 and 487 of Fig. 5.1. For each of these two occurrences, a random choice is made either to leave the occurrence of ADF0 unchanged or to replace it with the newly created ADF1. For the first invocation of ADF0 at 481 of Fig. 5.1, the choice is randomly made to replace ADF0 481 of Fig. 5.1 with the ADF1 581 in Fig. 5.2. The arguments for the invocation of ADF1 581 in Fig. 5.2 are D1 582 and D2 583 (i.e., they are identical to the arguments D1 482 and D2 483 for the invocation of ADF0 at 481 of the original program in Fig. 5.1). For the second invocation of ADF0 at 487 of Fig. 5.1, the choice is randomly made to leave ADF0 unchanged in Fig. 5.2.

Because the duplicated new function-defining branch is identical to the previously existing function-defining branch (except for the name ADF1 at 541 in Fig. 5.2) and because ADF1 is invoked with the same arguments as ADF0 had been invoked, this operation is a semantics-preserving operation in that the operation does not affect the value returned by the overall program.

The operation of branch duplication can be interpreted as a *case splitting*. After the branch duplication, the result-producing branch invokes ADF0 at 587 and ADF1 at 581 of Fig. 5.2. ADF0 and ADF1 can be viewed as separate procedures for handling the two separate newly created subproblems (cases).

Subsequent genetic operations may alter one or both of these two presently identical function-defining branches and these subsequent changes to lead to a divergence in structure and behavior. This divergence may be interpreted as a *specialization* or *refinement*. That is, once ADF0 and ADF1 diverge, ADF0 can be viewed as a specialization for handling the subproblem (case) associated with its invocation by the result-producing branch. Similarly, ADF1 can be viewed as a specialization for handling its subproblem (case).

The operation of branch duplication as defined above (and all the other new operations described herein) always produce a syntactically valid program. Analogs of the naturally occurring operation of gene duplication have been previously used with genetic algorithms operating on character strings and with other evolutionary algorithms. Holland (1975) suggested that intrachromosomal gene duplication might provide a means of adaptively modifying the effective mutation rate by making two or more copies of a substring of adjacent alleles. Cavicchio (1970) used intrachromosomal gene duplication in early work on pattern recognition using the genetic algorithm. Gene duplication is implicitly used in the messy genetic algorithm (Goldberg, Korb, & Deb, 1989). Lindgren (1991) analyzed the prisoner's dilemma game using an evolutionary algorithm that employed an operation analogous to gene duplication applied to chromosome strings. Gruau (1994) used genetic programming to develop a clever and innovative technique to evolve the architecture of a neural network at the same time as the weights are being evolved.

Argument Duplication

The operation of *argument duplication* duplicates one of the dummy arguments (format parameters) in one of the automatically defined functions of a program in the following way:

1. Select a program from the population to participate in this operation.
2. Pick one of the function-defining branches of the selected program.
3. Choose one of the arguments of the picked function-defining branch of the selected program as the argument-to-be-duplicated.
4. Add a uniquely-named new argument to the argument list of the picked function-defining branch of the selected program, thus increasing, by one, the number of arguments in its argument list.
5. For each occurrence of the argument-to-be-duplicated anywhere in the body of picked function-defining branch of the selected program, randomly choose either to leave that occurrence unchanged or to replace that occurrence with the new argument.
6. For each occurrence of an invocation of the picked function-defining branch anywhere in the selected program, identify the argument subtree corresponding to the argument-to-be-duplicated and duplicate that argument subtree in that invocation, thereby increasing, by one, the number of arguments in the invocation.

Because the function-defining branch containing the duplicated argument is invoked with an identical copy of the previously existing argument, the effect of this operation is to leave unchanged the value returned by the overall program.

Just as the operation of branch duplication was interpreted as a case splitting, the operation of argument duplication can be similarly interpreted. After the argument duplication, the result-producing branch invokes ADF0 with a new third argument. The particular instantiations of the second and third arguments in each invocation of ADF0 provide potentially different ways of handling the two separate subproblems (cases). Once the second and third arguments diverge, this divergence may be interpreted as a specialization or refinement.

Branch Deletion

The operation of branch deletion deletes one of the automatically defined functions of a program in the following way:

1. Select a program from the population to participate in this operation.
2. Pick one of the function-defining branches as the branch-to-be-deleted.
3. Delete the branch-to-be-deleted from the selected program, thus decreasing, by one, the number of branches in the selected program.
4. For each occurrence of an invocation of the branch-to-be-deleted anywhere in the selected program, replace the invocation of the branch-to-be-deleted with an invocation of a surviving branch (described next).

When a function-defining branch is deleted, the question arises as to how to modify invocations of the branch-to-be-deleted by the other branches of the overall program. One alternative (called *branch deletion by consolidation*) involves identifying a suitable second function-defining branch of the overall program as the surviving branch and replacing (consolidating) the branch-to-be-deleted with the surviving branch in each invocation of the branch-to-be-deleted. Branch deletion by consolidation can be interpreted as a way to achieve generalization in a problem-solving procedure. A second alternative (called *branch deletion with random regeneration*) is to randomly generate new subtrees composed of the available functions and terminals in lieu of an invocation of the branch-to-be-deleted. A third alternative (called *branch deletion by macro expansion*) is a semantics-preserving approach that involves inserting the entire body of the branch-to-be-deleted for each instance of an invocation of that branch.

Both the argument duplication and the branch duplication operations create larger programs. The operations of argument deletion and branch deletion (described below) can create smaller programs and can balance the growth that would otherwise occur (provided the alternative of argument deletion by macro expansion is not used).

Argument Deletion

The operation of *argument deletion* deletes one of the arguments to one of the automatically defined functions of a program in the following way:

1. Select a program from the population to participate in this operation.
2. Pick one of the function-defining branches of the selected program.
3. Choose one of the arguments of the picked function-defining branch of the selected program as the argument-to-be-deleted.
4. Delete the argument-to-be-deleted from the argument list of the picked function-defining branch of the selected program, thus decreasing, by one, the number of arguments in the argument list.
5. For each occurrence of an invocation of the picked function-defining branch anywhere in the selected program, delete the argument subtree in that invocation corresponding to the argument-to-be-deleted, thereby decreasing, by one, the number of arguments in the invocation.
6. For each occurrence of the argument-to-be-deleted in the body of the picked function-defining branch, replace the argument-to-be-deleted with a surviving argument.

The operation of argument deletion may be viewed as a *generalization* in that some information that was once considered in executing a procedure is now ignored. When an argument is deleted, references to the argument-to-be-deleted are modified by using *argument deletion by consolidation*, *argument deletion with random regeneration*, or *argument deletion by macro expansion*.

Branch Creation

The operation of branch creation creates a new automatically defined function (ADF) within an overall program in the following way:

1. Select a program from the population to participate in this operation.
2. Pick a point in the body of one of the function-defining branches or result-producing branches of the selected program. This picked point will become the top-most point of the body of the branch-to-be-created.
3. Starting at the picked point, begin traversing the subtree below the picked point in a depth-first manner.
4. As each point below the picked point in the selected program is encountered during the traversal, make a determination as to whether to designate that point as being the top-most point of an argument subtree for the branch-to-be-created. If such a designation is made, no traversal is made of the subtree below that designated point. The depth-first traversal continues and this Step 4 is repeatedly applied to each point encountered during the traversal so that when the traversal of the subtree below the picked point is completed, zero points, one point, or more than one point are so designated during the traversal.

5. Add a uniquely named new function-defining branch to the selected program. The argument list of the new branch consists of as many consecutively numbered dummy variables (formal parameters) as the number of points that were designated during the depth-first traversal. The body of the new branch consists of a modified copy of the subtree starting at the picked point. The modifications to the copy are made in the following way: For each point in the copy corresponding to a point designated during the traversal of the original subtree, replace the designated point in the copy (and the subtree in the copy below that designated point in the copy) by a unique dummy variable. The result is a body for the new function-defining branch that contains as many uniquely named dummy variables as there are dummy variables in the argument list of the new function-defining branch.

6. Replace the picked point in the selected program by the name of the new function-defining branch. If no points below the picked point were designated during the traversal, the operation of branch creation is now completed.

7. If one or more points below the picked point were designated during the traversal, the subtree below the just-inserted name of the new function-defining branch will be given as many argument subtrees as there are dummy arguments in the new function-defining branch in the following way: For each point in the subtree below the picked point designated during the traversal, attach the designated point and the subtree below it as an argument to the function whose name was just inserted in the new function-defining branch.

Several different methods may be used to determine how to designate a point below the picked point during the depth-first traversal described previously. The operation of branch creation is similar to, but different than, the compression (module acquisition) operation described by Angeline and Pollack (1994).

Argument Creation

The operation of argument creation creates a new dummy argument (formal parameter) within a function-defining branch of an overall program in the following way:

1. Select a program from the population to participate in this operation.

2. Pick a point in the body of a function-defining branch of the selected program.

3. Add a uniquely named new argument to the argument list of the picked function-defining branch for the purpose of defining the argument-to-be-created.

4. Replace the picked point (and the entire subtree below it) in the picked function-defining branch by the name of the new argument.

5. For each occurrence of an invocation of the picked function-defining branch anywhere in the selected program, add an additional argument subtree to that invocation. In each instance, the added argument subtree consists of a modified copy of the picked point (and the entire subtree below it) from the picked function-defining branch. The modification is made in the following way: For each dummy argument in a particular added argument subtree, replace the dummy argument with the entire argument subtree of that invocation corresponding to that dummy argument.

IMPLICATIONS OF THE ARCHITECTURE-ALTERING OPERATIONS

The six new architecture-altering operations can be viewed from five perspectives.

First, the new architecture-altering operations provide a new way to solve the problem of determining the architecture of the overall program in the context of genetic programming with automatically defined functions.

Second, the new architecture-altering operations provide an automatic implementation of the ability to specialize and generalize in the context of automated problem solving.

Third, the new architecture-altering operations, in conjunction with automatically defined functions, provide a way to automatically and dynamically change the representation of the problem while simultaneously solving the problem.

Fourth, the new architecture-altering operations, in conjunction with automatically defined functions, provide a way to decompose problems automatically and dynamically into subproblems and then automatically solve the overall problem by assembling the solutions of the subproblems into a solution of the overall problem.

Fifth, the new architecture-altering operations, in conjunction with automatically defined functions, provide a way to discover useful subspaces automatically and dynamically (usually of lower dimensionality than that of the overall problem) and then automatically assemble a solution of the overall problem from solutions applicable to the individual subspaces.

In addition, the new architecture-altering operations affect the implementation of genetic programming with regard to the creation of the initial random population and the crossover operation, as described next.

Creation of the Initial Population

When automatically defined functions are being used, the initial random generation of the population must be created so that each individual overall program in the population has the intended constrained syntactic structure. For example, Fig. 5.1 shows a program for which the constrained syntactic structure calls for one result-producing branch and one function-defining branch. The function-defining branch for ADF0 is a random composition of functions from the function set, F_{adf}, and terminals from the terminal set, T_{adf}. Here the function set, F_{adf}, consists of the two-argument Boolean primitive functions AND, OR, NAND, and NOR. The terminal set, T_{rpb}, of the function-defining branch consists of the two dummy arguments (formal parameters), ARG0 and ARG1. The result-producing branch is a random composition of functions from the function set, F_{rpb}, and terminals from the terminal set, T_{rpb}. In the result-producing branch of Fig. 5.1, the function set, F_{rpb}, consists of the two-argument Boolean primitive functions AND, OR, NAND, and NOR as well as the now-defined automatically defined function, ADF0. The terminal set, T_{rpb}, of the result-producing branch consists of the five actual variables (i.e., D0, D1, D2, D3, D4).

When the architecture-altering operations are used, the initial population of programs may be created in any one of three ways. One possibility (called the *minimalist approach*) is that each multipart program in the population at Generation 0 has a uniform architecture with exactly one automatically defined function possessing a minimal number of arguments appropriate to the problem. A second possibility (called *the big bang*) is that each program in the population has a uniform architecture with no automatically defined functions (i.e., only a result-producing branch). This approach relies on the operation of branch creation to create multipart programs in such runs. A third possibility is that the population at Generation 0 is architecturally diverse (as described in Koza, 1994a).

Structure-Preserving Crossover

In the crossover operation in genetic programming, a crossover point is randomly and independently chosen in each of two parents and genetic material from one parent is then inserted into a part of the other parent to create an offspring. A population may be architecturally diverse either because it was initially created with architectural diversity (as described above) or because the six new architecture-altering genetic operations (described below) create a diversity of new architectures during the run. Structure-preserving crossover with point typing (as described in Koza, 1994a) permits robust recombination while simultaneously guaranteeing that any pair of architecturally different parents will produce syntactically and semantically valid offspring.

If the population is architecturally diverse, the parents selected to participate in the crossover operation will often possess different numbers of automatically defined functions. Moreover, an automatically defined function with a certain name (e.g., ADF2) belonging to one parent will often possess a different number of arguments than the same-named automatically defined function belonging to the other parent (if indeed ADF2 is present at all). After a crossover is performed, each call to an automatically defined function actually appearing in the crossover fragment from the contributing parent will no longer refer to the automatically defined function of the contributing parent, but instead will refer to the same-named automatically defined function of the receiving parent. Thus, we must redefine the crossover operation when it is employed in an architecturally diverse population.

When automatically defined functions are involved, each program in the population conforms to a more complex constrained syntactic structure (such as shown in Fig. 5.1). The initial random population is created in accordance with this constrained syntactic structure. Crossover must be performed in a structure-preserving way to preserve the syntactic validity of all offspring. In structure-preserving crossover, the architecture-defining points of an overall program are never eligible to be chosen as crossover points and are never altered by crossover. Instead, structure-preserving crossover is restricted to the work-performing points. In structure-preserving crossover, the work-performing points in the overall program are partitioned into a certain number of types.

The basic idea of structure-preserving crossover is that any work-performing point anywhere in the overall program is randomly chosen, without restriction, as the crossover point of the first parent. That point has a type assigned to it. Then, once the crossover point of the first parent has been chosen, the crossover point of the second parent is randomly chosen from among points of the same type.

The typing of the work-performing points of an overall program constrains the set of subtrees that can potentially replace the chosen crossover point and the subtree below it. This typing is done so that the structure-preserving crossover operation will always produce valid offspring.

There are several ways of assigning types to the work-performing points of an overall program.

1. *Branch typing* assigns the same type to all the work-performing points of each separate branch of an overall program (but a different type to each different branch). There are as many types of work-performing points as there are branches in the overall program.

2. *Like-Branch Typing* assigns the same type to all the work-performing points of each separate branch of an overall program and assigns a different type to each different branch, except that if the function sets and terminal sets of two branches are identical, all the points of both such branches are assigned the same type.

3. *Point typing* assigns a type to each individual work-performing point in the overall program reflective of both the branch where the point is located and the contents of the subtree starting at the point. The characteristics of the branch where the point is located is relevant in determining whether a subtree from another program may be inserted at the point. The contents of the subtree starting at the point are relevant in determining if the subtree may be inserted at a particular point of another program.

If a program is subject to any additional problem-specific constrained syntactic structure, that additional structure, if any, must also be considered in typing. When all the programs in the population have a common architecture, any of the three methods of typing may be used. In practice, branch typing is most commonly used. The crossover operation starts with two parents and produces two offspring when either branch typing or like-branch typing is being used.

Point typing is used for architecturally diverse populations. If, for the sake of argument, branch typing or like-branch typing were to be used on an architecturally diverse population, the crossover operation would be virtually hamstrung; hardly any crossovers could occur. The types produced by branch typing or like-branch typing are insufficiently descriptive and overly constraining in an architecturally diverse population.

When point typing is used, the crossover operation acquires a directionality that did not exist with branch typing or like-branch typing. A distinction must be made between the contributing (first) parent and the receiving (second) parent. Consequently, the crossover operation starts with two parents, but produces only one offspring.

The crossover point (called the *point of insertion*) of the receiving (second) parent must be chosen from the set of points for which the crossover fragment from the contributing (first) parent has meaning if the crossover fragment were to be inserted at the point. When genetic material is inserted into the receiving parent during structure-preserving crossover with point typing, the offspring inherits its architecture from the receiving parent (the maternal line) and is guaranteed to be syntactically and semantically valid.

Point typing is governed by three general principles.

1. Every terminal and function actually appearing in the crossover fragment from the contributing parent must be in the terminal set or function set of the

branch of the receiving parent containing the point of insertion. This first general principle applies to actual variables of the problem, dummy variables, random constants, primitive functions, and automatically defined functions.

2. The number of arguments of every function actually appearing in the crossover fragment from the contributing parent must equal the number of arguments specified for the same-named function in the argument map of the branch of the receiving parent containing the insertion point. This second general principle governing point typing applies to all functions. However, the emphasis is on the automatically defined functions because the same function name is used to represent entirely different functions with differing number of arguments for different individuals in the population.

3. All additional problem-specific syntactic rules of construction, if any, must be satisfied.

Structure-preserving crossover with point typing (Koza, 1994a) permits robust recombination while simultaneously guaranteeing that any pair of architecturally different parents will produce syntactically and semantically valid offspring. In addition, structure-preserving crossover with point typing enables the architecture appropriate for solving the problem to be *evolutionarily selected* during a run while the problem is being solved. In addition, when the six new architecture-altering operations are being used, structure-preserving crossover with point typing enables the architecture appropriate for solving the problem to be *evolved* during a run while the problem is being solved in the sense of *actually changing* the architecture of programs dynamically during the run.

Steps for Executing Genetic Programming

Execution of genetic programming consists of the following steps. The six new architecture-altering operations appear as items (2)(c)(iii) through (2)(c)(ix).

The steps for executing genetic programming are as follows:

1. Generate an initial random population (Generation 0) of computer programs.
2. Iteratively perform the following substeps until the termination criterion of the run has been satisfied:
   a. Execute each program in the population and assign it (explicitly or implicitly) a fitness value according to how well it solves the problem.
   b. Select program(s) from the population to participate in the genetic operations in (c).
   c. Create new program(s) for the population by applying the following genetic operations.

i.   *Reproduction*: Copy an existing program to the new population.
ii. *Crossover*: Create new offspring program(s) for the new population by recombining randomly chosen parts of two existing programs.
iii. *Mutation*: Create one new offspring program for the new population by randomly mutating a randomly chosen part of one existing program.
iv. *Branch duplication*: Create one new offspring program for the new population by duplicating one function-defining branch of one existing program and making additional appropriate changes to reflect this change.
v.   *Argument duplication*: Create one new offspring program for the new population by duplicating one argument of one function-defining branch of one existing program and making additional appropriate changes to reflect this change.
vi. *Branch deletion*: Create one new offspring program for the new population by deleting one function-defining branch of one existing program and making additional appropriate changes to reflect this change.
vii. *Argument deletion*: Create one new offspring program for the new population by deleting one argument of one function-defining branch of one existing program and making additional appropriate changes to reflect this change.
viii. *Branch creation*: Create one new offspring program for the new population by adding one new function-defining branch containing a portion of an existing branch and creating a reference to that new branch.
ix. *Argument creation*: Create one new offspring program for the population by adding one new argument to the argument list of an existing function-defining branch and appropriately modifying references to that branch.

3. After satisfaction of the termination criterion (which usually includes a maximum number of generations to be run as well as a problem-specific success predicate), the single best computer program in the population produced during the run (the best-so-far individual) is designated as the result of the run. This result may (or may not) be a solution (or approximate solution) to the problem.

EXAMPLE OF AN ACTUAL RUN

The architecture-altering operations described herein are now illustrated by showing an actual run of the problem of symbolic regression of the even-5-

parity function. The Boolean even-k-parity function takes k Boolean arguments, D0, D1, D2, and so forth (up to a total of k arguments). The
even-k-parity function returns T (true) if an even number of its Boolean
arguments are T, but otherwise returns NIL (false). Boolean parity functions
are often used as benchmarks for experiments in machine learning because a
change in any one input (environmental sensor) toggles the outcome. The
problem is to discover a computer program that mimics the behavior of the
Boolean even-k-parity problem for every one of the 2^k combinations of its k
Boolean inputs. The primitive functions for this problem are AND, OR,
NAND, and NOR.

Example with a Complete Genealogical Audit Trail

The run starts with the random creation of a population of 1,000 individual
programs. The minimalist approach is used herein. That is, each program in
the initial random population at Generation 0 consists of one result-producing
branch and 1 one-argument function-defining branch and has an argument map
of {1}.

Thus, the terminal set for the result-producing branch, T_{rpb}, for a program
in the population for the Boolean even-3-parity problem is:

$$T_{rpb} = \{D0, D1, D2\}.$$

The function set for the result-producing branch, F_{rpb}, is

$$F_{rpb} = \{AND, OR, NAND, NOR, ADF0\},$$

with an argument map of

$$\{2, 2, 2, 2, 1\}.$$

The terminal set for the automatically defined function, ADF0, is:

$$T_{adf0} = \{ARG0\}.$$

The function set, F_{adf0}, for ADF0 is:

$$F_{adf0} = \{AND, OR, NAND, NOR\},$$

with an argument map for this function set of

$$\{2, 2, 2, 2\}.$$

After creating the 1,000 programs for the initial random population, each
program in the population is evaluated as to how well it solves the problem at
hand. The fitness of a program in the population of 1,000 programs is measured
according to how well that program mimics the target function for all eight
combinations of three Boolean arguments. The raw fitness of a program is the
number of matches.

In one particular run, the best program from among the 1,000 randomly created programs in Generation 0 has the function-defining branch (defining ADF0) shown as:

(OR (AND (NAND ARG0 ARG0) (OR ARG0 ARG0)) (NOR (NOR
      ARG0 ARG0) (AND ARG0 ARG0))).

The behavior of this function-defining branch is the Boolean constant function zero (called *Always False*).

The result-producing branch of this best-of-generation program from Generation 0 ignores ADF0 and is shown as:

(NOR (AND D0(NOR D2 D1)) (AND (AND D2 D1))).

Of course, it should be no surprise that the function-defining branch of even the best program of the initial random generation is not particularly useful or that this branch is ignored by the result-producing branch. The minimalist approach is not intended to provide a highly useful function-defining branch, but rather merely to provide a starting point for the evolutionary process.

Table 5.5 shows the behavior of this program from Generation 0. The first three columns show the values of the three Boolean variables, D0, D1, and D2. The fourth column shows the value produced by the overall program. The fifth column shows the value of the target function, the even-3-parity function. The last column shows how well the program performed at matching the behavior of the target function. As is shown, the program was correct for six of the eight possible combinations (fitness cases). Thus, the program scored a raw fitness of 6 (out of a possible 8).

A new population of 1,000 programs is then created from the existing population of 1,000 programs. Each successive generation of the population is created from the existing population by applying various genetic operations. Reproduction and crossover are the most frequently performed genetic opera-

TABLE 5.5.
Operation of the Best-of-Generation Program From Generation 0.

D0	D1	D2	Best-of-Generation Program for Generation 0	Even-3-Parity Function	Score
0	0	0	1	1	Correct
0	0	1	1	0	Wrong
0	1	0	1	0	Wrong
0	1	1	1	1	Correct
1	0	0	0	0	Correct
1	0	1	1	1	Correct
1	1	0	1	1	Correct
1	1	1	0	0	Correct

tions. In addition, the architecture-altering operations described herein are used on this run. Mutation and other previously described genetic operations may also be used in the process (although they are not used here).

The raw fitness of the best-of-generation program for Generation 5 improves to 7. That is, this program correctly mimics the behavior of the target even-3-parity function for seven of the eight fitness cases. The program achieving this new and higher level of fitness has a total of four branches (i.e., one result-producing branch and three function-defining branches). The change in the number of branches from 1 at Generation 0 to 4 at Generation 5 is the consequence of the architecture-altering operations. In addition to its one result-producing branch, this best-of-generation program for Generation 5 has branches defining ADF0 (taking two arguments), ADF1 (taking two arguments), and ADF2 (taking three arguments), so that its argument map is {2, 2, 3}. The result producing branch of this program is shown as:

(NOR (ADF2 D0 D2 D1) (AND (ADF1 D2 D1)D0)).

The first function-defining branch (defining ADF0) of the best-of-generation program for Generation 5 takes two dummy arguments, ARG0 and ARG1, and is shown next. The existence of two dummy arguments in this function-defining branch is a consequence of an argument duplication operation. As it happens, the behavior of this ADF0 is not important because ADF0 is not referenced by the result-producing branch:

(OR (AND (NAND ARG0 ARG0) (OR ARG1 ARG0)) (NOR (NOR
ARG1 ARG0) (AND ARG0 ARG1))).

The second function-defining branch (defining ADF1) of the best-of-generation program for Generation 5 also takes two dummy arguments, ARG0 and ARG1, and is shown next. The existence of this second function-defining branch is a consequence of a branch duplication operation:

(OR (AND ARG0 ARG1) (NOR ARG0 ARG1)).

Table 5.6 shows the behavior of ADF1 of the best-of-generation program for Generation 5 that, as can be seen, is equivalent to the even-2-parity function.

TABLE 5.6.
ADF1 of the Best-of-Generation Program of Generation 5

ARG0	ARG1	ADF0
0	0	1
0	1	0
1	0	0
1	1	1

The function-defining branch for ADF2 of this best-of-generation program for Generation 5 takes three dummy arguments, ARG0, ARG1, and ARG2, and is shown next. This third function-defining branch exists as a consequence of yet another branch duplication operation:

(AND ARG1 (NOR ARG0 ARG2)).

Table 5.7 shows that the behavior of ADF2 consists of returning 1 only when ARG0 and ARG2 are 0 and ARG1 is 1.

The raw fitness of the best individual program in the population remains at a value of 7 for Generations 6, 7, 8, and 9; however, the average fitness of the population as a whole improves during these generations.

On Generation 10, the best program in the population of 1,000 perfectly mimics the behavior of the even-3-parity function. This 100%-correct solution to the problem has a total of six branches (i.e., five function-defining branches and one result-producing branch). The argument map of this program is {2, 2, 3, 2, 2}. This multiplicity of branches is a consequence of the repeated application of the branch duplication operation and the branch creation operation. The function-defining branches of this program each have more than one dummy argument. All of these additional arguments exist as a consequence of the repeated application of the argument duplication operation.

The result-producing branch of this best-of-generation program for Generation 10 is shown as:

(NOR (ADF4 D0(ADF1 D2 D1)) (AND (ADF1 D2 D1) D0)).

The function-defining branch for ADF0 of this best-of-generation program for Generation 10 takes two dummy arguments, ARG0 and ARG1, and follows next. The behavior of ADF0 is equivalent to the odd-2-parity function:

(OR (AND (NAND ARG0 ARG0) (OR ARG1 ARG0)) (NOR (NOR
ARG1 ARG0) (AND ARG0 ARG1))).

TABLE 5.7.
ADF2 of the Best-of-Generation Program of Generation 5

ARG0	ARG1	ARG2	ADF2
0	0	0	0
0	0	1	0
0	1	0	1
0	1	1	0
1	0	0	0
1	0	1	0
1	1	0	0
1	1	1	0

The function-defining branch for ADF1 of the best-of-generation program for Generation 10 takes two dummy arguments, ARG0 and ARG1, and is shown next. ADF1 is equivalent to the even-2-parity function:

(OR (AND ARG0 ARG1) (NOR ARG0 ARG1)).

The function-defining branch for ADF2 takes three dummy arguments, ARG0, ARG1, and ARG2, and is seen next. ADF2 returns 1 only when ARG0 and ARG2 are 0 and ARG1 is 1. However, ADF2 is ignored by the result-producing branch:

(AND ARG1 (NOR ARG0 ARG2)).

The function-defining branch for ADF3 is the one-argument identity function. This relatively useless branch is ignored by the result-producing branch. The function-defining branch for ADF4 of the best-of-generation program for generation 10 takes two dummy arguments, ARG0 and ARG1, and is seen next. ADF4 is equivalent to the even-2-parity function:

(OR (AND ARG0 ARG1) (NOR ARG0 ARG1)).

Because both ADF1 and ADF4 are both even-2-parity functions, the result-producing branch can be simplified to the next expression. This expression is equivalent to the even-3-parity function:

(NOR (EVEN-2-PARITY D0(EVEN-2-PARITY D2 D1)) (AND (EVEN-2-PARITY D2 D1) D0)).

An examination of the genealogical audit trail shows the interplay between the Darwinian reproduction operation, the one-offspring crossover operation using point typing, and the new architecture-altering operations.

Figure 5.3 shows all of the ancestors of the just-described 100%-correct solution from Generation 10 of the run in Example 1 of the problem of symbolic regression of the even-3-parity problem. The generation numbers (from 0 to 10) are shown on the left edge of Fig. 5.3. Figure 5.3 also shows the sequence of reproduction operations, crossover operations, and architecture-altering operations that gave rise to every program that was an ancestor to the 100%-correct program in Generation 10. The 100%-correct solution from Generation 10 is represented by the box labeled M10 at the bottom of the figure. The argument map of this solution, namely {2, 2, 3, 2, 2}, is shown in this box.

The two lines flowing into the box M10 indicate that the solution in Generation 10 was produced by a crossover operation acting on two programs from the previous generation (Generation 9). Figure 5.3 uses the convention of placing the Mother M9 (the receiving parent) on the right and Father P9 (the contributing parent) on the left. Recall that, in a one-offspring, crossover

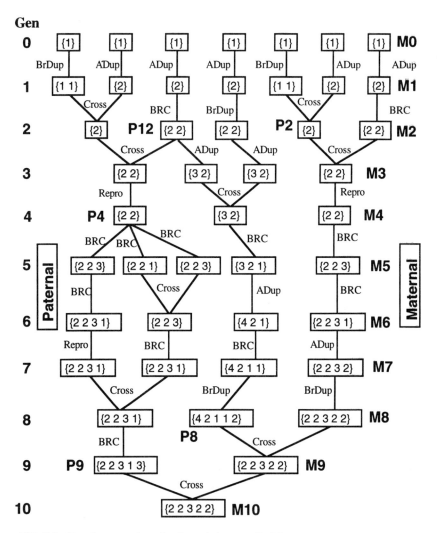

FIG. 5.3. Complete genealogical audit trail showing all of the ancestors in Generations 0 through 9 for the ultimate solution M10from Generation 10 for the run in Example 1. The maternal line is shown at the right and the paternal line at the left.

operation using point typing, the bulk of the structure of a multipart program comes from the mother since the father contributes only one subtree into only one of the many branches of the mother. Thus, the 11 boxes on the right side of this figure (consecutively numbered from M0 to M10) represent the maternal genetic lineage (from Generation 0 through Generation 10) of the

100%-correct solution M10 that emerged in Generation 10. The 100%-correct solution M10 in Generation 10 has the same argument map, {2, 2, 3, 2, 2}, as the Mother M9 because the crossover operation is not an architecture-altering operation and does not change the architecture (or argument map) of the offspring (relative to the mother).

The maternal lineage is now reviewed in detail to illustrate the overall process of evolving the architecture of a solution to a problem while evolving the solution to the problem.

The Mother M9 from Generation 9 (shown on the right side of Fig. 5.3) has an argument map of {2, 2, 3, 2, 2}, has a raw fitness of 7, was itself the result of a crossover of two parents from Generation 8. The grandfather of the 100%-correct solution M10 in Generation 10 (and the father of M9) was P8. The Grandmother of the 100%-correct solution M10 in Generation 10 (and the Mother of M9) was M8.

The Grandmother M8 from Generation 8 of the 100%-correct solution M10 in Generation 10 (and the Mother of M9) has an argument map of {2, 2, 3, 2, 2}, has a raw fitness of 7, and was the result of a branch duplication from a single ancestor M7 from Generation 7. Because of the branch duplication operation, the program M7 from Generation 7 of the maternal lineage at the far right of Fig. 5.3 has one fewer branch than its offspring M8. Program M7 has an argument map of {2, 2, 3, 2}. Program M7 was the result of an argument duplication from a single ancestor from Generation 6.

Because of the argument duplication operation, the fourth function-defining branch of the program M6 from Generation 6 of the maternal lineage at the far right of Fig. 5.3 has one less argument than its offspring M7. Program M6 from Generation 6 has an argument map of {2, 2, 3, 1}, whereas program M7 from Generation 7 has an argument map of {2, 2, 3, 2}. Program M6 was the result of an branch creation from a single ancestor M5 from Generation 5. Because of the branch creation operation, the program M5 from Generation 5 shown on the right side of Fig. 5.3 has one fewer function-defining branch than program M6. Program M5 has an argument map of {2, 2, 3}. In turn, program M5 was the result of a branch creation from a single ancestor, M4 from Generation 4.

Program M4 from Generation 4 shown on the right side of Fig. 5.3 has one less function-defining branch than its offspring program M5. Program M4 has an argument map of {2, 2}. Program M4 was the result of a reproduction operation from a single ancestor, M3 from Generation 3.

Program M4 from Generation 3 (shown on the right side of Fig. 5.3) has an argument map of {2, 2} and was the result of a crossover involving Father P2 and Mother M2 from Generation 2. Program M2 from Generation 2 (shown on the right side of Fig. 5.3) has an argument map of {2, 2} and was the result of a branch creation from a single ancestor M1 from Generation 1.

Program M1 from Generation 1 has an argument map of {2} and was the result of an argument duplication of a single ancestor, M0 from Generation 0.

Program M0 from Generation 0 at the upper right corner of Fig. 5.3 has an argument map of {1} and has a raw fitness of 6. It has an argument map of {1} because all programs at Generation 0 consist of one result-producing branch and a single one-argument, function-defining branch when the minimalist approach is being used. The sequence of genetic operations and architecture-altering operations of this run shows the simultaneous evolution of the architecture while solving the problem.

Example with Even-5-Parity Problem

For the problem of symbolic regression of the even-5-parity function, a population size, M, of 96,000 was used. The targeted maximum number of generations, G, was set at 76. The run used the *minimalist approach* in which each program in Generation 0 consists of one result-producing branch and a single two-argument function-defining branch. Branch deletion and argument deletion with random regeneration were used. The percentage of operations at each processing node on each generation was 74% crossovers; 10% reproductions; 0% mutations; 5% branch duplications, 5% argument duplications; 0.5% branch deletions; 0.5% argument deletions; 5% branch creations; and 0% argument creations. Other minor parameters were chosen as in Koza (1994a).

The problem was run on a home-built, medium-grained parallel computer system. In the so-called *distributed genetic algorithm* or *island model* for parallelization (Tanese, 1989), different semiisolated subpopulations (called *demes* after Wright, 1943) are situated at the different processing nodes of the system. The system consisted of a host PC 486-type computer running Windows and 64 Transtech TRAMs (containing one INMOS T805 transputer and 4 megabytes of RAM memory) arranged in a toroidal mesh. There were $D = 64$ demes, a population size of $Q = 1,500$ per deme, and a migration rate (boatload size) of $B = 8\%$ (in each of four directions on each generation for each deme). Generations were run asynchronously. Additional details of the parallel implementation of genetic programming on a network of transputers can be found in Koza and Andre (1995).

On Generation 13 of one run, a 100%-correct solution to the even-5-parity problem emerged in the form of a computer program with one 3-argument automatically defined function and one 2-argument automatically defined function.

Three-argument ADF0 (which had only two arguments in Generation 0) performed Boolean rule 106, a non-parity rule, and is:

```
(NOR (OR (AND (OR (OR ARG0 ARG2) (NAND ARG0 ARG2))
(AND (NAND ARG0 ARG1) (NOR ARG2 ARG0))) (AND (AND
```

(NOR ARG1 ARG0) (OR ARG2 ARG0)) (OR (NAND ARG0 ARG1)
(NOR ARG2 ARG0)))) (NAND (NAND (AND (NOR ARG0 ARG2)
(NAND ARG0 ARG0)) (NOR (NAND ARG0 ARG0) (NOR ARG2
ARG1))) (OR (NAND (AND ARG1 ARG0) (OR ARG1 ARG0)) (OR
(NAND ARG2 ARG2) (NOR ARG0 ARG0))))).

Two-argument ADF1 (which did not exist at all in Generation 0) is
equivalent to the odd-2-parity function and is:

(NOR (OR (AND (OR (OR ARG0 ARG1) (NAND ARG0 ARG1))
(AND (NAND ARG0 ARG1) (NOR ARG1 ARG0))) (AND (AND
(NOR ARG1 ARG0) (OR ARG1 ARG0)) (OR (NAND ARG0 ARG1)
(NOR ARG1 ARG0)))) (NAND (NAND (AND (NOR ARG0 ARG1)
(NAND ARG0 ARG0)) (NOR (NAND ARG0 ARG0) (NOR ARG1
ARG1))) (OR (NAND (AND ARG1 ARG0) (OR ARG1 ARG0)) (OR
(NAND ARG1 ARG1) (NOR ARG0 ARG0))))).

The result-producing branch of this program invokes both ADF0 and ADF1
and is:

(AND (OR (ADF0 (NAND D1 D2) (ADF0 D2 D0 D0) (ADF0 D2 D0
D0)) (NAND (OR D3 D1) (ADF1 D3 D3))) (ADF0 (ADF1 (NAND
D1 D2) (NOR D4 D4)) (ADF1 (ADF1 D3 D0) (NOR D1 D2)) (ADF1
(ADF1 D3 D0) (NOR D1 D2))))).

PERFORMANCE CHARACTERISTICS
OF THE NEW OPERATIONS

We now use the Boolean even-5-parity problem to compare, over a series of
runs, three performance characteristics of the architecture-altering operations
for the following five approaches:

1. Without automatically defined functions (corresponding to the style of
   runs discussed throughout most of *Genetic Programming*),
2. With automatically defined functions, evolutionary selection of the
   architecture (corresponding to the style of runs discussed in chapters
   21 to 25 of *Genetic Programming II* on the evolutionary selection of the
   architecture), an architecturally diverse initial population, and struc-
   tureipreserving crossover with point typing,
3. With automatically defined functions, the architecture-altering operations, an architecturally diverse population (after Generation 0), and
   structure-preserving crossover with point typing,

4. With automatically defined functions, a fixed, user-supplied architecture (i.e., an argument map of {3, 2} that is known to be a good choice of architecture for this problem), and structure-preserving crossover with point typing, and
5. With automatically defined functions, the fixed, known-good, user-supplied {3, 2} architecture, and structure-preserving crossover with branch typing (corresponding to the style of runs discussed throughout most of *Genetic Programming II*).

The comparisons are made for the following three performance characteristics: computational effort, E (with 99% probability); the wallclock time, $W(M,t,z)$ in seconds (with 99% probability); and the average structural complexity, $\overline{S}$. These three measures are described in detail in Koza (1994a).

The comparisons in Table 5.8 all used a common population size, M, of 96,000. All runs solved well before the targeted maximum number of generations, G, of 76.

As can be seen from the table, all four approaches (2, 3, 4, or 5) employing automatically defined functions required less computational effort than not using them (Approach 1). Approach 5 (which benefits from the most user-supplied information) required the least computational effort. At the other extreme, Approach 1 required the most computational effort.

Approach 3 (using the architecture-altering operations) required less computational effort than solving the problem without automatically defined functions (Approach 1), but more computational effort than with the fixed, known-good, user-supplied architecture (Approach 5).

Approach 2 required greater computational effort than Approach 3, but less than that for Approach 1.

Approach 4 isolated the additional computational effort required by point typing (relative to Approach 5). Greater computational effort was required by Approach 4 than Approach 5. Because the computational effort for

TABLE 5.8.
Comparison of the Five Approaches

Approach	Runs	Computational Effort, E	Wallclock Time, W(M,t,z)	Average Size of Solution $\overline{S}$
A – No ADFs	14	5,025,000	36,950	469.1
B – ADFs Evolutionary Selection of Architecture	14	4,263,000	66,667	180.9
C – ADFs Architecture-Altering Operations	25	1,789,500	13,594	88.8
D – ADFs Point Typing-Fixed, Known-Good Architecture	25	1,705,500	14,088	130.0
E – ADFs Branch Typing-Fixed, Known-Good Architecture	25	1,261,500	6,481	112.2

Approach 3 was virtually tied with Approach 4, the cost of architecture-altering operations for this problem was not much greater than the cost of point typing.

Approach 5 consumed less wallclock time than Approach 3 (using the architecture-altering operations), which, in turn, consumed less wallclock time than Approach 1 (without automatically defined functions).

The average structural complexity, s, for all four approaches (2, 3, 4, or 5) employing automatically defined functions was less than that for Approach 1 (without automatically defined functions). Approach 3 (using the architecture-altering operations) has the lowest value of $\overline{S}$.

Note also that all four approaches (2, 3, 4, or 5) employing ADFs required less computational effort, require less wallclock time, and produce smaller solutions (i.e., are more parsimonious) than the ADF-less approach (Approach 1).

Additional work in this area is described in Koza (1995a, 1995b, 1995c) and Koza and Andre (1996). Application of the architecture-altering operations to the automated design of analog electrical circuits using genetic programming with automatically defined functions was described in Koza, Andre, Bennett, and Keane (1996).

CONCLUSIONS

This chapter describes how the biological theory of gene duplication described in Ohno's 1970 book was applied to a vexatious problem of architecture discovery for automated machine learning. The resulting biologically motivated approach enables genetic programming to automatically discover the size and shape of the solution at the same time as genetic programming is evolving a solution to the problem. This is accomplished using six biologically motivated, architecture-altering operations that provide a way to automatically discover, during a run of genetic programming, both the architecture and the sequence of steps of a multipart computer program that will solve the given problem.

ACKNOWLEDGMENTS

David Andre and Walter Alden Tackett wrote the program in C for the architecture-altering operations used in this chapter. The midge comes from *Destructive and Useful Insects* by C. L. Metcalf and W. P. Flint.

REFERENCES

Angeline, P. J., & Kinnear, K. E. Jr. (Eds.). (1996). *Advances in genetic programming 2.* Cambridge, MA: MIT Press.

Angeline, P. J., & Pollack, J. B. (1994). Coevolving high-level representations. In C. G. Langton (Ed.), *Artificial life III, SFI studies in the sciences of complexity* (Vol. XVII, pp. 55–71). Redwood City, CA: Addison-Wesley.

Brooks Low, K. (1988). Genetic recombination: A brief overview. In K. Brooks Low (Ed.), *The recombination of genetic material* (pp. 1–21). San Diego: Academic Press.

Cavicchio, D. J. (1970). *Adaptive search using simulated evolution.* Unpublished doctoral dissertation, University of Michigan, Ann Arbor.

Darwin, C. (1859). *On the origin of species by means of natural selection.* John Murray.

Dyson, P., & Sherratt, D. (1985). Molecular mechanisms of duplication, deletion, and transposition of DNA. In T. Cavalier-Smith (Ed.), *The evolution of genome size.* Chichester, England: Wiley.

Galli, J., & Wislander, L. (1993). Two secretary protein genes in *Chironomus tentans* have arisen by gene duplication and exhibit different developmental expression patterns. *Journal of Molecular Biology, 231,* 324–334.

Galli, J., & Wislander, L. (1994). Structure of the smallest salivary-gland secretory protein in *Chironomus tentans. Journal of Molecular Evolution, 38,* 482–488.

Go, M. (1991). Module organization in proteins and exon shuffling. In S. Osawa & and T. Honjo (Eds.), *Evolution of life.* Tokyo: Springer-Verlag.

Goldberg, D. E. (1989). *Genetic algorithms in search, optimization, and machine learning.* Reading, MA: Addison-Wesley.

Goldberg, D. E., Korb, B., & Deb, K. (1989). Messy genetic algorithms: Motivation, analysis, and first results. *Complex Systems, 3*(5), 493–530.

Gruau, F. (1994). Genetic micro programming of neural networks. In K. E. Kinnear Jr. (Ed.), *Advances in genetic programming* (pp. 495–518). Cambridge, MA: MIT Press.

Holland, J. H. (1975). *Adaptation in natural and artificial systems: An introductory analysis with applications to biology, control, and artificial intelligence.* Ann Arbor, MI: University of Michigan Press.

Hood, L., & Hunkapiller, T. (1991). Modular evolution and the immunoglobin gene superfamily. In S. Osawa & T. Honjo (Eds.), *Evolution of life.* Tokyo: Springer-Verlag.

Kinnear, K. E. Jr. (Ed.). (1994). *Advances in genetic programming.* Cambridge, MA: MIT Press.

Koza, J. R. (1989). Hierarchical genetic algorithms operating on populations of computer programs. In *Proceedings of the 11th international joint conference on artificial intelligence* (Vol I, pp. 768–774). San Mateo, CA: Kaufmann.

Koza, J. R. (1992). *Genetic programming: On the programming of computers by means of natural selection.* Cambridge, MA: MIT Press.

Koza, J. R. (1994a). *Genetic programming II: Automatic discovery of reusable programs.* Cambridge, MA: MIT Press.

Koza, J. R. (1994b). *Genetic programming II videotape: The next generation.* Cambridge, MA: MIT Press.

Koza, J. R. (1994c). *Architecture-altering operations for evolving the architecture of a multi-part program in genetic programming* (Tech. Rep. No. STAN-CS-TR-94-1528). Palo Alto, CA: Stanford University.

Koza, J. R. (1995a). Evolving the architecture of a multi-part program in genetic programming using architecture-altering operations. In J. R. McDonnell, R. G.

Reynolds, & D. B. Fogel (Eds.), *Evolutionary programming IV: Proceedings of the fourth annual conference on evolutionary programming* (pp. 695–717). Cambridge, MA: MIT Press.

Koza, J. R. (1995b). Gene duplication to enable genetic programming to concurrently evolve both the architecture and work-performing steps of a computer program. In *Proceedings of the 14th international joint conference on artificial intelligence* (pp. 734–740). San Francisco, CA: Kaufmann.

Koza, J. R. (1995c). Two ways of discovering the size and shape of a computer program to solve a problem. In L. J. Eshelman (Ed.). *Proceedings of the Sixth International Conference on Genetic Algorithms* (pp. 287–294). San Francisco, CA: Kaufmann.

Koza, J. R., & Andre, D. (1995). *Parallel genetic programming on a network of transputers* (Tech. Rep. No. STAN-CS-TR-95-1542). Palo Alto, CA: Stanford University.

Koza, J. R., & Andre, D. (1996). Classifying protein segments as transmembrane domains using architecture-altering operations in genetic programming. In P. J. Angeline & K. E. Kinnear Jr. (Eds.), *Advances in genetic programming 2* (pp. 155–176). Cambridge, MA: MIT Press.

Koza, J. R., Andre, D., Bennett, F. H., III, & Keane, M. A. (1996). Use of automatically defined functions and architecture-altering operations in automated circuit synthesis with genetic programming. In J. R. Koza, D. E. Goldberg, D. B. Fogel, & R. L. Riolo (Eds.), *Genetic programming 1996: Proceedings of the first annual conference, July 28–31, 1996, Stanford University* (pp. 132–140). Cambridge, MA: MIT Press.

Koza, J. R., Goldberg, D. E., Fogel, D. B., & Riolo, R. L. (Eds.). (1996). *Genetic programming 1996: Proceedings of the first annual conference, July 28–31, 1996, Stanford University.* Cambridge, MA: MIT Press.

Koza, J. R., & Rice, J. P. (1992). *Genetic programming: The movie.* Cambridge, MA: MIT Press.

Lazcano, A., & Miller, S. L. (1994). How long did it take for life to begin and evolve to cyanobacteria? *Journal of Molecular Evolution, 39*, 546–554.

Lindgren, K.. (1991). Evolutionary phenomena in simple dynamics. In C. Langton, C. Taylor, J. D. Farmer, & S. Rasmussen (Eds.), *Artificial life II, SFI studies in the sciences of complexity* (Vol. X, pp. 295–312). Redwood City, CA: Addison-Wesley.

Maeda, N., & Smithies, O. (1986). The evolution of multigene families: Human haptoglobin genes. *Annual Review of Genetics 20*, 81–108.

Mitchell, M. (1996). *An introduction to genetic algorithms.* Cambridge, MA: MIT Press.

Nei, M. (1987). *Molecular evolutionary genetics.* New York: Columbia University Press.

Ohno, S. (1970). *Evolution by gene duplication.* New York: Springer-Verlag.

Patthy, L. (1991). Modular exchange principles in proteins. *Current Opinion in Structural Biology, 1*, 351–361.

Riley, M. (1993). Functions of the gene products of *Escherichia coli. Reviews of Microbiology, 32*, 519–560.

Samuel, A. L. (1959). Some studies in machine learning using the game of checkers. *IBM Journal of Research and Development, 3*(3), 210–229.

Smith, T. F., & Waterman, M. S. (1981). Identification of common molecular subsequences. *Journal of Molecular Biology, 147*, 195–197.

Stryer, L. (1988). *Biochemistry.* San Francisco: Freeman.

Tanese, R. (1989). *Distributed genetic algorithm for function optimization.* Unpublished doctoral dissertation, University of Michigan, Ann Arbor.

Wright, S. (1943). Isolation by distance. *Genetics, 28*, 114–138.

6

The Frightening Complexity of Avoidance: A Neural Network Approach

Nestor A. Schmajuk
David W. Urry
B. Silvano Zanutto
Duke University

In spite of the vast behavioral data collected on avoidance over several decades, the question of how animals learn to avoid dangerous stimuli remains largely unanswered. Although several theories have been proposed to explain avoidance—notably Mowrer's (1947) two-factor theory, Herrnstein's (1969) one-factor theory, and Seligman and Johnston's (1973) cognitive theory—none of them was able to provide a completely successful account of the experimental data accumulated over time.

The failure of earlier theories might be attributed to what Kamin (1957) called the "frightening complexity of the theoretical aspects of avoidance" (p. 445). A common problem of the older theories of avoidance is that they did not provide accurate, unequivocal descriptions of behavior, thereby making it difficult to evaluate their theoretical constructs. Recently, computational models have been developed to describe classical conditioning, avoidance, spatial learning, and cognitive mapping. Unlike verbal theories, these models provide precise quantitative descriptions that can be contrasted with experimental data. Some of these computational models are real-time models and, therefore, their output can be compared to behavior as it unfolds in real time. Furthermore, the dynamics of their intervening variables can be contrasted with neural activity, providing a basis for the study of the physiological foundations of behavior.

Given the success of computational models in generating explicitly testable theories and offering guidance for the investigation of the physiological basis of behavior, the present chapter describes avoidance in terms of a neural network. This neural network theory depicts classical and operant conditioning processes assumed to underlie escape and avoidance. The theory successfully accounts for many of the features that characterize both behaviors.

A NEURAL NETWORK MODEL OF ESCAPE
AND AVOIDANCE

In 1947, Mowrer proposed a two-process theory that appealed to classical and operant conditioning to describe avoidance behavior. The classical conditioning process consists of the association between the warning stimulus (WS) and the shock unconditioned stimulus (US), and the consequent generation of a fear conditioned response (CR) when WS is presented. According to drive–reduction theories popular at the time (Hull, 1951), Mowrer suggested that the operant conditioning process consists of the reinforcement of the avoidance response (Ra) by the reduction of fear as the animal's response terminates the fear-eliciting WS. Recently, Zhuikov, Couvillon, and Bitterman (1994) presented and tested a quantitative restatement of Mowrer's two-process avoidance theory.

A two-process neural network theory of avoidance is presented here. In cognitive terms, the model assumes that through classical conditioning animals build an internal model of their environment (see Sokolov, 1960), and that through operant conditioning animals select from alternative behavioral strategies (see Hull, 1951). The internal model provides *predictions* of what environmental events precede other environmental events, such as the US. Behavioral strategies refer to the *prescription* of the responses to be generated in different circumstances. Whenever there is a mismatch between predicted and actual environmental events, the internal model is modified and the behavioral strategies are adjusted.

In simple terms, the network intimately combines classical and operant conditioning principles. The classical conditioning process involves the formation of associations between different environmental stimuli and the animal's responses with the US. These associations are used to predict the presence or absence of the US. Classical conditioning is regulated by the mismatch between the actual and the predicted intensity of the US: When the US is underpredicted, classical associations increase, and decrease otherwise. The operant conditioning process entails the formation of associations between environmental stimuli with the escape or the avoidance response. These associations are used to select the adequate response in each case. Operant conditioning is controlled by a novel algorithm that mirrors the classical conditioning algorithm: When the US is underpredicted operant associations decrease, and increase otherwise.

Figure 6.1 shows a real-time neural network that describes escape and avoidance. The network is a real-time mechanism that describes behavior as a moment-to-moment phenomenon. Nodes in the network represent neural populations, rather than individual neurons. Appendix A presents a formal description of the network as a set of differential equations that depict changes in the values of neural activities and connectivities as a function of time. Table 6.1 summarizes the intervening variables used in the model. The model has eight variables and nine parameters.

Real-Time Internal Representations

Environmental events are internally represented by real-time variables in the model. We assume that both the WSs and the animal's responses (R) can be regarded as biologically neutral CSs and might become associated with the US. In addition, following Konorski (1967),we assume that the US itself has both sensory (the US as a CS) and emotional (the US as a biologically meaningful stimulus) representations. Hull (1929) proposed that CSs give rise to short-term memory (STM) traces, $\tau_{cs}(t)$, in the central nervous system that increase over time to a maximum and then gradually decay back to zero. In the same vein, we assume that WS, US, and Rs activate different neural populations whose activity constitute STM traces $\tau_{ws}(t)$, $\tau_{us}(t)$, and $\tau_R(t)$. The use of STM traces allows the model to describe paradigms, such as trace conditioning, in which the STM traces (but not the physical stimuli them-selves) temporally overlap with the US.

Gormezano, Kehoe, and Marshall (1983) suggested that the curve repre-senting the strength of the CS–US association as a function of the CS–US interstimulus interval (ISI) reflects the variation in the intensity of trace $\tau_{cs}(t)$

TABLE 6.1
Glossary of Intervening Variables

Symbol	Variable
τ_i	STM trace
V_i	Classical association of τ_i with the US
$\Sigma_j V_j \tau_j$	Aggregate prediction of the US
CR	Strength of the conditioned response
Z_{ik}	Operant association of τ_i with the R_k
R_k	Strength of alternative response k
r_k	Random number for response R'_k
R_k'	Strength of operant response R_k

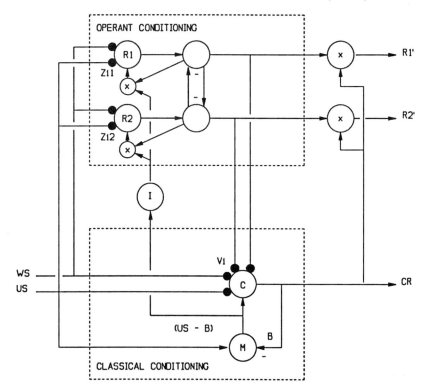

FIG. 6.1. Neural network for escape and avoidance learning.
Classical conditioning processes build an internal model of the environment used to generate predictions of the US. Operant conditioning processes establish the animal's behavioral strategy by selecting the adequate avoidance or escape responses. Mismatch between the actual and predicted intensity of the US modulates changes in both classical and operant associations. The prediction of the US also serves to regulate the strength of the operant response. WS: Warning stimulus. US: Unconditioned stimulus. *Classical conditioning.* CR: Conditioned response. Solid circles represent V, the associations of the US, WS, and Rs with the US [V(US,US), V(WS,US), V(R1,US), and V(R2,US)]. B: Aggregate prediction of the US. C: Classical conditioning element. M: Mismatch between actual and predicted values of the US. *Operant conditioning.* Solid circles represent Z, the associations of WS with R [Z(WS, R1) and Z(WS,R2), and of the US with R [Z(US, R1) and Z(US, R2)]. Open circles with an X inside represent neural populations that compute the product of the inputs. Inhibitory connections indicate competition between alternative responses R1 and R2. I: Neural population converting (US - B) into (B - US). R'1 and R'2: Strength of the output of the operant responses.

over time. In the case of the rabbit's nictitating membrane conditioned response, it has been consistently reported (Schneiderman, 1966; Smith, 1968) that CS–US associations are negligible at zero ISI, present a peak at ISIs of 200 msec, and gradually decrease for longer ISIs. Based on the shape of the ISI curve, Schmajuk and Moore (1990) suggested that $\tau_{cs}(t)$ grows fast after a 80 msec delay following the CS presentation, reaches its maximum value at around 200 msec, and then decays to zero in approximately 2,000 msec. In the case of fear conditioning in rats, Davis, Schlesinger, and Sorenson (1989) reported that CS–US associations are negligible at zero ISI and monotonically increase from 50 msec to 50 sec ISIs. The shape of the ISI curve in fear conditioning suggests that $\tau_{ws}(t)$ grows fast after a 50 msec delay following the CS presentation, reaches its maximum value at around 50 sec, and then decays to zero.

Real-time internal representations participate in two processes: classical and operant conditioning. The classical conditioning process describes the acquisition of WS–US, R–US, and US–US associations. The operant conditioning process describes US–Re and WS–Ra associations.

Classical Conditioning.   The Classical Conditioning block of Fig. 6.1 shows that node C receives inputs from the US (as a stimulus), WS, Rl', and R2'. Solid circles V_i represent the associations of these inputs with the US. Classical associations, regarded as a prediction of the US by the US, WS, or R, are changed according to a real-time version of the Rescorla–Wagner rule (or least mean-squares (LMS) algorithm), described by Schmajuk and DiCarlo (1992). LMS algorithms minimize the mean square difference between the actual and predicted values of the US. Independently introduced by Widrow and Hoff (1960) and Rescorla and Wagner (1972), LMS algorithms were applied by Pearce and Hall (1980), Sutton and Barto (1981), and Kehoe (1988) to classical conditioning.

Node M computes the mismatch (US - B) between the actual US intensity and its aggregate prediction, $B = \Sigma_i \tau_i(t) V_i$ (where t$\tau_i(t)$ represents the trace of either a WS, the US, or an R). Classical associations, V_i, increase when (a), $\tau_{ws}(t)$, $\tau_{us}(t)$, or $\tau_R(t)$ is active, and (b) (US - B) is greater than zero. Classical associations, V_i, decrease when (a) $\tau_{ws}(t)$, $\tau_{us}(t)$, or $\tau_R(t)$ is active, and (b) (US - B) is less than zero. Because no learning occurs when US = B, USs, WSs, and Rs compete with each other in order to gain association with the US. When the US is predicted but is not presented, other WSs or Rs present at that time might acquire inhibitory associations with the US, that is, predict the absence of the US. The algorithm provides real-time descriptions of classical conditioning paradigms that include acquisition and extinction of delay and trace conditioning, blocking, overshadowing, conditioned inhibition, and discrimination acquisition and reversal.

The aggregate prediction of the US, B, is interpreted as *fear* of the aversive US in the context of the model. Figure 6.1 shows that B is used: (a) to define the strength of the conditioned response, $CR = f [\Sigma_i \tau_i(t) V_i]$, (b) to compute (US - B) at node M, and (c) to regulate the strength of the operant response (R1' or R2').

Operant Conditioning. The Operant Conditioning block in Fig. 6.1 shows that nodes R1 and R2 receive inputs from WS and the US (as a stimulus). Solid circles Z_{i1}, and Z_{i2}, represent the associations of these inputs with alternative responses R1 and R2. In Fig. 6.1, node I inverts (US–B), computed by node M and used by the classical conditioning block, to compute (B–US), used by the operant conditioning block. Operant associations, Z_{ik}, increase when (a) $\tau_{WS}(t)$ or $\tau_{US}(t)$ are active together with $\tau_{Rk}(t)$, and (b) (B–US) is greater than zero. Operant associations, Z_{ik}, decrease when (a) $\tau_{WS}(t)$ or $\tau_{US}(t)$ are active together with $\tau_{Rk}(t)$, and (b) (B–US) is less than zero. Therefore, the model selects the adequate response by first increasing Z_{ik} of all responses active in the presence of fear alone (i.e., increasing variability) and then by decreasing the Z_{ik}, of those responses active when the US is present (i.e., discarding the wrong responses). Notice that the US and WS accrue excitatory associations with responses that gain inhibitory associations with the US, that is, with responses, such as Ra or Re, that predict the absence of the US.

Output of nodes R1 and R2, proportional to $\Sigma_i Z_{ik}\tau_i$ compete to decide which alternative response will be generated. Noise is added to the intensity of each alternative response in order to provide initial random responses at the beginning of training. The response, R1 or R2, most strongly activated becomes selected and is executed by the system. This response is the one that, at given time, predicts the minimal amount of US (either generates the strongest prediction of the absence of the US or generates the weakest prediction of its presence). Figure 6.1 shows that the selected response is combined with the error signal arriving from node I to modify its association with the active $\tau_{WS}(t)$ or $\tau_{US}(t)$.

Figure 6.1 shows that the intensity of the selected operant response, R1' or R2', is proportional to the product of the aggregate prediction of the US, B, and the STM trace of the maximal response (R1 or R2). Because the strength of the output response is proportional to the product of WS–US and WS–R associations, and therefore decreases in one are compensated by increases in the other, similar output response strength can be obtained with multiple combinations of classical and operant associations.

Escape and Avoidance

This section illustrates how the model processes information during acquisition of avoidance in a shuttle box. Figure 6.2 shows a graphic depiction of escape and avoidance in a two-way shuttle-box. The shuttle-box is a chamber with two compartments separated by a barrier with a door. Each compartment has a metal-grid floor that can deliver a shock, US. Lights above the chambers are used to signal the US. The experiment may start with both compartments being dark (Fig. 6.2A). At time zero, the light above the compartment where

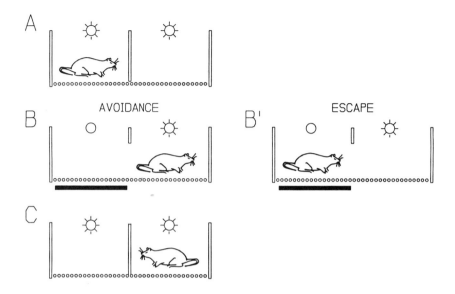

FIG. 6.2. Sequence of events during escape or avoidance in a shuttle-box. A: The experiment starts with both compartments being illuminated. B: At time zero, the light above the compartment where the animal is located turns off (WS) and the door separating both compartments opens. If the animal has not crossed to the opposite side before a given time, it escapes the US. B': If the animal has crossed to the opposite side after a given time, it avoids the US. C: After a constant or an average intertrial interval, the whole sequence restarts. The black solid line underneath the grid represents the US.

the animal is located turns on (WS) and the door separating both compartments opens. If the animal has not crossed to the opposite side after a given time (that may vary between 2 and 40 sec), the shock US is applied (Fig. 6.2B). If the animal has crossed to the opposite side before that time, it avoids the US (Fig. 6.2B'), and the separating door closes behind it. After a constant or an average intertrial interval (ITI), that varies from 15 sec to 4 min, the whole sequence restarts (Fig. 6.2C).

Figure 6.3 shows the temporal arrangement of WS, Re, Ra, US, and their respective STM traces, $\tau(t)$, during escape and avoidance. Figure 6.3 shows that when WS is on, $\tau_{WS}(t)$ increases toward one, and decreases toward zero when WS turns off. Normally, when the animal produces the avoidance response, Ra, WS turns off. Notice that the more rapidly the animal generates the Ra, the shorter the $\tau_{WS}(t)$. Also notice that $\tau_{WS}(t)$ is greater than zero even when the animal has generated the Ra. These temporal properties of $\tau_{WS}(t)$ are essential to the functioning of the model. When the animal generates the avoidance response Ra, its trace $\tau_{Ra}(t)$ increases toward one. As the response is terminated, $\tau_{Ra}(t)$ decreases back to zero.

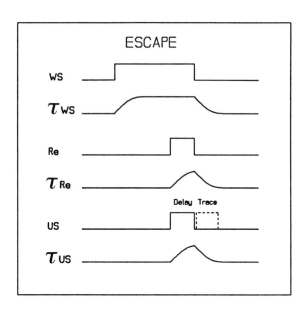

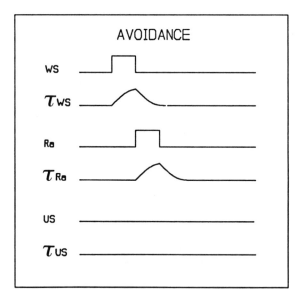

FIG. 6.3. Real-time variables used in the model during escape and avoidance. WS: Warning signal, τ_{WS}: Trace of the WS, Re: Escape response, τ_{Re}: Trace of the escape response, Ra: Avoidance response, τ_{Ra}: Trace of the avoidance response, US: Shock unconditioned stimulus, τ_{US}: Trace of the US. The US represented by a solid line corresponds to delay avoidance. The US represented by a dashed line corresponds to trace avoidance.

At the beginning of training, when the WS is on, $\tau_{WS}(t)$ is active in the presence of the US. Therefore, $\tau_{WS}(t)$ accrues an excitatory association with the US. When the animal emits the correct avoidance response, Ra, crossing to the opposite side and terminating the WS, $\tau_{WS}(t)$ and $\tau_{Ra}(t)$ are active in the absence of the predicted US. At this time, the prediction of the US exceeds its actual value (US < B). Consequently, in the Classical Conditioning block, $\tau_{WS}(t)$ partially decreases its excitatory association and $\tau_{Ra}(t)$ acquires an inhibitory association with the US (see Equation 2 in Appendix A). Simultaneously, in the Operant Conditioning block, $\tau_{WS}(t)$ accrues an excitatory association with Ra (see Equation 3 in Appendix A).

We assume that the animal is able to avoid the US when the strength of the operant response, Ra′ = $\tau_{Ra}(t)$ B, exceeds an arbitrary threshold value. Ra′ is a real-time function whose temporal topography is basically defined by the shape of $\tau_{Ra}(t)$, which exponentially grows and decays. The latency of the avoidance response is the time when Ra′ reaches threshold. Therefore, latency decreases for a fast-growing $\tau_{Ra}(t)$ and large Bs. Because $\tau_{Ra}(t)$ grows faster with increasing values of $R_a = K_6 f[\Sigma_i \tau_i Z_{ia}]$, latency decreases with increasing values of Z_{ia}, (WS–R_a and US–R_a operant associations). Also, because B = $\Sigma_j \tau_j, V_j$ latency decreases with increasing values of V_j (WS–US and US–US classical associations). Notice that the same amplitude and latency of the output response Ra′ can be achieved by various combinations of operant and classical associations.

As response latency decreases, $\tau_{WS}(t)$ becomes increasingly shorter, and its shorter duration partially prevents the extinction of its excitatory association with the US. In addition, because inhibitory associations do not extinguish with WS presentations (Zimmer-Hart & Rescorla, 1974), $\tau_{Ra}(t)$ does not extinguish its inhibitory association with the US (see Equation 2 in Appendix A). Furthermore, because $\tau_{WS}(t)$ and $\tau_{Ra}(t)$ are simultaneously active, the inhibitory association of $\tau_{Ra}(t)$ retards the extinction of the excitatory association of $\tau_{WS}(t)$ (see Equation 2 in Appendix A). In Soltysik's (1985) terms, $\tau_{Ra}(t)$ partially protects to $\tau_{WS}(t)$ from extinguishing its excitatory association with the US.

COMPUTER SIMULATIONS

The present section contrasts experimental data with computer simulations carried out with the model presented in Fig. 6.1. All simulations were carried out with identical parameter values. Parameter values used in the simulations are presented in Appendix B.

In Figs. 6.4 to 6.14 (except for Fig. 6.5 and Fig. 6.12), Panel A displays the percentage of avoidance responses, Panel B displays response latency (as the number of simulated time units), Panel C displays Ri–US, US–US, WS–US and associations, and Panel D represents WS–Ri and US–Ri associations. In Figs. 6.15 and 6.16, Panel A displays the average number of responses over a 100 time-unit period, whereas Panel B displays the number of USs received during a 100 time-unit period. Ri represent responses such as running, turning, rearing, or pressing. Any of these responses can be chosen to be the escape or avoidance response.

Acquisition of Delay Avoidance

Experimental Data. During acquisition of avoidance, fear decreases as the animal masters avoidance. Although initially subjects show intense fear to the US, this fear decreases as the animal learns the avoidance response (Kamin, Brimer, & Black, 1963; Solomon & Wynne, 1953; Starr & Mineka, 1977). Subjects have been shown to substitute fear responses with stereotyped behavior: immediately after jumping to the opposite side of the shuttle-box, dogs position themselves in a specific part of the apparatus, body and head facing a fixed direction, and maintain this position until the next WS presentation (Solomon & Wynne, 1953).

Interestingly, even after the avoidance response is established, the response latency still decreases with increasing number of trials (Mowrer, 1947; Schoenfeld, 1950; Solomon & Wynne, 1953). Therefore, although acquisition of avoidance is a relative rapid process, the gradual decrease in response latency suggests that learning continues after the initial avoidance behavior has been acquired.

Computer Simulations. Figure 6.4 shows computer simulations of delay avoidance. During the first trials, simulations showed alternated escape and avoidance responses, followed by a long period of uninterrupted avoidance behavior (over 800 trials). Latency time decreased over the first 200 trials and the subject showed no sign of extinction, even after 650 trials. These simulated results agreed with Solomon and Wynne's (1953, 1954) earlier data.

Panel A in Fig. 6.4 shows that, in agreement with empirical results (Solomon, Kamin, & Wynne, 1953; Solomon & Wynne, 1954), the percentage of avoidance responses did not decrease over trials, that is, avoidance has almost negligible extinction. This negligible extinction is explained as follows. Although protected by shorter response latencies and the inhibitory association of Ra with the US, the WS–US association, and therefore fear, decreases over trials. As WS–US decreases, WS–Ra increases. Because the intensity of the output response is proportional to the product of the WS–US and WS–Ra associations, decreases in WS–US are compensated by increases in WS–Ra,

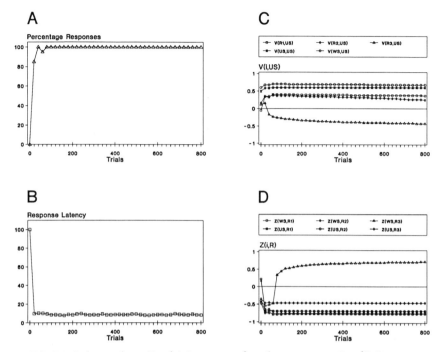

FIG. 6.4. Delay avoidance. *Panel A*: Percentage of avoidance responses. *Panel B*: Response latency. *Panel C*: Associations of the US, WS, and Rs with the US (V(US, US), V(WS, US), V(R1, US), V(R2, US), and V(R3, US)). *Panel D*: Associations of WS and US with the Rs (Z(WS, R1), Z(WS, R2), Z(WS, R3), Z(US, R1), Z(US, R2), Z(US, R3)). R3 is the avoidance response.

and therefore, the percentage of avoidance responses remains constant for a long number of trials. The model's description of resistance to extinction is compatible with Baum's (1970) data, showing that response extinction is independent of fear extinction.

Panel B in Fig. 6.4 shows that in agreement with experimental data (Mowrer, 1947; Schoenfeld, 1950; Solomon & Wynne, 1953), response latency decreased after avoidance response is acquired. According to the model, response latency decreased because the WS–Ra association, driven by the absence of the predicted US, keeps increasing over trials. Panel C shows that in agreement with experimental results (Kamin et al.; 1963; Solomon & Wynne, 1953; Starr & Mineka, 1977), the WS–US association, and therefore fear, continued to gradually decrease after the animal mastered avoidance. Panel C shows that whereas WS, R1, and R2 accrued an excitory association with the US, R3 (the avoidance response, Ra) acquired an inhibitory association. Panel D shows that the WS–R3 association increases over trials.

Fear Modulates Avoidance Behavior

Experimental Data.   Whereas techniques used to decrease fear decrease avoidance responding, techniques used to increase fear increase ongoing avoidance responses (Rescorla, 1967; Rescorla & LoLordo, 1965; Weisman & Litner, 1969).

Computer Simulations.   Figure 6.5 shows that when a CS previously associated with the US (a conditioned excitor) is presented together with the WS in an avoidance paradigm, the latency of avoidance responses decreases and the percentage of avoidance responses increase. Conversely, when a CS previously associated with the absence of the US (a conditioned inhibitor) is presented together with the WS, the latency of avoidance responses increase and the percentage of avoidance responses decreases. These results are explained in terms of the modulation of the avoidance response by the intensity of the prediction of the US, $CR = f[\ B\ ] = f[\ \Sigma_i\ V_i\ X_i\ ]$, by all CSs present at a given time (Equation 5 in Appendix A).

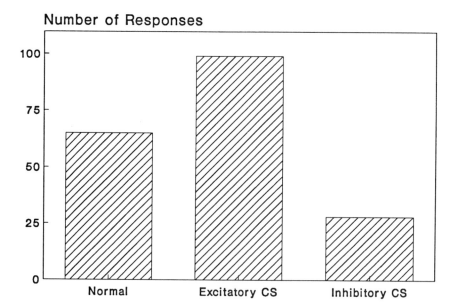

FIG. 6.5. Effect of introducing an excitatory and an inhibitory CS on the rate of responding during Sidman avoidance. Average number of avoidance responses per 100 time units during Trials 100 to 200.

Transfer of Control

Experimental Data. Overmier and Bull (1969) trained an animal to generate an avoidance response at the presentation of WS1. After the animal was restricted so that avoidance responses could not occur, it was presented with unreinforced WS1 presentations and WS2 reinforced presentations. Later, whereas the WS1 did not elicit avoidance responses, WS2 elicited the avoidance response. The transfer of control to WS2 was independent of the continued responding to WS1. According to Overmier and Bull (1969), the results supported the views that: (a) WS1 controls fear and fear controls the avoidance response, and (b) classical fear conditioning and operant conditioning are independent sequential links.

Computer Simulations. Assuming that WS1 and WS2 share a common element, WG, that represents 1% of the total value of both WS1 and WS2 traces, the model is able to reproduce Overmier and Bull's (1969) experimental results. Figures 6.6a and 6.6b show computer simulations of a transfer of control experiment. During the first 200 trials, simulations showed avoidance acquisition to WS1. On Trial 201, the simulated animal was presented with the WS1, the avoidance response was blocked, and no US was delivered. From Trials 400 to 450, the simulated animal was presented with WS2, the avoidance response was blocked, and the US was delivered. Panel A in Fig. 6.6a shows that when the barrier was removed on Trial 451, the simulated animal did not generate avoidance responses to WS1. However, Panel A in Fig. 6.6b shows that when the barrier was removed on Trial 451, the simulated animal did generate avoidance responses to WS2. These results were explained in terms of the WG–Ra associations that remained unchanged during the extinction of the WS1–US association and the acquisition of the WS2–US association. During extinction of the WS1–US associations, WG–US associations were also extinguished, and animals did not avoid WS1 (Fig. 6.6a). When WS2 was presented together with the US, both WS2–US and WG–US associations increased. When animals were given the opportunity to avoid the US, WS2–US and WG–US associations modulated the WG–Ra associations (see Equation 6), and the simulated animal avoided WS2 (Fig. 6.6b). In contrast to this explanation, Overmier and Bull (1969) applied Mowrer's view and interpreted the results in terms of both WS1 and WS2 activating fear and fear controlling the avoidance response.

Extinction of Avoidance

Experimental Data. An enigmatic feature of avoidance is that in some cases it shows very slow extinction. For example, Solomon, Kamin, and Wynne

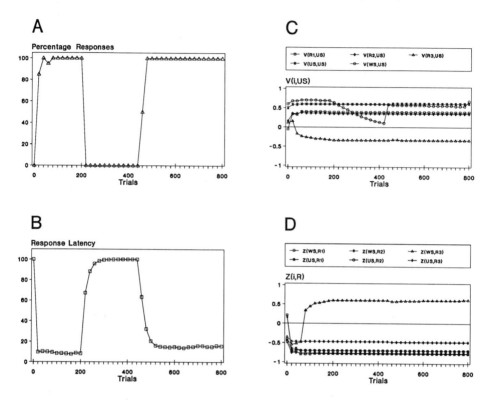

FIG. 6.6a. Transfer of control: Responses to WS1. *Panel A*: Percentage of avoidance responses. *Panel B*: Response latency. *Panel C*: Associations of the US, WS, and Rs with the US (V(US, US), V(WG, US), V(WS1, US), V(R1, US), V(R2, US), and V(R3, US)). *Panel D*: Associations of WG and WS1 with the Rs (Z(WG, R1), Z(WG, R2), Z(WG, R3), Z(WS1, R1), Z(WS1, R2), Z(WS1, R3)). R3 is the avoidance response.

(1953) reported that latency times kept decreasing even after 200 trials without shock, and Solomon and Wynne (1954) reported no sign of extinction in experiments with up to 650 trials.

Several factors might influence the resistance to extinction of avoidance. Whereas Solomon et al. (1953) and Solomon and Wynne (1954) reported insignificant extinction using a strong (10.0 to 12.5 mA) US, Bolles, Moot, and Grossen (1971) found extinction of avoidance using a weaker (1.2 mA) US. In the same vein, Brush (1957) reported that resistance to extinction tended to increase with increasing shock intensities ranging from 0.7 mA to 3.1 mA. In addition to US intensity, decreasing WS–US intervals also resulted in an increased resistance to extinction (Kamin, 1954). In agreement with these experimental data, computer simulations showed that resistance to extinction increased with increasing US intensity and decreasing WS–US interval. In terms of the model, all these procedures resulted in increased WS–US associations.

Although under certain conditions avoidance might show little ordinary extinction, avoidance can be extinguished through various alternative procedures. One procedure, called response blocking or flooding, refers to blocking the capability of an animal to generate the avoidance response without presenting the US (Baum, 1966, 1976; Page & Hall, 1953; Solomon, Kamin, & Wynne, 1952). A second procedure consists of making the avoidance response ineffective by shocking both sides of the shuttle-box (Davenport & Olson, 1968). Finally, a third procedure consists of shocking the animal when it emits the avoidance response (Seligman and Campbell, 1965; Solomon, et al., 1953).

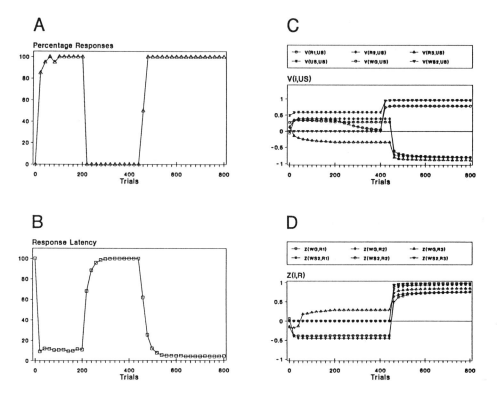

FIG. 6.6b. Transfer of control: Responses to WS1 (trials 1–200) and WS2 (trials 451–800). *Panel A*: Percentage of avoidance responses. *Panel B*: Response latency. *Panel C*: Associations of the US, WS, and Rs with the US (V(US, US), V(WG, US), V(WS2, US), V(R1, US), V(R2, US), and V(R3, US)). *Panel D*: Associations of WG and WS2 with the Rs (Z(WG, R1), Z(WG, R2), Z(WG, R3), Z(WS2, R1), Z(WSZ, R2), Z(WS2, R3)). R3 is the avoidance response.

Computer Simulations

Blocking the Avoidance Response. Figure 6.7 shows computer simulations of blocking. During the first 200 trials, simulations showed avoidance acquisition. On Trial 201, the simulated animal was presented with the WS, the avoidance response was blocked, and no US was delivered. Panel A shows that when the barrier was removed on Trial 401, the simulated animal did not generate avoidance responses. These simulated results are in agreement with Baum's (1966, 1969, 1976) and Page and Hall's (1953) data. Panel A in Fig. 6.7 also shows simulations of the effect of reinstating an avoidance protocol on Trial 450. Panel A in Fig. 6.7 shows that animals reacquired avoidance in few (two) shocked trials. This savings effect is based on the preserved WS–Ra association (see Panel D). This prediction of the model has not been experimentally tested.

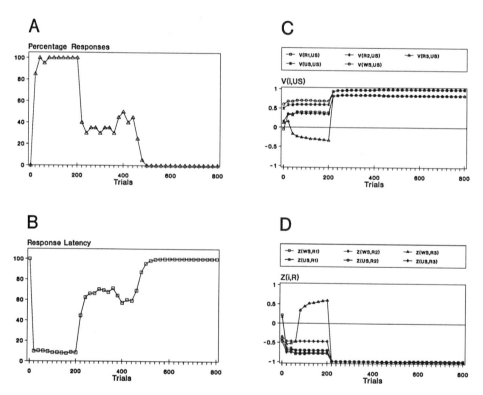

FIG. 6.7. Blocking the avoidance response. *Panel A*: Percentage of avoidance responses. *Panel B*: Response latency. *Panel C*: Associations of the US, WS, and Rs with the US (V(US, US), V(WS, US), V(R1, US), V(R2, US), and V(R3, US)). *Panel D*: Associations of WS and US with the Rs (Z(WS, R1), Z(WS, R2), Z(WS, R3), Z(US, R1), Z(US, R2), Z(US, R3)). R3 is the avoidance response.

Panel C in Fig. 6.7 shows that on Trial 400 the WS–US association (aversion or fear to the WS) was almost entirely extinguished. According to the model, extinction of the WS–US association occurred at a faster rate than during avoidance because: (a) $\tau_{WS}(t)$ was long enough to decrease its excitatory association with the US, and (b) $\tau_{Ra}(t)$ was absent and therefore did not provide protection from extinction. Based on these principles, the model is able to explain experimental data showing that weak responding (which in the model causes $\tau_{WS}(t)$ to become longer and, therefore, to decrease the WS–US association) lead to even weaker responding (Beecroft, 1967).

Shocking All Responses. Figure 6.8 shows computer simulations of the effect of presenting the US 10 secs after the beginning of the trial, regardless of the animal's response. During the first 200 trials, simulations showed acquisition of delay avoidance. From Trials 201 to 600, the animal always received the shock US independently of the response. Panel C in Fig. 6.8 shows that from Trial 201 to Trial 600, as all responses were shocked, WS–US and R3–US associations became excitatory. Therefore, the model explains extinction of avoidance because all responses became associated with the US and all WS–R associations became too small to activate a response (Panel D). These simulated results are in agreement with Davenport and Olson (1968) and Kamin (1957).

Panel A in Fig. 6.8 also shows simulations of the effect of reinstating an avoidance protocol on Trial 601. Panel A of Fig. 6.8 shows that when a normal avoidance protocol was reinstated, simulated animals did not generate any response and, therefore, avoidance was never reinitiated. This predicted behavior might be related to learned helplessness (Maier & Seligman, 1976; Overmier & Seligman, 1967; Seligman, 1975), a phenomenon by which animals exposed to an inescapable shock do not attempt to escape the shock even when later the shock becomes escapable. Panel D in Fig. 6.8 suggests that learned helplessness was the consequence of all $Z(US,Re)$ being very low for the system to try any response. Furthermore, Seligman and Maier (1967) reported that learned helplessness can be prevented by training the animal to escape–avoid. Panel D in Fig. 6.8 suggests that training the animal in avoidance could immunize against inescapable shocks by increasing the value of $Z(US,Re)$, and thereby increasing the number of inescapable trials needed to decrease $Z(US,Re)$ to a helpless value.

Shocking the Avoidance Response. Figure 6.9 shows computer simulations of the effect of presenting a US only when the animal generates Ra. During the first 200 trials, simulations showed acquisition of delay avoidance. From Trials 201 to 400, the simulated animal received a US when generating Ra. From Trials 201 to approximately 320, the animal decreased its average

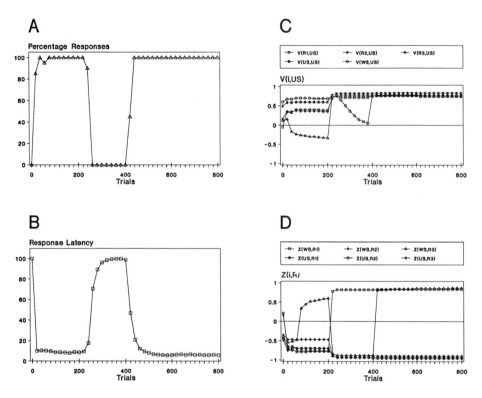

FIG. 6.8. Shocking all responses. *Panel A*: Percentage of avoidance responses. *Panel B*: Response latency. *Panel C*: Associations of the US, WS, and Rs with the US (V(US, US), V(WS, US), V(R1, US), V(R2, US), and V(R3, US)). *Panel D*: Associations of WS and US with the Rs (Z(WS, R1), Z(WS, R2), Z(WS, R3), Z(US, R1), Z(US, R2), Z(US, R3)). R3 is the avoidance response.

response latency, and after that it rapidly extinguished Ra. These simulated results are in agreement with Solomon, Kamin, and Wynne (1953). Panel A in Fig. 6.9 also shows simulations of the effect of reinstating an avoidance protocol on Trial 401. Panel A in Fig. 6.9 shows that animals reacquired avoidance in eight shocked trials. As in the flooding case, this savings effect was based on the preserved WS–US association. This prediction of the model has not been experimentally tested.

Panels C and D in Fig. 6.9 show that from Trial 201 to Trial 240, WS–US increased, WS–R3 (Ra) decreased and WS–R1 and WS–R2 increased. The model explains extinction of avoidance in terms of the decrease in WS–R3 and increase in WS–R1 and WS–R2 associations. In agreement with Bolles et al. (1971) simulated results showed that shocking the avoidance response produced faster extinction than shocking all responses.

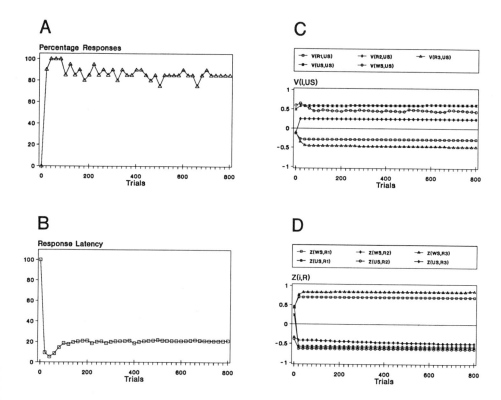

FIG. 6.9. Shocking the avoidance response. *Panel A*: Percentage of avoidance responses. *Panel B*: Response latency. *Panel C*: Associations of the US, WS, and Rs with the US (V(US, US), V(WS, US), V(R1, US), V(R2, US), and V(R3, US)). *Panel D*: Associations of WS and US with the Rs (Z(WS, R1), Z(WS, R2), Z(WS, R3), Z(US, R1), Z(US, R2), Z(US, R3)). R3 is the avoidance response.

Effects of Extending the WS, Terminating the WS, and Classical Conditioning—Experimental Data

Kamin (1957) studied the effect that different procedures have on acquisition of avoidance. In normal avoidance, the avoidance response was followed by the immediate termination of the WS, and no US was delivered on that trial. In a *terminate WS* situation, the avoidance response was followed by the immediate termination of the WS but a US was delivered to either side of the box. In an *extended WS* situation, the avoidance response prevented the US presentation but did not terminate the WS. Finally, in a classical conditioning situation, the avoidance response had no effect on either the WS duration or the delivery of the US. Kamin's (1957) data showed that whereas in the classical conditioning and terminate WS cases animals did not acquire consistent avoidance, the extended WS case resulted in a slower acquisition and lower asymptotic levels of avoidance.

Computer Simulations

Extending the WS. Figure 6.10 shows computer simulations of delay avoidance when the WS is not terminated even after the animal has generated the Ra. In agreement with Kamin (1957), when the WS was extended after the Ra, consistent avoidance was attained but reinforcement was intermittently needed. This alternating behavior is explained in terms of the extinction of the aversive association between the WS and the US. Panel C in Fig. 6.10 shows that the WS–US association decreased when the WS was extended. In contrast, WS–Ra is largely increased as Ra accrued an inhibitory association

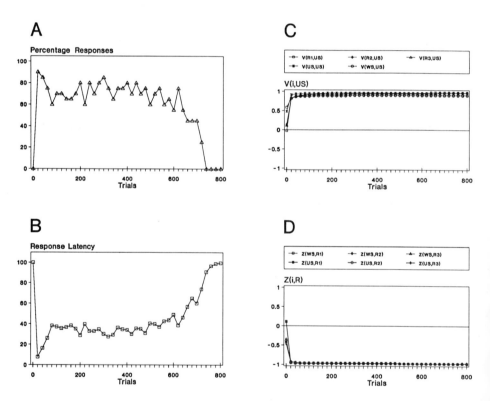

FIG. 6.10. Extending the WS. *Panel A:* Percentage of avoidance responses. *Panel B:* Response latency. *Panel C:* Associations of the US, WS, and Rs with the US (V(US, US), V(WS, US), V(R1, US), V(R2, US), and V(R3, US)). *Panel D:* Associations of WS and US with the Rs (Z(WS, R1), Z(WS, R2), Z(WS, R3), Z(US, R1), Z(US, R2), Z(US, R3)). R3 is the avoidance response.

with the US. As mentioned, extinction of the WS–US association occurred during: (a) the period when the WS was on and the US was absent, and (b) the period when WS was off but $\tau_{WS}(t)$ was still active. When the WS terminated as the animal generated Ra, $\tau_{WS}(t)$ was protected from extinction by $\tau_{Ra}(t)$ (see Equation 2 in Appendix A). By extending the WS, the value of $\tau_{WS}(t)$ was large even when the animal avoided the US, and therefore extinction of WS–US increased, and periodical reinforcement was needed in order to maintain the generation of avoidance responses.

Terminating the WS. Figure 6.11 shows computer simulations of delay avoidance when the WS is terminated after Ra but the US is still delivered. When the WS was terminated after the Ra but the US was still delivered, consistent avoidance was never attained. In agreement with Kamin (1957), simulations show alternating avoidance and escape behavior. This behavior is

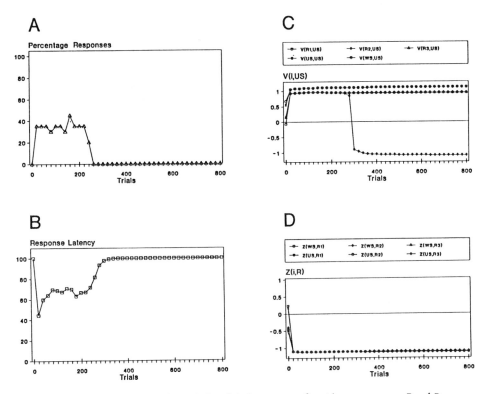

FIG. 6.11. Terminating the WS. *Panel A*: Percentage of avoidance responses. *Panel B*: Response latency. *Panel C*: Associations of the US, WS, and Rs with the US (V(US, US), V(WS, US), V(R1, US), V(R2, US), and V(R3, US)). *Panel D*: Associations of WS and US with the Rs (Z(WS, R1), Z(WS, R2), Z(WS, R3), Z(US, R1), Z(US, R2), Z(US, R3)). R3 is the avoidance response.

explained in terms of the extinction of the WS–Ra association. As mentioned, acquisition of the WS–Ra association occurs during the period when the WS and Ra are active and the predicted US exceeds the value of the actual US. When the US is delivered as the animal generates Ra, the actual US exceeds the predicted US and, therefore, WS–Ra association decreases. Panel C in Figure 6.11 shows that the WS–US association increased as the US is consistently presented. Panel D shows that all WS–R associations are greatly decreased.

Classical Conditioning. Figure 6.12 shows computer simulations of delay avoidance when the US follows the WS at a fixed time interval, independently of the animal's response. When the WS was always followed by the US, consistent avoidance was never attained. Panel A shows that, in agreement with Kamin (1957), Ra increased at the beginning of training and then

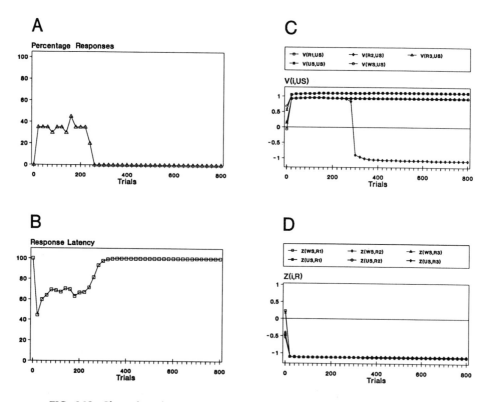

FIG. 6.12. Classical conditioning. *Panel A*: Percentage of avoidance responses. *Panel B*: Response latency. *Panel C*: Associations of the US, WS, and Rs with the US (V(US, US), V(WS, US), V(R1, US), V(R2, US), and V(R3, US)). *Panel D*: Associations of WS and US with the Rs (Z(WS, R1), Z(WS, R2), Z(WS, R3), Z(US, R1), Z(US, R2), Z(US, R3)).

gradually decreased to zero. At the beginning of training, animals generated avoidance responses on a random basis, and the intensity of the responses was magnified by an increased WS–US association. As all Rs, including Ra, became increasingly aversive, the animal stopped responding. This result is similar to the paradigm in which all Rs are punished, and therefore, avoidance is extinguished. Panel C in Fig. 6.12 shows that the WS–US association increased as the US was consistently presented, as in classical conditioning. In contrast, Panel D shows that all WS–R associations were greatly decreased.

Trace Avoidance

Experimental Data.   Whereas the WS and the US temporally overlap in delay avoidance, WS offset precedes US onset in trace avoidance. Trace avoidance seems to be less resistant to extinction than delay avoidance (Brush, 1957; Church, Brush, & Solomon, 1956; Kamin, 1954). In contrast to delay avoidance, trace avoidance deteriorates with increasing interstimulus interval (ISI). Church et al. (1956) studied the effect of varying the ISI on the acquisition of avoidance under trace and delay conditions. Church et al. (1956) reported that with the trace procedure, acquisition was faster and extinction slower with decreasing WS–US intervals. Interestingly, Brush (1957) reported that avoidance learning was not characterized by learning the timing of the US, as is the case in classical conditioning (Smith, 1968). However, Davis et al. (1989) reported that fear conditioning was also time specific, that is, it is maximal at the time that matches the CS–US interval used in training.

Computer Simulations.   Figure 6.13 shows the total exposure to the US in delay and trace avoidance with different ISIs. Figure 6.13 shows that the total exposure to the US in trace avoidance rapidly grew with increasing ISIs, whereas the total exposure to the US in delay avoidance decreased toward an asymptotic value with increasing ISIs. In trace avoidance, the intensity of Ra was determined by the classical WS–US association, which decreased with increasing ISIs. In delay avoidance, the intensity of Ra was also determined by the classical WS–US association, which also decreased with increasing ISIs. However, in delay avoidance, the trace of WS reached a higher asymptotic value than in the trace avoidance case, thereby strongly activating the WS–Ra association, and this increased Ra compensated for decreases in the WS–US associations. This is in agreement with experimental data (Church et al., 1956). In addition, Figure 6.13 shows that, also in agreement with experimental data (Brush, 1957; Church, et al., 1956; Kamin, 1954), trace avoidance was less resistant to extinction than delay avoidance and, therefore, US reinforcement was intermittently needed.

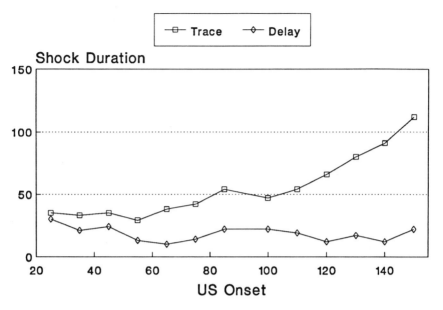

FIG. 6.13. Total shock duration during delay and trace avoidance for different WS–US interstimulus intervals (ISIs).

Different Escape and Avoidance Responses

Experimental Data. Mowrer and Lamoreaux (1946) and Bolles (1969) reported that, although animals can learn an avoidance response different from that required to escape the US, in most cases avoidance performance was better if the responses were identical.

Computer Simulations. Figure 6.14 shows that when Ra (R3) is different from Re (R2), acquisition of delay avoidance proceeded at a slower pace than when Ra was identical to Re (see Fig. 6.4). At the beginning of training, the animal generated Re either in the presence of WS or the US because both WS–Re and US–Re were the dominant responses. After 25 trials, WS–Ra became stronger than WS–Re, and the animal correctly discriminated avoidance from escape, that is, in the presence of WS the animal avoided, and if WS was omitted, the animal escaped when the US was presented. Because it took more trials for the animal to learn the correct discrimination, the model predicts that WS–US and US–US associations are stronger with identical avoidance and escape responses than with different ones.

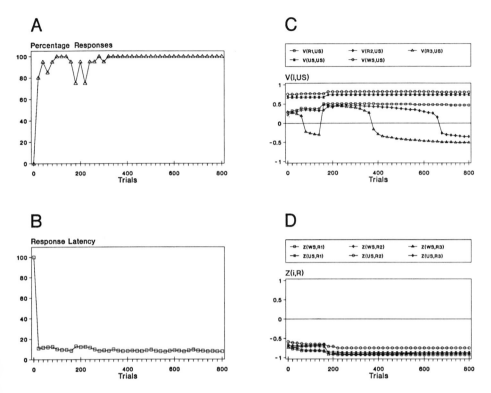

FIG. 6.14. Different escape and avoidance responses. *Panel A*: Percentage of avoidance responses. *Panel B*: Response latency. *Panel C*: Associations of the US, WS, and Rs with the US (V(US, US), V(WS, US), V(R1, US), V(R2, US), and V(R3, US)). *Panel D*: Associations of WS and US with the Rs (Z(WS, R1), Z(WS, R2), Z(WS, R3), Z(VS, R1), Z(US, R2), Z(US, R3)). R3 is the avoidance response. R2 is the escape response.

Discriminative Avoidance

Experimental Data.   In a discriminative avoidance procedure, animals are taught to avoid the shock by performing one response in the presence of a warning stimulus and a different response in the presence of another warning stimulus. For instance, Overmier, Bull, and Trapold (1971) trained dogs to avoid shock by pressing a panel on the left in the presence of one stimulus and pressing a panel on the right in the presence of another stimulus. Similarly, Young (1976) trained rats to avoid shock by pressing a lever in the presence of one stimulus and licking in the presence of another stimulus.

Computer Simulations.   Figure 6.15 shows that when two different avoidance responses, R1 and R3, were respectively required in the presence of two different warning stimuli, WS2 and WS1, acquisition proceeded at a slower pace than when only one response was needed (see Fig. 6.4). After respectively, 150 and 200 trials, WS2–R1 and WS1–R3 became strong, and the animal reliably generated the correct response in the presence of the corresponding WS. After 400 and 500 trials, both R1 and R3 became predictors of the absence of the US.

Sidman Avoidance

Experimental Data.   In a Sidman (1953) avoidance task, subjects were taught to avoid a 5-second shock interval by pressing a lever that gave them a 30-second shock-free period. In Hernnstein and Hineline's (1966) experi-

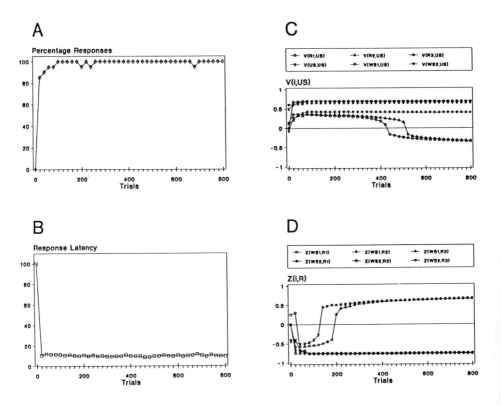

FIG. 6.15. Discriminative avoidance. *Panel A*: Percentage of avoidance responses. *Panel B*: Response latency. *Panel C*: Associations of the US, WS, and Rs with the US (V(US, US), V(WS1, US), V(WS2, US), V(R1, US), V(R2, US), and V(R3, US)). *Panel D*: Associations of WS and US with the Rs (Z(WS1, R1), Z(WS1, R2), Z(WS1, R3), Z(WS2, R1), Z(WS2, R2), Z(WS2, R3)). R1 is the correct the avoidance response when WS2 is present. R3 is the correct the avoidance response when WS1 is present.

ment, subjects learned to choose between high and low random chances of being shocked in the absence of external or internal WSs.

Computer Simulations. Panel A in Figs. 6.16 and 6.17 displays the number of responses as a percentage over 100 time units. Panel B in Figs. 6.16 and 6.17 displays the number of USs as a percentage over 100 time units. Figure 6.16 shows that percentage of responses increased and the percentage of USs decreased with increasing number of trials. Over trials, US–US associations increase (the US predicted itself) as well as US–Ra associations (the US became associated with the Ra response). According to the model, animals learn the Sidman avoidance task because Ra becomes associated to the US trace. In addition, Fig. 6.16 shows that in agreement with Hernnstein and Hineline's (1966) experimental data, Ra increased after the subject received the US and gradually decreased until the next US was presented.

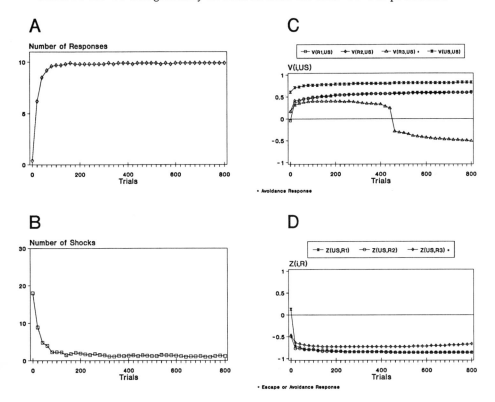

FIG. 6.16. Sidman avoidance. *Panel A*: Average number of responses over a 100 time-unit period. *Panel B*: Number of US received during a 100 time-unit period. *Panel C*: Associations of the US, WS, and Rs with the US (V(US, US), V(WS, US), V(R1, US), V(R2, US), and V(R3, US)). *Panel D*: Associations of the US with the Rs (Z(US, R1), Z(US, R2), Z(US, R3)). R3 is the escape or avoidance response.

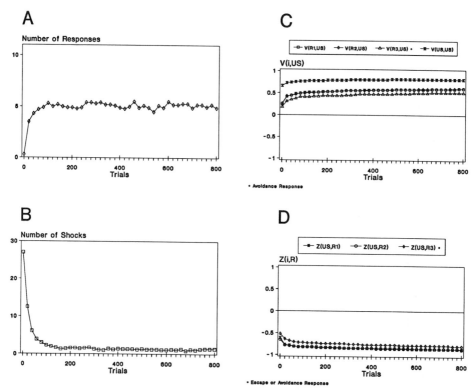

FIG. 6.17. Reduction of the rate of aversive stimulation. *Panel A*: Average number of responses over a 100 time-unit period. *Panel B*: Number of US received during a 100 time-unit period. *Panel C*: Associations of the US, WS, and Rs with the US (V(US, US), V(WS, US), V(R1, US), V(R2, US), and V(R3, US)). *Panel D*: Associations of the US with the Rs (Z(US, R1), Z(US, R2), Z(US, R3)). R3 is the escape or avoidance response.

DISCUSSION

This chapter introduced a novel, real-time, two-process theory of avoidance that combines elements of classical and operant conditioning. The classical conditioning process generates associations between the WS, the US, and Rs generated by the animal with the US. Whereas WS and the US become predictors of the US, escape and the avoidance response, Re and Ra, become predictors of the absence of the US. The operant conditioning process associates WS and US with those Rs that predict the absence of the US (avoidance and escape responses). The classical conditioning process: (a) provides predictions of the presence or absence of the US used by the operant conditioning process to generate WS–R associations, and (b) controls the strength of the operant behavior. Because the warning WS and the US can become associated with different available responses, animals can learn different escape and avoidance responses. Because the model describes behavior in real time, it is able to capture the effects of using different temporal WS and US arrangements and to describe the latency of avoidance and escape responses.

The network comprises two processes, classical and operant conditioning. The classical conditioning process assumes that US–US, WS–US, and R–US associations increase whenever the traces of US, WS, or R are active and the *actual* intensity of the US *exceeds* its *predicted* intensity. The operant conditioning process assumes that US–Re and WS–Ra associations increase whenever the traces of US and Re or WS and Ra are active together and the *predicted* intensity of the US *exceeds* its *actual* intensity. Whereas classical conditioning is regulated by a LMS algorithm, operant conditioning is regulated by a novel algorithm that mirrors the classical conditioning algorithm. In other words, by computing the mismatch between the actual and predicted intensity of the US, and using this mismatch in opposite ways to regulate classical and operant conditioning, the network is able to describe escape and avoidance in a unified manner. Importantly, responses that acquire an excitatory association with a stimulus gain an inhibitory association with the US, that is, responses activated by a stimulus in order to escape or avoid the US come to predict the absence of the US. Furthermore, Gabriel and Schmajuk (1990) suggested that the mismatch between the actual and predicted intensity of the US might also control behavioral inhibition during avoidance: When a mismatch is detected the avoidance response is inhibited. Gabriel and Schmajuk (1991) showed through computer simulations that behavioral inhibition retarded acquisition and accelerated extinction of avoidance in a running wheel paradigm.

The operant algorithm offered in this chapter differs from Mowrer's two-factor theory in several important ways. First, Mowrer suggested that Fear–Ra associations were reinforced by the *reduction of fear* as Ra terminated the fear-eliciting WS. In contrast, our model assumes that WS–R associations for *all responses are reinforced by fear*, and WS–R associations for *responses other than the avoidance response are weakened by the presentation of the US*. In both cases, however, when fear decreased ((US - B) is less than zero)) operant associations increased ((B - US) is greater than zero)). Second, Mowrer assumed that classical and operant processes were organized in a serial fashion, that is, a WS was associated with the US and activates fear, and fear was associated with the escape response and activated the (now) avoidance response. The strength of the operant response is proportional to the product of the strength of classical and operant associations. In contrast, our model assumes that classical associations modulate the operant responses in a parallel fashion, that is, a WS is associated both to the US and to the avoidance response, and WS–US associations multiplicatively modulate the strength of the operant response. No associations between fear with the operant responses are assumed. Third, Mowrer assumed that the US and fear were associated with the same response, and therefore escape and avoidance responses were identical. In contrast, our model establishes independent US–Re and WS–Ra associations. Fourth, whereas Mowrer's two-factor theory explained transfer of control experiments in terms of both WS1 and WS2 activating fear and fear activating the avoidance response, our model describes the results in terms of a common element in WS1 and WS2.

Some empirical evidence that presents difficulties for Mowrer's two-factor theory is correctly addressed by our model. For instance, two-factor theory has problems explaining trace avoidance, in which a temporal delay is introduced between the WS offset and the presentation of the US. Although in trace avoidance the association between a WS temporal trace and the US can be explained in the same terms as classical trace conditioning, WS termination reduces fear only slightly. Our model does not need WS termination to create the reduction of fear that reinforces the avoidance response. Instead, the model explains trace avoidance in terms of the association between the WS and the avoidance response in the presence of the fear generated by the temporal trace of the WS. Another major difficulty for Mowrer's theory is the sometimes reported slow extinction of avoidance. According to the theory, WS–US associations extinguish as the animal avoids the US and, therefore, animals should cease avoiding the WS. In contrast to the two-factor theory, our model adequately addresses extinction of avoidance: extinction of the WS–US association (supported by experiments showing a decrease in fear during learning) is perfectly compensated by the increase in the WS–R association (corroborated by data showing that response latency decreases as the animal increasingly masters the avoidance response).

Also in contrast to Mowrer's theory, our neural network contributes to the understanding of learned helplessness (Maier & Seligman, 1976; Overmier & Seligman, 1967; Seligman, 1975), a phenomenon by which animals exposed to an inescapable shock do not attempt to escape the shock even when later the shock becomes escapable. In terms of the model, learned helplessness is the consequence of all $Z(US,Re)$ being very low for the system to try any response. Furthermore, Seligman and Maier (1967) reported that learned helplessness can be prevented by training the animal to escape–avoid. The model suggests that training the animal in avoidance could immunize against inescapable shocks by increasing the value of $Z(US,Re)$, and thereby increasing the number of inescapable trials needed to decrease $Z(US,Re)$ to a helpless value.

The network was applied to the description of escape and avoidance behavior in a shuttle-box and Sidman avoidance. Computer simulations demonstrate that the model describes many of the features that characterize avoidance behavior:

1. Fear of the US decreases as the animal masters the response,
2. Techniques that decrease fear decrease, and techniques that increase fear increase, ongoing avoidance responses,
3. When WS–US classical associations are extinguished, the avoidance response can be elicited by another WS classically conditioned to the US (transfer of control),
4. The amount of time needed to generate the avoidance response decreases with increasing number of trials,
5. In some situations avoidance has negligible extinction,
6. Trace avoidance is less resistant to extinction than delay avoidance,
7. Acquisition of trace avoidance is slower and extinction faster with increasing WS–US intervals,
8. Extinction of avoidance is obtainable by blocking the animal's capability to elicit the avoidance response without delivering the US, by shocking the animal when it emits the avoidance response, or by shocking the animal whether or not it emits the avoidance response,
9. When the avoidance response terminates the WS but does not prevent the US presentation animals show a slower acquisition and lower asymptotic levels of avoidance,
10. When the avoidance response prevents the US presentation but does not terminate the WS, animals learn less than in the normal avoidance situation,
11. When the avoidance response has no effect either in the WS duration or the delivery of the US (classical conditioning) animals do not acquire consistent avoidance,
12. The avoidance response may be different from that required to escape the US, and
13. Different avoidance responses may be associated with different discriminative stimuli.

The Frightening Complexity of Avoidance

Recently, Schmajuk and DiCarlo (1992) presented a neural network model that extends the Rescorla–Wagner model to the description of stimulus configuration in classical conditioning. The model (a) describes behavior in real time, (b) incorporates hidden units that code *configural stimuli,* (c) includes inputs directly and indirectly connected to the output, and (d) employs a biologically plausible backpropagation procedure.

In a similar vein, the present study attempts to extend the Rescorla–Wagner model to the description of avoidance. The network intimately combines classical and operant conditioning principles. The classical conditioning process involves the formation of WS–US and R–US associations. These associations are used to predict the presence or absence of the US. Classical conditioning is regulated by a real-time version of the Rescorla–Wagner algorithm that computes the mismatch between the actual and the predicted intensity of the US: When the US is underpredicted, classical associations increase, and decrease otherwise. The operant conditioning process entails the formation of WS–a and US–Re associations. These associations are used to select the adequate response in each case. Operant conditioning is controlled by a novel algorithm that mirrors the Rescorla–Wagner classical conditioning algorithm: When the US is underpredicted, operant associations decrease, and increase otherwise. Interestingly, the elegant and simple combination of these algorithms results in the simultaneous development of WS–Ra excitatory operant associations and Ra–US inhibitory classical associations. Importantly, the real-time properties of the model are critical for the description of the changing durations of the WS during avoidance learning.

In general terms, the model assumes that through classical conditioning animals build an internal model of their environment and through operant conditioning animals select from alternative behavioral strategies. Whenever there is a mismatch between predicted and actual environmental events (a) the internal model is modified and (b) the behavioral strategies are adjusted. This approach can be easily extended to appetitive operant conditioning. As in the aversive case, in the appetitive condition classical associations, V_i, increase when (US–B) is greater than zero and decrease when (US–B) is less than zero. In contrast to the aversive case, in the appetitive condition operant associations, Z_{ik} increase when (US–B) is greater than zero and decrease when (US–B) is less than zero.

To the extent that the model improves our understanding of a large collection of experimental data and the problematic question of how animals learn avoidance is answered in terms of already familiar algorithms, the complexity of the theoretical aspects of avoidance looks less frightening and simply challenging.

APPENDIX A

A FORMAL DESCRIPTION OF THE MODEL

This section formally describes the model depicted in Fig. 6.1. The network presented is able to describe in real-time many classical conditioning paradigms, such as acquisition, extinction, blocking, overshadowing, and conditioned inhibition. As shown by Schmajuk and DiCarlo (1992), with the addition of a configural system, it also describes positive and negative patterning. Importantly, in agreement with experimental data (Zimmer-Hart & Rescorla, 1974), the network characterizes conditioned inhibition as not extinguishable by nonreinforced presentations of a conditioned inhibitor.

Classical Conditioning Process

We assume that $WS_i(t)$, $R_i(t)$, and the $US(t)$ (generically represented by $X_i(t)$) generate a STM trace, $\tau_i(t)$, according to

$$d(\tau_i(t))/d\tau = -K_1\,\tau_i + K_2(K_3 - \tau_i(t))X_i(t), \quad (1)$$

where $-K_i\tau_i(t)$ represents the passive decay of the STM of $WS_i(t)$, $R_j(t)$, or the $US(t)$. K_2 represents the rate of increase of $\tau_i(t)$, and constant K_3 is the maximum possible value of $\tau_i(t)$ (see Grossberg, 1975).

Changes in the association of $\tau_i(t)$ with the US, are given by:

$$d(V_i(t))/d\tau = K_4\tau_i(US(t) - B(t))(1 - |V_i(t)|). \quad (2)$$

where $\tau_i(t)$ represents the trace of $WS_i(t)$, $R_i'(t)$, or the $US(t)$. $B(t) = \Sigma_j V_j(t) \tau_j(t)$ represents the aggregate prediction of the US by all τs active at a given time. $K_4 = K_4'$ if $US\,(t) > \Sigma_j V_j(t)\,\tau_j(t)$, $K_4 = K_4''$ if $US < \Sigma_j V_j(t)\,\tau_j(t)$. By Equation 2, $V_i(t)$ increases whenever $\tau_i(t)$ is active and $US(t) > \Sigma_j V_j(t)\,\tau_j(t)$ and decreases when $US(t) < \Sigma_j V_j(t)\,\tau_j(t)$. In order to prevent the extinction of conditioned inhibition or the generation of an excitatory CS by presenting a neutral CS with an inhibitory CS, we assume that when $\Sigma_j V_j(t)\,\tau_j(t) < 0$ then $\Sigma_j V_j(t)\,\tau_j(t) = 0$. The term $(1 - |V_j(t)|)$ bounds $V_i(t)$ $(-1 \leq V_i(t) \leq 1)$.

Operant Conditioning Process

We adopt a response–selection view of operant conditioning by which WS or the US become associated to different alternative responses. Changes in the association of τ_i of WS_i with R_j, are given by:

$$d(Z_{ik}(t))/d\tau = K_5\tau_i(t)\,\tau_k(t)\,(B(t) - US(t))\,(1 - |Z_{ik}(t)|) \quad (3)$$

where $\tau_i(t)$ is the STM trace of $WS(t)$ or the $US(t)$, and $\tau_k(t)$ is the STM trace of the response that has the maximum output at a given time. $K_5 = K_5'$ if $US(t) < \Sigma_j V_j(t)\,\tau_j(t)$, $K_5 = K_5''$ if $US > \Sigma_j V_j(t)\,\tau_j(t)$. According to Equation 3, $Z_{ik}(t)$ increases whenever $\tau_i(t)$ and $\tau_k(t)$ are active simultaneously and $\Sigma_j V_j(t)\,\tau_j(t) > US(t)$, and decreases if $\Sigma_j V_j(t)\,\tau_j(t) < US(t)$. Because WS and the US may become associated with different responses, Ra and Re might be different. More generally, the system can learn operant discriminations by generating R_1 in the presence of WS_1 and R_2 in the presence of WS_2. The term $(1 - |Z_{ik}(t)|)$ bounds $Z_{ik}(t)$ $(-1 \leq Z_{ik} \leq 1)$.

Performance Rules

The strength of the conditioned response is given by:

$$CR(t) = f[B(t)] = f[\Sigma_j V_j(t) \tau_j(t)]. \quad (4)$$

The strength of $R_k(t)$ is given by:

$$R_k(t) = K_6 f[\Sigma_i Z_{ik}(t) \tau_i(t) + r_i(t)], \quad (5)$$

where $f[x] = 1 / 1 + e^{-(x)}$ and $r_i(t)$ is a random number. $R_k(t)$ is always positive, even when x may vary between -1 and 1. We assume that the animal selects and tries the response with the maximum $R_k(t)$, denoted by $R_{max}(t)$, every 5 time units. When $R_{max}(t)$ is selected, it activates $\tau_{max}(t)$ according to Equation 1.

The amplitude of the operant response, $R'(t)$, is given by

$$R'(t) = \tau_{max}(t) B(t), \quad (6)$$

where $B(t) = \Sigma_j V_j(t) \tau_j(t)$. Equation 6 implies that the same operant response $R'(t)$ can be achieved by various combinations of $\tau_{max}(t)$ and $B(t)$.

Let $t_{latency}$ denote the time at which the animal crosses to the opposite side in a shuttle-box and thereby avoids or escapes the shock Then $t_{latency}$ is the earliest time such that:

$$R'(t_{latency}) = \tau_{max}(t_{latency}) B(t_{latency}) \geq K_7, \quad (7)$$

where K_7 is the threshold to escape or avoid. Equation 7 implies that latency decreases for a fast-growing $\tau_{max}(t)$ and large Bs. Because $\tau_{max}(t)$ grows faster with increasing values of $R_{max}(t) = K_6 f[\Sigma_i Z_{ik}(t) \tau_i(t)]$, latency decreases with increasing values of $Z_{imax}(t)$ (WS–R_{max} and US–R_{max} operant associations). Also, because $B(t) = \Sigma_j V_j(t) \tau_j(t)$, latency decreases with increasing values of $V_j(t)$ (WS–US, US–US, and R–US classical associations).

APPENDIX B

Simulation Procedures and Parameters

In our computer simulations, each trial is divided into 250 time units for shuttlebox avoidance and 100 time units for Sidman avoidance. Each time unit represents approximately 500 msec. At time zero, the WS is presented. If the avoidance response has not been performed after 25 time units, a shock US is applied. Three alternative responses are considered and no initial hierarchy (see Hull, 1951) is assumed, that is, $Z_{ik} = 0$.

Table 6.2 summarizes the parameters values used in all simulations. K_4' and K_4'' are selected to yield normal rates of acquisition and extinction of classical conditioning.

TABLE 6.2
Parameter Values

Parameters	Meaning	Value
K_1	Passive decay of τ_i	2×10^{-2}
K_2	Rate of increase of τ_i	25×10^{-3}
K_3	Maximum value of τ_i	1
K_4'	Acquisition rate (classical conditioning)	1×10^{-5}
K_4''	Extinction rate (classical conditioning)	6×10^{-4}
K_5'	Acquisition rate (operant conditioning)	5×10^{-3}
K_5''	Extinction rate (operant conditioning)	3×10^{-1}
K_6	Mechanic constant	5×10^{4}
K_7	Threshold to escape or avoid	5×10^{-2}
r_k	Random number for response R_k	2×10^{-4}

ACKNOWLEDGMENTS

The authors are thankful to Tad Blair, Jeffrey Gray, Mike Gabriel, Win Hill, John Moore, John Staddon, Aaron Thieme, and two anonymous reviewers for their helpful comments on previous versions of the manuscript. Aaron Thieme and Elise Axelrad assisted with the graphics. This project was supported in part by Contract N00014-91-J-1764 from the Office of Naval Research.

REFERENCES

Baum, M. (1966). Rapid extinction of an avoidance response following a period of response prevention in the avoidance apparatus. *Psychological Reports, 18*, 59–64.

Baum, M. (1969). Extinction of avoidance response following response prevention: Some parametric investigations. *Canadian Journal of Psychology, 23*, 1–10.

Baum, M. (1970). Extinction of avoidance responding through response prevention (flooding). *Psychological Bulletin, 74*, 276–284.

Baum, M. (1976). Instrumental learning: comparative studies. In M. P. Feldman & A. Broadhurst (Eds.), *Theoretical and experimental bases of the behavior therapies* (pp. 113–131). New York: Wiley.

Beecroft, R. S. (1967). Near-goal punishment of avoidance running. *Psychonomic Sciences, 8*, 109–110.

Bolles, R. C. (1969). Avoidance and escape learning: Simultaneous acquisition of different responses. *Journal of Comparative and Physiological Psychology, 68,* 355–358.

Bolles, R. C., Moot, S. A., & Grossen, N. E. (1971). The extinction of shuttlebox avoidance. *Learning and Motivation, 2,* 324–333.

Brush, E. S. (1957). Traumatic avoidance learning: The effects of conditioned stimulus length in a free responding situation. *Journal of Comparative and Physiological Psychology, 50,* 541–564.

Church, R. M., Brush, F. R., & Solomon, R. L. (1956). Traumatic avoidance learning: The effects of CS–US interval with a delayed-conditioning procedure in a free-responding situation. *Journal of Comparative and Physiological Psychology, 49,* 301–308.

Davenport, D. G., & Olson, R. D. (1968). A reinterpretation of extinction in discriminated avoidance. *Psychonomic Science, 13,* 5–6.

Davis, M., Schlesinger, L. S., & Sorenson, C. A. (1989). Temporal specificity of fear conditioning: Effects of different conditioned stimulus–unconditioned stimulus intervals on the fear-startled reflex. *Journal of Experimental Psychology: Animal Behavior Processes, 15,* 295–310.

Gabriel, M., & Schmajuk, N. A. (1991). Neural substrate of avoidance learning in rabbits. In J. W. Moore & M. Gabriel (Eds.), *Neurocomputation and Learning: Foundations of Adaptive Networks* (pp. 143–170). Cambridge, MA: MIT Press.

Gormezano, I., Kehoe, E. J., & Marshall, B. S. (1983). Twenty years of classical conditioning research with the rabbit. *Progress in Psychobiology and Physiological Psychology, 10,* 197–275.

Grossberg, S. (1975). A neural model of attention, reinforcement, and discrimination learning. *International Review of Neurobiology, 18,* 263–327.

Herrnstein, R. J. (1969). Method and theory in the study of avoidance. *Psychological Review, 76,* 49–69.

Herrnstein, R.J., & Hineline, P.N. (1966). Negative reinforcement as shock-frequency reduction. *Journal of the Experimental Analysis of Behavior, 9,* 421–430.

Hull, C. L. (1929). A functional interpretation of the conditioned reflex. *Psychological Review, 36,* 498–511.

Hull, C. L. (1951). *Essentials of Behavior.* Westport, CT: Greenwood Press.

Kamin, L. J. (1954). Traumatic avoidance learning: The effects of CS–US interval with a trace-conditioning procedure. *Journal of Comparative and Physiological Psychology, 47,* 65–72.

Kamin, L. J. (1957). The gradient of delay of secondary reward in avoidance learning. *Journal of Comparative and Physiological Psychology, 50,* 445–449.

Kamin, L. J., Brimer, C. J., & Black, A. H. (1963). Conditioned suppression as a monitor of fear of the CS in the course of avoidance training. *Journal of Comparative and Physiological Psychology, 56,* 497–501.

Kehoe, E. J. (1988). A layered network model of associative learning: Learning to learn and configuration. *Psychological Review, 95,* 411–433.

Konorski, J. (1967). *Integrative activity of the brain.* Chicago: University of Chicago Press.

Maier, S. F., & Seligman, M. E. P. (1976). Learned helplessness: Theory and evidence. *Journal of Experimental Psychology: General, 105,* 3–46.

Mowrer, O. H. (1947). On the dual nature of learning—A reinterpretation of conditioning and problem solving. *Harvard Educational Review, 17,* 102–148.

Mowrer, O. H., & Lamoreaux, R. R. (1946). Fear as an intervening variable in avoidance conditioning. *Journal of Comparative Psychology, 39*, 29–50.

Overmier, J. B., & Bull, J. A. (1969). On the independence of stimulus control of avoidance. *Journal of Experimental Psychology, 79*, 464–467.

Overmier, J. B., Bull, J. A., III, & Trapold, M. A. (1971). On instrumental response interaction as explaining the influences of Pavlovian CSs upon avoidance behavior. *Learning and Motivation, 2*, 103–112.

Overmier, J. B., & Seligman, M. E. P. (1967). Effects of inescapable shock upon subsequent escape and avoidance responding. *Journal of Comparative and Physiological Psychology, 63*, 28–33.

Page, H. A., & Hall, J. F. (1953). Experimental extinction as a function of the prevention of response. *Journal of Comparative and Physiological Psychology, 46*, 33–34.

Pearce, J. M., & Hall, G. (1980). A model for Pavlovian learning: Variations in the effectiveness of conditioned but not unconditioned stimuli. *Psychological Review, 87*, 532–552.

Rescorla, R. A. (1967). Inhibition of delay in Pavlovian fear conditioning. *Journal of Comparative and Physiological Psychology, 64*, 114–120.

Rescorla, R. A., & LoLordo, V. M. (1965). Inhibition of avoidance behavior. *Journal of Comparative and Physiological Psychology, 59*, 406–412.

Rescorla, R. A., & Wagner, A. R. (1972). A theory of Pavlovian conditioning: Variations in the effectiveness of reinforcement and nonreinforcement. In A. H. Black & W. F. Prokasy (Eds.), *Classical conditioning II: Current research and theory* (pp. 64–99). New York: Appleton-Century-Crofts.

Schmajuk, N. A., & DiCarlo, J. J. (1992). Stimulus configuration, classical conditioning, and the hippocampus. *Psychological Review, 99*, 268–305

Schmajuk, N. A., & Moore, J. W. (1988). The hippocampus and the classicaly conditioned nictitating membrane response: A real-time attentioal–associative model. *Psychobiology, 46*, 20–35.

Schneiderman, N. (1966). Interstimulus interval function of the nictitating membrane response of the rabbit under delay versus ace conditioning. *Journal of Comparative and Physiological Psychology, 62*, 397–402.

Schoenfeld, W. N. (1950). An experimental approach to anxiety, escape, and avoidance behavior. In P. H. Hoch & J. Zubin (Eds.), *Anxiety* (pp. 70–99). New York: Grune & Stratton.

Seligman, M. E. P. (1975). *Helplessness: On depression, development, and death.* San Francisco: Freeman.

Seligman, M. E. P., & Campbell, B. A. (1965). Effects of intensity and duration of punishment on extinction and avoidance response. *Journal of Comparative and Physiological Psychology, 59*, 295–297.

Seligman, M. E. P., & Johnston, J. C. (1973). A cognitive theory of avoidance learning. In F. J. McGuigan & D. B. Lumsden (Eds.), *Contemporary approaches to conditioning and learning* (pp. 69–110). Washington, DC: V. H. Winston.

Seligman, M. E. P., & Maier, S. F. (1967). Failure to escape traumatic shock. *Journal of Experimental Psychology, 74*, 1–9.

Sidman, M. (1953). Two temporal parameters of the maintenance of avoidance behavior by the white rat. *Journal of Comparative and Physiological Psychology, 46*, 253–261.

Smith, M. C. (1968). CS–US interval and US intensity in classical conditioning of the rabbit's nictitating membrane response. *Journal of Comparative and Physiological Psychology, 66,* 679–687.

Sokolov, E. N. (1960). Neuronal models and the orienting reflex. In M. A. B. Brazier (Ed.), *The central nervous system and behavior* (pp. 31–36). New York: Macy Foundation.

Solomon, R. L., & Wynne, L. C. (1953). Traumatic avoidance learning: Acquisition in normal dogs. *Psychological Monographs, 67,* 354.

Solomon, R. L., & Wynne, L. C. (1954). Traumatic avoidance learning: The principles of anxiety conservation and partial reversibility. *Psychological Review, 61,* 353–385.

Solomon, R. L., Kamin, L. J., & Wynne, L. C. (1953). Traumatic avoidance learning: The outcomes of several extinction procedures with dogs. *Journal of Abnormal and Social Psychology, 48,* 291–302.

Soltysik, S. (1985). Protection from extinction: New data and a hypothesis of several varieties of conditioned inhibition. In R. R. Miller & N. E. Spear (Eds.), *Information Processing in Animals: Conditioned Inhibition* (pp. 369–394). Hillsdale, NJ: Lawrence Erlbaum Associates.

Starr, M. D., & Mineka, S. (1977). Determinants of fear over the cause of avoidance learning. *Learning and Motivation, 8,* 332–350.

Sutton, R. S., & Barto, A. G. (1981). Toward a modern theory of adaptive networks: Expectation and prediction. *Psychological Review, 88,* 135–170.

Weisman, R. G., & Litner, J. S. (1969). Positive conditioned reinforcement in Sidman avoidance rats. *Journal of Comparative and Physiological Psychology, 68,* 598–603.

Widrow, B., & Hoff, M. E. (1960). Adaptive switching circuits. *1960 IRE WESCON Convention Record,* 96–104.

Young, G. A. (1976) Electrical activity of the dorsal hippocampus in rats operantly trained to lever press and to lick. *Journal of Comparative and Physiological Psychology, 90,* 78–90.

Zimmer-Hart, C. L., & Rescorla, R. A. (1974). Extinction of Pavlovian conditioned inhibition. *Journal of Comparative and Physiological Psychology, 86,* 837–845.

Zhuikov, A. Y., Couvillon, P. A., & Bitterman, M. E. (1994). Quantitative two-process analysis of avoidance conditioning in goldfish. *Journal of Experimental Psychology: Animal Behavior Processes, 20,* 32–43.

7

In Praise of Parsimony

J. E. R. Staddon
Duke University

B. Silvano Zanutto
Universidad de Buenos Aires

Although some contributors to this book are concerned with neural mechanisms, the primary focus of all is how behavior depends on the organism's stimulus–response history. Given that the focus even of the neural chapters is on environment–behavior relations—rather than brain–behavior or brain–environment relations—we are all doing what used to be called (somewhat disparagingly) *black-box* psychology. But now *all* psychology is black-box psychology, if only because brain–behavior relations are the special province of a new science, behavioral neuroscience. In this chapter we ponder some implications for the new psychology of models for action.

Three Distractions

A theme that runs through several chapters is that apparently intelligent behavior may often be the outcome of simple processes. This chapter argues that simplicity is fundamental, a principle to which the science of psychology should pay more than lip-service. Parsimony is not just a good thing, it may be the *only* thing. Yet, it has been relegated to an insignificant role in psychology. Why? Perhaps the reason is that psychology is uniquely subject to three distractions that make parsimony seem less than critical. In psychology, we seem to have special access to at least two sources of knowledge about behavior: the mind, and the brain. Together with another distraction, the

239

behavioristic reaction against *mind*, these three—mind, the reaction against mind, and the brain—have sidelined parsimony in the search for explanations of behavior.

Because we are thinking beings and think that we have direct access to some of our own mental processes—and because we are largely mistaken in this—our attempt to use our own introspections as a guide is almost invariably a distraction. Cognitive psychology, for example, has spent much of its time trying to explain "mental states," sometimes explicitly eschewing parsimony if it seems to get in the way (Johnson-Laird, 1988). It is perhaps too early to pass a conclusive judgement, but our guess is that this project has stalled, if not failed. Equally distracting are the various more or less violent reactions against mentalism. The several behaviorisms have tried to substitute behavior, defined in different and often mutually incompatible ways, for mind or cognition, but this is little improvement, for reasons that have been repeatedly spelled out (e.g., Dennett, 1978; Smith, 1986; Staddon, 1993a, 1993b). Unless defined in such an inclusive way that the word retains almost nothing of its usual meaning, behavior alone is not sufficient. Explanations that invoke internal states of some kind cannot be avoided (Staddon, 1997).

The third distraction is biological plausibility—brain–behavior relationships. Most neurobiologists, for example, regard it as both certain and obvious that the only *real* explanation for behavior (or mind) is the underlying physiology. Yet this is a distraction also, because we have no guarantee that the way neurophysiology works to make behavior is simple or even always comprehensible. In fact, what we know of the process of evolution through natural selection suggests that brain–behavior relations may routinely be mind-numbingly complex—so complex, that we may not understand interesting behavior—behavior strongly dependent on history—at the neurophysiological level any time in the foreseeable future. Natural selection is a notoriously messy engineer. Even human engineering, when the construction is complex, is more art than science. Anyone who has worked and reworked a long computer program, or attempted to maintain a large program written by others, knows how quickly it escapes the understanding of any individual. Pretty soon, maintenance reduces to trying *ad hoc* fixes rather than deducing the causes of malfunction from first principles. Each fix only increases the incomprehensibility of the whole. Natural selection, of course, is nothing but *ad hoc* fixes.

This is not to say that either the cognitive project or the neurophysiological approach should be abandoned. It *is* to say that neither is without its difficulties. Neither is so successful as to rule out other approaches. In this chapter we advocate an alternative: Psychology as the search for parsimonious mechanism. But to be really persuasive, our argument must do more than point to the failings of other methods. Lacking the intuitive plausibility of cognitive psychology, the attractive (if illusory) directness of radical behaviorism, and the concreteness of neurobiology, the new view—which has elsewhere been termed *theoretical behaviorism* (Staddon, 1993b)—must prove itself through successful example. We describe one such example in this chapter.

PARSIMONIOUS MECHANISM

We argue for a particular view of what scientific psychology should be. That view is resolutely *mechanical*. Animals and human beings are to be studied as *machines*, formed through the tendentious haphazard of evolution. These natural machines seem to be quite different from machines made by humans,[1] just because the evolutionary process *is* so accidental. Redundancy and awkwardness in design seem to be eliminated only slowly and imperfectly by the process of natural selection. Any paleontology text gives dozens of examples of organs evolving (often not very efficiently) from one function into organs that serve a very different one: Legs into wings (bats), lateral line into inner ear (fish to mammals), wrist bones into thumbs (panda) and so on. The brain also has evolved in this highly contingent fashion, and all evidence is that the links between brain structure and function and the behavior (or mind) of the whole animal are often going to be very difficult to unravel. Even apparently simple behavior patterns, like the swimming behavior of the leech, are still far from completely understood at the neural level (e.g., Churchland & Sejnowski, 1992).

Neuroscientists aim at higher things than leeches, of course.[2] Learning, preferably in human beings, is a major focus. But learning, because of the very long time delays involved (what is learned today may not show up in behavior until next year), undoubtedly affects very many brain areas. Consequently a full account of the neural basis[3] for associative learning may have to wait until

---

[1]Perhaps they are not so different these days. As microchips become smaller, the function even of human-made machines is less easily discerned from their structure than used to be the case when machines had a few, large parts. Indeed, large computer programs, and the newest microchips, are no longer comprehensible by any single individual—as the apparently unavoidable recurrence of bugs proves.

[2]We take the liberty of stating for the first time in print a new law in the fashionable new discipline of *science studies*. This principle is well-known to working psychobiologists but has hitherto eluded publication. The *Law of Aesthetic Conservation* is: "The cuter the animal, the worse the research" (or, more formally, $C_{ute} + Q_{uality} = K_{onstant}$). Like all principles in evolutionary biology (Bergmann's Law, Cope's Law, etc.) there are many individual exceptions. But, by and large, the principle is true, as can easily be seen by studying the research literature on babies, dogs, and chimps, on the one hand, and research on rats, leeches, flies, and worms, on the other.

[3]*Neural basis* means more than (for example) the fact that brain area X is involved in processing task Y at time t. Structure-function-time maps of this sort are common, and becoming commoner as brain-imaging technology becomes more widely available. The complete story involves an understanding of the real-time functioning of the neural and biochemical *circuitry* in the brain as it produces the muscular movements we identify as behavior. We are light years from understanding interesting behavior at this fundamental level.

we have acquired a largely complete understanding of the structure and function of the entire brain. Even then, when all the parts of the puzzle are known, the number of parts is so large, the types of interaction so varied, and the number of possible interconnections so astronomical that the task may turn out to be a practical impossibility.

Just because the natural machines that are brains are so different from familiar human-made machines, the obvious strategy for understanding them, taking them apart and trying to understand how the pieces work together to produce behavior—the so-called bottom-up approach—has turned out to be extraordinarily difficult. There is an alternative, and it works because brains have evolved not for convenience of neurophysiological analysis but for effectiveness in action. The function of the brain is to produce behavior. The function of behavior is to promote the Darwinian fitness of the behaver. Because behavior has evolved in an orderly world, it is itself orderly, even though the underlying neural processes are often dazzlingly complex and obscure. Theoretical behaviorism is just the idea that the psychologist's task is *to discover the simplest possible process* that can explain observed behavior.[4] The keys to this approach are *parsimony*—our theories should be as simple as possible and explain as much as possible—and *mechanism*—our theories should be real theories not verbal glosses that inhibit inquiry without augmenting understanding. Our theories should *do* something: be real processes acting in real time, the kind of thing that might usefully animate a robot. After all, what use is a theory that purports to explain adaptive behavior if, when left to itself, it can do nothing at all? Just try and do something useful with dissonance theory, or social learning theory, or memory interference theory, or any theory at that purely verbal level. In this chapter, we illustrate how this strategy of looking for the most parsimonious mechanism has helped us to understand the dynamics of feeding and foraging.

FEEDING AND REINFORCEMENT LEARNING

Animal learning was once the dominant field in experimental psychology. Even now, it is pretty lively—and theorists continue to mine its decades of research reports for data to model (e.g., Commons, Grossberg, & Staddon, 1991; Grossberg, 1982; Klopf, 1988; Schmajuk & DiCarlo, 1991; Sutton & Barto, 1981). The vast majority of animal-learning experiments are on what is now termed *reinforcement learning*. (Animals cannot follow instructions and with-

---

[4]Anderson (1978) reminded psychologists that there are an indefinite number of "black box" explanations for a given set of input–output relations. His proposal for reducing this number to one was to use biological constraints—but this method may sometimes be impossible because of the intrinsic complexity of brain mechanisms. Theoretical behaviorism emphasizes another method for achieving uniqueness: parsimony, which is the method used in all other sciences.

out the inducement of explicit rewards or punishments, little useful behavior can be got from most of them.) Most animal-learning experiments involve hungry animals and food reinforcement. But feeding behavior is also involved in basic mechanisms of *homeostasis*, the maintenance of constancy in the *milieu intérieur* made famous by Claude Bernard (1865/1927). Feeding is usually considered as the prototypical *motivational* system. Yet, motivation is largely ignored in most animal-learning experiments and also in most of the theoretical schemes that have been developed to explain them. The animals are kept hungry in these experiments so (it is assumed) their motivational level may be assumed to be high—and constant.

How reasonable is this assumption? First, it is unlikely to hold in so-called *closed-economy* experiments; that is, experiments in which the animal is not explicitly food-deprived, but lives in the experimental cage and must get all his food via some kind of reinforcement schedule (see, for example, Collier, Hirsch, & Hamlin, 1972, and many subsequent experiments from the laboratories of Collier and others). The schedule may be relatively simple—the animal must complete a fixed number of lever presses, but then gets essentially unlimited access to food—or more complex, if a second consumption response is required for each bit of food (see Fig. 7.9). Animals in closed-economy experiments are close to satiated most of the time, but presumably at those times when they actually work to get food, they are hungrier than at other times. Yet data from these experiments are usually discussed not in motivational or regulatory terms, but in terms of operant learning and planned economic strategies (e.g., Collier, 1986).

Second, even in *open-economy* experiments, with very hungry animals and short (a few hours or less) experimental sessions, there is reason to question the assumption that motivational factors are irrelevant. In the first place, we do not know much about the dynamics of motivation, so we have no principled basis for excluding motivational factors, a priori. Also, a substantial body of recent experimental work from the laboratory of McSweeney has shown that there are changes, sometimes quite substantial changes, in the rate at which rats and pigeons will work for food reinforcement within short open-economy experimental sessions (McSweeney, Hinson, & Cannon, in press; McSweeney, Weatherly, & Swindell, 1996). The reinforcement schedule is constant, the animals are clearly very hungry, yet the response rate varies—perhaps because of motivational dynamics or some other nonassociative process. Even in open-economy experiments, therefore, motivational effects cannot be ignored.

It seems that the lowly topic of motivation has been slighted in research on reinforcement learning. In their eagerness to study *learning*, researchers have neglected to explore the motivational processes without which they would have no behavior to study. Yet, without understanding motivation, how can they assess the role of learning? Or, to put the same thing in an older way, "In no case may we interpret an action as the outcome of the exercise of a higher psychological faculty, if it can be interpreted as the outcome of the exercise of one which stands lower in the psychological scale" (p. 163).[5] Lloyd Morgan's canon is often mentioned, but rarely obeyed, in psychology. We would like to elevate it to a primary principle. Next we show how this approach has helped us to understand the elementary motivational dynamics that underlie feeding behavior as well as some operant (i.e., apparently learned) behavior in rats.

A MODEL FOR FEEDING DYNAMICS[6]

Beginning in the early 1970s, numerous theorists attempted to construct computer models to explain the essentials of feeding behavior in rats: Regulation in response to different kinds of challenge, the temporal pattern (eating in meals) and the relation between eating patterns and body weight (Booth, 1978; Guillot & Meyer, 1987; McFarland, 1974; Toates & Halliday, 1980; Toates & Rowland, 1987). The effort peaked in the early 1980s, but then fizzled to the point that an eating model rarely makes its way into print anymore, despite the great boost in computer power and the growth of computer modeling in general. Why this decline? Our guess is that it is because essentially all the models focused on physiology; every element of every model had to *represent something physiological*: gut distention, fat stores, glucose concentration in the blood, metabolic rate, and so forth. But we did not know then, nor do we know now, *all* the physiology that is relevant for feeding.[7] If we do not know *all* the relevant physiology, a model built exclusively out of the physiology we *do* know is very likely to fail.

On the other hand, if you just look at feeding *behavior*, it *is* possible to come up with a very simple model that can duplicate the regulation of feeding behavior and the basic patterns of feeding under challenge. The model is parsimonious: It has very few assumptions and explains many facts. But the ingredients of the model are designed to explain the behavioral data, not to match up with known physiology. It is not contrary to physiology—as we shall see, this model is perfectly consistent with known feeding physiology. But it

---

[5]Apparently *process*—or *model* as we might say now—was actually Morgan's preferred term—see Boakes (1984). Morgan's canon is not the same as the principle of parsimony (Thomas, in press), but our attempt to account for as much as possible of learned behavior with the simplest possible model for a motivational (i.e., lower) process is consistent both with Morgan's canon and the principle of parsimony.

[6]A more complete account of this model of feeding dynamics appears in Staddon & Zanutto (1997).

[7]It's interesting that as we write, the press is full of a new study that has discovered hitherto unsuspected neural structures directly associated with the digestive tract. What do they do? We do not know.

is *agnostic* about what bits of the model might correspond to what bits of physiology.

First, we summarize the salient behavioral facts. We then consider what these imply for the nature of the underlying mechanism. Finally, we propose a mechanism and see how well it does in explaining the original facts plus some new ones.

The Data

The large number of experiments on feeding in rats lead to three important generalizations:

1. *Eating in Meals*: Rats (and many other animals, including people) do not eat either at fixed intervals of time, or at random times. They eat in bouts, called meals.

2. *Regulation*: Rats regulate their rate of eating. If eating is interrupted for a time, the rat will make up the food it has lost when eating is allowed to resume—so long as the interruption is not too long. This regulation of eating rate occurs at several time scales, from interruptions of a few minutes to a day or so.

3. *First-Meal Effect*: When eating is interrupted for a few hours, the first (and only the first) meal is extra long. Eating rate and meal size thereafter both revert to normal values.

The aim of this section is to show how these well-known properties of feeding behavior can be explained by a dynamic model—providing we ignore at this stage, how the bits of the model relate to bits of physiology. Our contention is that the *behavior* itself is orderly and can be explained—whereas its relation to physiology is imperfectly known and possibly very complex. The second point remains to be seen, but the first point can be proved by example, by finding a simple process that does the job of explanation.

Past attempts to model feeding behavior have tended to focus on the regulatory property (see earlier references). A few theorists have addressed the *eating-in-meals* problem. So far as we know, no one has discussed the theoretical significance of the first-meal effect of interrupting feeding. We begin with eating in meals.

First, we assume that feeding behavior is turned off and on by some kind of *satiation signal*; and that this satiation signal is an aftereffect of eating. The major task for modeling is to define how eating affects the satiation signal. Second, we will assume that the satiation signal affects eating behavior via a *threshold*: when the satiation signal rises above a threshold, θ, eating ceases; when it falls below a threshold, eating begins. What is this threshold (or thresholds) and how does eating determine the satiation signal?

The fact that eating is episodic poses a problem. Either different processes must turn eating on and off, or the feedback from food must be delayed in some way. The first possibility was explored by Booth (1978) and Guillot & Meyer (1987) as well as Staddon (1988). It implies two thresholds, a *start* threshold and a second, *stop* threshold. The idea is that eating begins when the satiation signal falls below a start threshold; but once eating has begun, the satiation signal must fall below a second, lower, stop threshold, before eating ceases. The two-threshold idea works surprisingly well, but there are at least three objections to the idea:

1. The definition of an eating bout is arbitrary. In simulations of a two-threshold model, a transition from not eating, in time step t, to eating, in time step t + 1, must trigger a change in the effective threshold from θ_{start} to θ_{stop}. This seems very arbitrary because the threshold change depends only on the transition from eating to not eating (or vice versa). It does not depend on the size of the time step in the simulation. Yet, we know that if the time step is changed, such a model may produce a completely different pattern of real-time behavior. No model should be so sensitive to the size of time step. The two-threshold approach is arbitrary in a second way also.

2. It takes no account of the number of time steps in between eating bouts. Two eating episodes separated by one time step cause the same threshold change as two episodes separated by two or one hundred, which is not plausible.

3. The behavior is too perfect: every meal is uninterrupted, for example. The two-threshold assumption can never yield three-state intereating time distributions, which are frequently observed (e.g., Davis, 1989). These problems could be partially mitigated by an assumption relating the size of the difference between θ_{start} and θ_{stop} to the time-step size or to the length of an eating (or not-eating) bout—but the model would then become cumbersome and *ad hoc*—a violation of parsimony.

This leaves the second possibility: that eating in meals is a consequence of inhibitory feedback that is *delayed*. Delay accounts for eating in meals in the following way. In the absence of food, the satiation signal decreases with time. When the signal falls below a (single) threshold, the *set point*, eating begins (if food is available). But because of the lag between eating and its inhibitory feedback, the satiation signal continues to decline for a while during the meal. After a while, the delayed effects of the ingested food begin to catch up and the satiation signal begins to increase, eventually rising above the threshold, at which point eating ceases. Eating does not resume at once, because the lagged satiation signal continues to increase for a while even after eating has ceased. In this way, eating under ad lib conditions consists of temporally separated eating bouts.

The Model

These arguments have led us to a very simple model for feeding dynamics[8] that incorporates delay. Properties are shown in Fig. 7.1 as the hypothetical satiating effect of a single bit of food, at time 0. The satiating effect of the food takes some time after ingestion to reach a maximum: It increases and then decreases. (The effect of a single brief stimulus is termed the *impulse response* of the system.[9] The impulse response we need to get eating in meals is biphasic, like the one in the figure.) The horizontal line shows the eating

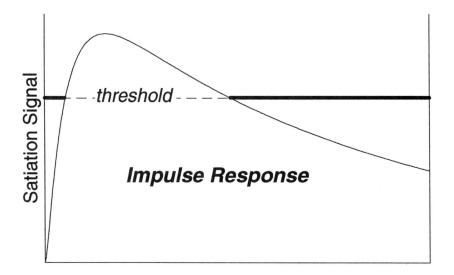

Time

FIG. 7.1 Biphasic impulse response of a system that will produce eating in meals. A single piece of food is delivered at time 0. The satiating effects rise to a maximum only after a delay. The height of the peak depends on the size of the piece of food; if it is small, the satiation signal will follow the same biphasic pattern, but may not reach the satiation threshold. This impulse response was obtained from the last three stages of the 6-stage system used to make Figs. 7.3, 7.5, 7.6, 7.7, 7.8, 7.10 and 7.11. The parameters were determined as described in the legend to Fig. 7.10. See Staddon and Zanutto (1997) for other details.

---

[8]Note that we discuss only feeding *behavior*, which does not translate directly into *body weight*, the aspect of feeding of most direct interest to many people. Feeding behavior is one component of body weight regulation, but not the only one (see Staddon & Zanutto, in press, for further discussion of this point).

[9]The satiation-signal model we describe is in fact linear (although the system as a whole is not, because of the threshold). The impulse response has a special significance for linear systems: it characterizes the system completely. Note that the impulse response in this model is always biphasic, but it will only cross the satiation threshold if the initial bit of food is large enough.

threshold. When the satiation signal is below the threshold (indicated by the heavy solid line), the system is ready to eat if food is available. When the satiation signal is above threshold (dashed line), the system will not eat. A system with a biphasic impulse response like this will produce eating in meals, for the reasons just described. What kind of process can produce a satiation signal with the properties shown in Fig. 7.1? Will such a system show regulation? And will it regulate over different time scales?

Figure 7.2 shows a physical model with the necessary properties. It is a series of "leaky buckets" with the output of the first being the input to the second and so on down the chain. The chain can be of any length—three buckets are shown in the figure. The water level in successive buckets represents the effects (neural as well as metabolic) of food at different stages in the chain of events initiated by ingestion. For the reasons just given, though, we imply no specific mapping between parts of the model and particular

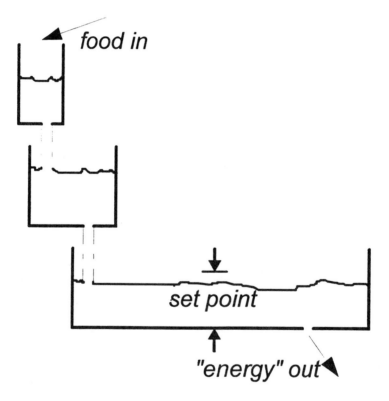

FIG. 7.2 A fluid-flow metaphor for feeding regulation. Energy (fluid) enters at one end of the chain and is depleted at the other. Intake (*food in*) is regulated by the fluid level(s) in the last compartment. When the level is above a threshold value, intake ceases; when the level falls below threshold, intake is resumed if food is available.

physiological variables. The satiation signal (the signal that shuts off fluid inflow) from such a system is not shown in the figure. As we will show in a moment, to get the required biphasic impulse response, all that is necessary is that the satiation signal be a weighted sum of all the water levels, *with levels later in the chain weighted more than earlier ones* (in the figure, the set-point is defined by the water level in the last compartment).

This system is intrinsically regulatory. The fluid-flow metaphor shows pretty clearly that it tends towards a stable fixed-point equilibrium. Look again at Fig. 7.2. Suppose we pour in fluid (food in = *eat*) only when the level in the last compartment exceeds a fixed threshold. It should be intuitively obvious that so long as inflow rate is not too large (so the top compartment overflows) or too small (lower than energy out), the average rate of inflow will tend towards a constant. In the steady state, total amount of fluid per unit time, averaged over a number of periods of inflow and no-inflow (termed *molar ingestion rate*), will equal energy outflow rate and be more or less independent of the local inflow rate.

It is not so obvious that a system like this will regulate at all the different time scales that feeding regulates: Within a meal, following short interruptions and after lengthy deprivation. To demonstrate this requires a formal model.

Figure 7.3 shows a mathematical model with the same dynamic properties as the physical model in Fig. 7.2. We term this the *cascaded-integrator* (CINT) model.[10] The leaky buckets in Fig. 7.1 are replaced by *leaky integrators* linked in a cascade. Each leaky integrator is simply the familiar linear operator: In discrete-time notation, $V(t+1) = aV(t) + x(t)$, where V corresponds to the water level in a given bucket in Fig. 7.2, x is the inflow rate, and a is a constant between 0 and 1 representing the rate of outflow in relation to the size of the bucket (the larger a, the smaller the change in V during each time step). Thus, at each time step, each bucket (integrator) loses an amount proportional to its current level, $(1-a)V(t)$, and gains an amount corresponding to the inflow rate, $x(t)$. In a cascade, the output, V_N, of integrator N is the input of integrator N+1. There is a set point or threshold, represented by a fixed value, θ. The strength of the tendency to eat, *eat command*, is simply the difference between the sum of integrator states, V_i, and θ.

Formally, the cascaded-integrator model works like this. The tendency to eat is simply the difference between a reference value, θ, and an aggregate satiation signal, V_S, which is the weighted sum of the states of all of the integrators:

---

[10]This cascaded structure is very similar to a class of models we have applied to data on simple habituation (Staddon, 1993; Staddon & Higa, 1996). The probable importance of time delays in eating dynamics was noted by Schilstra (1978). The general idea that both long- and short-term food stores exert an inhibitory effect on eating was briefly discussed by Bolles (1980) and explored somewhat more extensively by Panksepp (1974) and Schilstra (1978).

A 3-Stage Cascade Eating Model

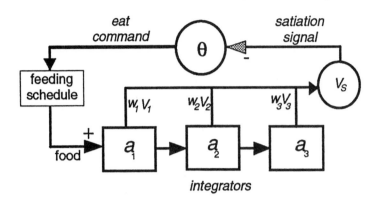

integrators

FIG. 7.3 A 3-stage cascaded-integrator feeding model. The effects of food are input to the first integrator (time constant: a_1), whose output is input to the second integrator, and so on. The weighted values (V_i) of integrators are summed together as the satiation signal, V_S, which is compared with a threshold, θ, the set point of the system. In the models discussed in the text, the weights for early integrators in the chain are 0, the weights for later stages, 1. The eating command is the difference between satiation signal and a fixed threshold (set point): Eating command = θ - V_S. If the eating command is positive, and the feeding schedule makes food available, the system will eat; each quantum of food, $x(t)$, is input to the first integrator in the chain.

$$V_S = w_1V_1 + w_2V_2 + ... + w_NV_N = \sum_{i=1}^{N} w_iV_i, \quad (1)$$

where w_i are the weights. Physiologically, Equation 1 reflects the well-established fact that the tendency to eat depends not only on the state of bodily deprivation (i.e., events late in the cascade), but also on taste factors and stomach fill (i.e., events at the beginning and in the middle of the cascade; Mook, 1987).

To translate V_S into an eating command (response strength) we need a reference (threshold), θ: When V_S falls below θ, eating begins and the tendency to eat (which we term *reflex strength*) is proportional to the difference between V_S and θ. Thus, reflex strength = θ - V_S. When eating actually occurs, *response strength*, the vigor of the measured response, is proportional to reflex strength. For simplicity, we assume initially that eating is all-or-none, occurring at a constant rate when reflex strength is above zero and not occurring when reflex strength is below zero. The model can only turn ingestion on and off; it has no control over the rate of ingestion in the *on* state. Our assumption resembles an experimental procedure used to study eating dynamics, the controlled-infusion-rate sucrose solution procedure of Seeley, Kaplan, and Grill (1993), but we are applying it to all eating. (The all-or-none assumption must be relaxed when we use the model to drive operant responding.)

The θ is similar to the set-point that figures prominently in most discussions of feeding regulation, but θ is different in three respects:

1. It corresponds to *no single physiological variable*, such as blood glucose level or the level of fat stores. The θ is a purely theoretical construct, which may reflect a whole constellation of physiological variables, represented by the V_i terms in Equation 1.

2. The θ does not necessarily correspond to a fixed settling point for eating rate.[11] As we will see, different feeding regimens can lead to different steady-state eating rates, even though θ is constant.

3. Moreover, as we demonstrate, the system as a whole shows regulation of eating rate—not just over the long term (molar regulation), but also more or less locally.

To obtain eating in meals, we need a biphasic impulse response like the one shown in Fig. 7.1. To get it, we need only order the weights to emphasize later integrators in the chain. For example, suppose the early weights are all zero. Then, the early integrators serve only as delays and have no direct inhibitory effect on eating. As we will see, this delay is sufficient to produce eating in meals and does not interfere with useful regulatory properties of the cascaded system. In most of our simulations, the early w values are all equal to zero and the last two or three equal to one.

Equations. For the simple case, we assume that eating is all-or-none: The system eats at a fixed rate during a time step if the satiation signal is below threshold, and not otherwise. The size of each eating input reflects the *satiating value* (SV) of each bit of food. *Satiating value* is presumably related to taste, texture, and bulk as well as nutrient value. These properties probably have separable effects on feeding, but it is simpler at this stage not to distinguish among them. The equations for this system are:

$$x(t) = \phi(t) \text{ if } V_s < \theta; \ 0 \text{ otherwise}, \quad (2)$$

$$V_1(t + 1) = a_1 V_1(t) + b_1 x(t), \quad (3)$$

$$V_i(t + 1) = a_i V_i(t) + b_i V_{i-1}(t), \quad 0 < i < N, \quad (4)$$

---

[11]Note the distinction between *set-point*, which is a fixed parameter of the (theoretical) feeding system, and *settling point*, which is the steady-state value of molar feeding rate (or some other variable, such as body weight) under fixed environmental conditions. Set-point is fixed, but settling point will vary as environmental regimens vary.

where $x(t)$ is the SV during each time step (i.e., the food actually eaten), $\phi(t)$ is the SV of each bit of food available (determined by the experimenter), V_i is the state of integrator i, a_i is the time constant of integrator i and b_i is the input weight. In all our simulations $b_i = 1 - a_i$ and the a_i values are linked as described in the legend to Fig. 7.10, limiting the number of free parameters to two, no matter how many integrators are in the model.

Equation 2 says that the system eats the food available (size: $\phi(t)$ in each time step) only if the satiation signal, V_S, is below threshold, θ. Equation 3 says that the state of the first integrator, $V_1(t+1)$, is determined by its previous state, $V_1(t)$, and the SV during that time step, $x(t)$. Equation 4 shows that the state of an integrator later in the cascade, $V_i(t+1)$, is determined by its previous state, $V_i(t)$, and the state of the integrator earlier in the chain, $V_{i-1}(t)$.

Predictions

This model is purely regulatory. It does not learn. It has no associative properties, and thus is insensitive to stimulus context. It takes no account of food quality (as opposed to quantity). Therefore, the model cannot explain everything about feeding and indeed *should not explain everything*. Rats do more than just regulate. They come to the experiment with preexisting feeding patterns—the Circadian pattern of eating, for example; and they do learn—even though our contention is that they learn less than is often assumed. They must learn at least some things even to eat at all in a new situation: What is food and what is not (not all foods are recognized instinctively by young rats) and where the food source is in the enclosure. Undoubtedly they learn other things as well, such as properties of the reinforcement contingencies. Because a purely regulatory model contains none of these things, we should not expect to explain every feature of the data. We should not expect, in other words, to fit a limited data set with high precision, in the fashion of traditional mathematical modeling. What we are looking for is a *pattern of results* in a broad data set that is consistent with an underlying regulatory process. What is left over—the details that cannot be explained by a regulatory model—is then a proper domain for "higher" processes like learning and cognition. As we pointed out earlier, the method is nothing more than a self-conscious application of Morgan's (1894) canon: Explain as much as you can at the lowest possible functional level.

The steady-state pattern of eating shown by the system in Fig. 7.1 is illustrated in Fig. 7.4, which shows 30,000 seconds of data after a long stabilization period. The heavy line shows the satiation signal, V_S, for a large bite size (1 unit), the light line is for a small bit size (0.5 unit). The two rows of blocks at the bottom show periods of eating. Meal size approximately doubles when bite size is halved, but meal frequency is little affected.

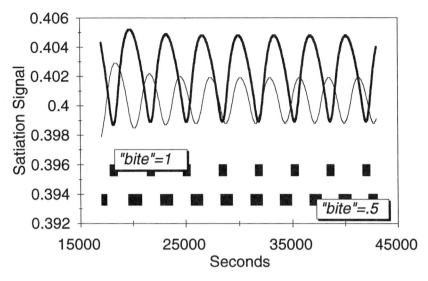

FIG. 7.4 Steady-state eating pattern for a 6-integrator version of the system illustrated in Fig. 7.3, with an eating threshold, $\theta = .4$. The wavy line is the satiation signal (i.e., $\sum_{i=M}^{N} V_i$, for $M = 4, N = 6$) for a bite size, $\phi(t) = .5$ (*light line*) or 1 (*heavy line*). The blips at the bottom show eating episodes (larger bite size above). Note eating in meals and similar meal frequency, but 2:1 meal duration, with the two bite sizes. Parameters as in Fig. 7.10.

Consequently, overall intake is regulated—as in rats (Adolph, 1947; Johnson, Ackroff, Peters, & Collier, 1986; Levitsky & Collier, 1968). Meal *duration* is not regulated, but because of our all-or-none eating assumption, the model has no control over the amount ingested in each bite (i.e., no control over within-meal eating rate) so cannot simultaneously regulate both meal duration and meal size.

Figure 7.5 shows the effect on the model of deprivation and interruption. The system begins with all V values set equal to zero—maximal deprivation. When eating is allowed to begin, the satiation signal increases almost linearly and eating occurs in a long initial meal (A in the figure), after which eating in the regular meal pattern resumes. A brief period when eating is interrupted (between the two vertical dashed lines) is followed by a single long meal (B) and resumption of the regular pattern. The effect of a relatively brief interruption is precisely the first-meal effect described earlier.

If the initial long meal in Fig. 7.5 is briefly interrupted, the meal resumes after the interruption and its total size remains the same—exactly duplicating the results of meal-interruption experiments by Seeley, Kaplan, & Grill

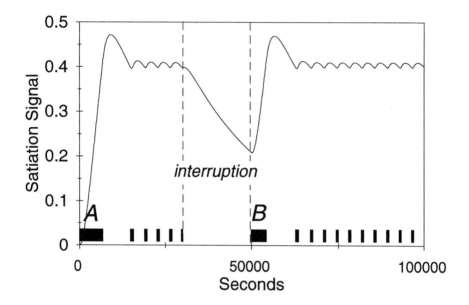

FIG. 7.5 Effects of deprivation and interruption on the eating pattern of the cascaded-integrator model. The system begins with all V values equal to 0 (infinite deprivation). The vertical dashed lines mark a brief period during which normal eating is interrupted. Blips at the bottom show eating episodes. Blocks A and B show extra-long meals after periods of deprivation. Parameters and other details as in Fig. 7.10.

(1993), who found that rats defend the size of their first meal following an intermediate (5 hours) period of food deprivation.

The CINT model regulates pretty well over the long term. After an interruption of feeding, postfast intake increases with deprivation up to a maximum after fasts of 24 hours (Le Magnen, 1985; Levitsky, Faust, & Glassman, 1976); longer fasts produce no further effect. Moreover, eating rate after an initial extra-large postfast meal very soon returns to normal. Figure 7.6 shows data from Levitsky et al. in which groups of rats were deprived of food for periods from 24 to 96 hours. The figure shows cumulative food intake (lines) for four groups of rats. The cumulative curves after the interruption show similar brief periods of acceleration and are thereafter approximately parallel, to each other and to the eating rate before the interruption. The symbols are predictions of the same 6-stage CINT model. The predictions are qualitatively the same as the data: Brief acceleration (similar after all four

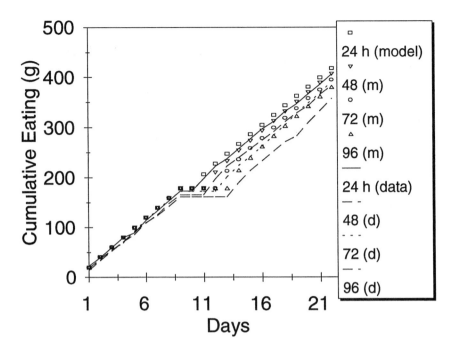

FIG. 7.6 The effects of 24- to 96-hour food deprivation on cumulative eating. Lines show group-average data from Levitsky, Faust, & Glassman (1976). Symbols are predictions of the 6-integrator CINT model with the parameter values shown in the legend to Fig. 7.10.

deprivation periods), followed by a return to the same slope as before. Note that the cumulative eating deficit caused by greater than 24 hours' deprivation is never made up, either in the data or in the model. The model is not a perfect fit to the data. It fits the 72-hour data well, the other data less well. The reason is that we used the same parameter values for all simulations (details in the legend to Fig. 7.10); a slight adjustment could have greatly improved the fits, but we wanted to see how well the model would do with no adjustments at all.

Figure 7.7 shows how overall eating rate is affected by the satiation value of each bite. When each bite is less than about 0.4, the system eats during every time step, so that eating rate is proportional to bite size. Above that point, eating rate is regulated at about 12,000 per 24 hours, with a slight tendency for the regulated level to increase with SV: Given a highly nutritious diet, an animal who behaved this way would slowly gain weight unless noningestive mechanisms act to prevent it (in practice, they seem not to: cf. Sclafani & Springer, 1976).

The CINT model also regulates in the face of imposed delays in between pellets (cf. Lucas & Timberlake, 1988). The heavy solid line in Fig. 7.8 shows

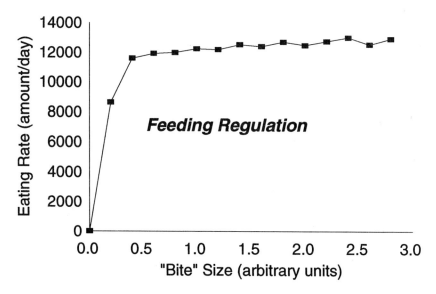

FIG. 7.7 Effect of bite size (or infusion rate, in a controlled-infusion rate procedure) on total ingestion rate (measured in satiation value units: $\sum_{t=0}^{T} x(t)$, where $x(t)$ is amount eaten, in SV units, and T = 24 hours) in the CINT model. Parameters as in Fig. 7.10.

the effect of increasing the time between permitted bites (interpellet interval, IPI) from 1 to 128, using the same parameters as we used for all the other simulations. Over a range of delays from 1 to 32, overall eating rate varies only about 10%; thereafter the decline is more substantial. The open squares are data from Lucas and Timberlake. The two curves show that data and model are very similar up to an IPI of about 64; thereafter, the model intake falls faster than the data.

The predictions of the CINT model are not very sensitive to the time constants of the integrator series. The qualitative predictions of the CINT model depend much more on its cascaded structure, and the delay imposed by the initial nonsatiating integrators, than on details of parameter values.

Foraging Behavior

There are many causes for behavior. The ethologist Niko Tinbergen (1963) discussed four, two of which are of interest in the present context: *final* causes and *proximal* causes. The final cause for behavior is its adaptive function: Wings are for flying, legs for walking, sexual behavior for reproduction, and so on. As a notable critic has pointed out, there is a *Just-So Stories* quality about some functional explanations. It is perhaps too easy to come up with a function for

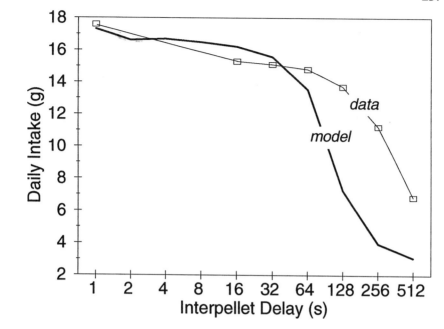

FIG. 7.8 Effect of imposing a minimum interpellet interval (IPI) on overall eating rate. Open squares: Data from Lucas & Timberlake (1988). Heavy solid line: Simulation of the 6-stage CINT model (data from last 6 days of 22 days at each IPI). Note the drop in intake at the 2 s IPI with approximately constant intake thereafter, until IPI > 32 s. Other details as in Fig. 7.10.

everything, even for things that are the result of developmental constraints or sheer accident.[12] Nevertheless, functional accounts can often be cast into precise, mathematical form using the techniques of optimization theory, and the result may be a powerful theoretical scheme that well satisfies the criterion of parsimony, explaining many things with few assumptions (see, for example, McNamara & Houston, 1986; Staddon, 1979). Yet, no matter how elegant these functional accounts, they necessarily lack a highly desirable feature: they provide no mechanism, no proximal cause, in Tinbergen's terms. In the case of feeding behavior, for example, we would like to know not just that the animal's pattern of feeding is in some sense optimal—because it minimizes effort, or maximizes safety in the presence of predators, or whatever—but what the process is. How does feeding actually work to satisfy these optimality criteria?

Notice that these two kinds of account are not in conflict. Feeding behavior is certainly the product of some kind of mechanism, a complex mechanism,

---

[12]There have been, and continue to be, extensive discussions in the biological literature of the legitimacy of functional explanations, with an emerging consensus in favor. See, for example, Borgia, 1994; Brandon, 1990; Gould & Lewontin, 1979; Maynard Smith, 1978; Queller, 1995.

perhaps, but a mechanism nevertheless. It is almost equally certain that feeding behavior is highly adaptive, it optimizes some Darwinian-fitness function, and the driving mechanism has evolved for this reason.

Bearing these considerations in mind, we conclude this chapter with a brief account of application of the CINT model to a sample experiment from the extensive and important work on feeding behavior of George Collier and his associates (see Staddon & Zanutto, 1997, for a complete account). Over a period of several decades, Collier has produced a series of careful, systematic laboratory studies of feeding behavior in rats under conditions that simulate natural foraging. These studies all tend to one conclusion: That rat eating behavior is close-to-optimal, highly tuned to the rat's ecological niche. For example, the rat tends to minimize time and effort expended in relation to food obtained. In a recent summary Collier and Johnson (in press) wrote "(1) Foraging costs are major determinants of [eating] bout patterns. Animals economize on foraging costs by initiating access to expensive resources less often and taking more on each occasion. . . . (2) Both time and effort are currencies of foraging cost. (3) The time window of foraging consequences is long" (p. 10).

For Collier, the key to feeding behavior is its adaptiveness. He argued that traditional methods of studying food-motivated behavior, using highly food-deprived animals in short experimental sessions (termed the *session paradigm* by Collier, and also called open economy experiments), yield misleading results; that the results of closed economy experiments (termed the *foraging paradigm* by Collier), where the rats are not explicitly deprived and live in the experimental apparatus 24 hours a day, more truly represent what animals do in nature. Most importantly for this chapter, Collier argued that "In the foraging paradigm the depletion: repletion hypothesis [i.e., homeostatic models of feeding] does not explain the bout patterns. For example, . . . when the price of access to the resource increases, animals initiate access less often and consume more on each occasion, conserving daily intake while restricting the increase in daily cost" (Collier & Johnson, in press b, p. 16).

Of course, as we have already pointed out, there is no contradiction between mechanism and optimization. Even if feeding behavior *is* adaptive, unless we are to believe in teleology, it still depends on *some* mechanism, and that mechanism must be fundamentally regulatory—homeostatic—because the behavior is regulated. When you perturb eating, behavior after the interruption is such as to reduce the effect of the interruption: Meal length and (up to a point) molar eating rate are, after all, conserved, as Collier pointed out. The issue, therefore, is not *whether* there is a homeostatic mechanism underlying these data, but *what* that mechanism is. More specifically, does it involve higher functions, like associative learning, or will something simpler do? In fact, it seems that the kind of nonassociative, regulatory process we have just described, the CINT model, can account for much of Collier's data on feeding.

Collier used procedures that differ in critical ways from standard operant conditioning studies: the animals are not food deprived (other than through limitations on their ability to get food imposed by the reinforcement schedule) and reinforcement is not a fixed amount of food, but a meal of indefinite duration. Collier's aim was to mimic natural foraging in which isolated meals (termed consumption) are separated by periods of searching (termed procurement) for the next meal. Experimental sessions are continuous throughout the 24-hour day, except for a brief daily maintenance period. Collier's basic procedure is diagrammed in Fig. 7.9. The idea is that the animal must make a large number of responses (usually lever-pressing, for rat subjects) for the opportunity to eat (this is termed procurement or *access* cost: P); but once

Collier Procedure

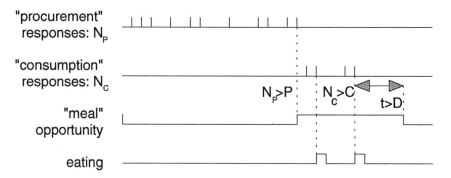

FIG. 7.9 Collier's procedure. The figure shows the procedure used in most of the single-choice, open-economy feeding experiments by Collier and his associates. Operant responses (usually, lever-pressing) are of two types: *procurement* or access responses, and *consumption* responses. Procurement responses (number = N_P) are made for the opportunity to engage in consumption responses (number = N_C). Consumption responses are made for the opportunity to eat. (This is called a *chain* schedule, in operant terminology.) The procedure works like this: After a meal (i.e., when the animal has ceased to make a consumption response [defined later] for experimenter-set time D—usually 10 min), further consumption responses are ineffective and the animal must make responses on the *procurement* or *access* lever to get further food. After he has made P procurement responses ($P = 11$ in the figure, but the usual requirement is much larger), a meal opportunity is offered. Now the animal must work on the *consumption* lever to get actual access to the food. After C consumption responses (C = 2 in the figure; usually C < P: Consumption cost is much less than procurement cost), a fixed amount of food is delivered. The animal can get additional food portions by paying the consumption cost again and this possibility remains available indefinitely—until the animal again ceases to make consumption responses for time D, at which point the procurement cost must be paid again and the cycle repeats. In some experiments, the only consumption response required is eating itself, detected by a photocell arrangement. See text for other details.

food is available, it remains available (at small, or no, *consumption* cost: C) until the animal ceases to make the consumption response for a substantial period (usually D = 10 min), at which point he must again pay the access cost.

If procurement cost is increased (and consumption cost is low), both meal frequency and meal size change. As we have seen, when procurement cost increases, meal frequency decreases, so that the animal needs to pay the cost less often. At the same time, meal size increases, so that the animal does not reduce its total food intake. Typical steady-state results are shown in Fig. 7.10

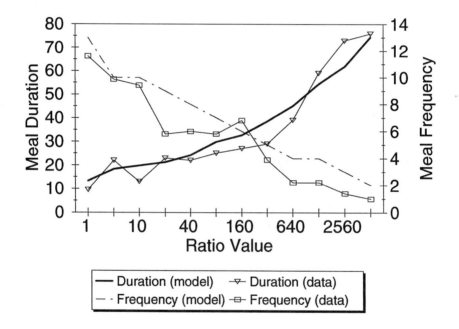

FIG. 7.10 The effect of increasing procurement cost (ratio value) in the Collier, Hirsch, & Hamlin (1972) experiment. *Data*: Open symbols: Mean meal duration (triangles) and meal frequency (squares). These are steady-state data from a single rat (the other two animals were similar) in a closed economy. *Simulation*: Lines: Predictions of a 6-stage CINT model. The parameter values (Equations 2–4) were determined as follows: $a_1 = \lambda_1$; $a_i = a_{i-1} + \lambda_2(1 - a_{i-1})$, $b_i = 1 - a_i$, that is, six stages, but only two free parameters. V_s was the sum of V-values for the last three stages. To deal with the fact that rats eat more in the 12-hour light phase of these experiments than in the dark phase (for the Collier simulations only), we set θ_{dark} as follows: $0.3\theta_{dark} = \theta_{light}$. Response probability was set by the equation $p = k[(\theta - V_s)/\theta] + (1 - k)$, with $k = 0.5$, which ensured a nonzero operant level of lever pressing of $p = k$. One time step = 1 s. To relate eating amount to Φ eating units/s (the output of the model) we assumed in all cases that 350 Φ units/time step = 1 g/s. To assure appropriate initial conditions, the model was run for 10 days under free-food conditions before imposing the first schedule constraint. For all the simulations the parameters were $\lambda_1 = 0.87$, $\lambda_2 = 0.93$, $\Phi = 2$ and $\theta_{light} = 0.4$.

(open squares and triangles) for an individual rat in a closed economy with food available according to the procedure diagrammed in the previous figure (Collier, Hirsch, & Hamlin, 1972, Fig. 7.5 & Fig. 7.7, Rat 1). Consumption cost was zero in this experiment. Food was made unavailable 10 minutes after the last eating episode, as measured by a photocell arrangement. The open squares show per diem frequency of meals (defined according to the 10-min criterion) at different procurement costs, from 1 (continuous reinforcement: CRF) through 5,120 lever-presses. The open triangles show average meal duration, in minutes.

Figure 7.11 shows how well the animal in Fig. 7.10 regulated its overall food intake. Overall ingestion rate (open squares) was approximately constant up to ratio values of 200 or so, and declined thereafter. The meal-frequency and meal-duration data in Fig. 7.10 show that this defense of overall eating rate is accomplished primarily by large increases in meal size. Thus, as Collier pointed out, the animal behaves in a wonderfully adaptive fashion: decreasing meal

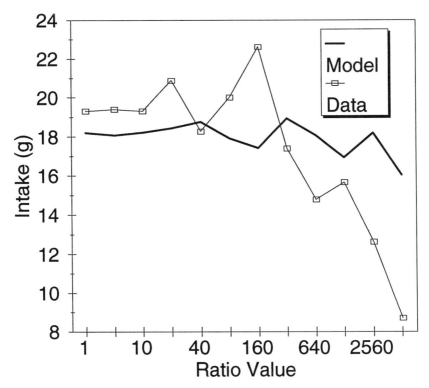

FIG. 7.11 Intake regulation in the Collier, Hirsch, & Hamlin (1972) experiment. *Open squares*: Daily food intake in g for the rat in Fig. 7.10. *Solid line*: Predictions of the CINT model. Other details as in Fig. 7.10.

frequency as procurement cost increases (so the cost need be paid less and less often), but compensating for the resulting drop in eating rate by enormous, largely costless, increases in meal size. Very similar effects have been found in cats and guinea pigs (see review in Collier, Hirsch, & Kanarek, 1977; Hirsch & Collier, 1974; Kanarek, 1975).

These results might be explained homeostatically in the following way. As procurement cost (fixed-ratio size) is increased over the very large range used in these experiments, the time between meals must also increase, because the rat does not increase its lever-press rate significantly at longer ratios. The question for the CINT model is whether this forced increase in intermeal interval is by itself sufficient to explain the effects of procurement cost. Mathis, Johnson, and Collier (1995) have shown that procurement time and procurement cost are roughly equivalent: "The functions relating meal frequency and size to the procurement interval were of the same shape as those seen when cost is the completion of the bar-press requirement" (p. 295). Thus, although the decline in meal frequency appears to be an adaptive, learned response to increased procurement cost, it may be largely an incidental by-product of the time taken to complete large procurement–cost ratios. The crucial question is: Can the CINT model (which is nonassociative, purely homeostatic, and takes no account at all of response cost) duplicate the effects of increased procurement cost?

Figures 7.10 and 7.11 show, along with the Collier et al. (1972) data, predictions[13] of a 6-stage CINT model, with the same parameters used earlier to fit the Levitsky et al. fasting data (Fig. 7.6). As you can see, the CINT model predicts the meal-duration and meal-frequency data reasonably well. The model regulates total intake better than the rat at high-ratio values, but otherwise model and data agree. Apparently the main effect of procurement cost on meal size can be explained by the same process that accounts for eating in meals and the first-meal effect of feeding interruption. The decrease in meal frequency with increasing procurement cost, and the maintenance throughout of close-to-control levels of daily ingestion, is also explained by the CINT model. This general pattern of agreement holds over a range of parameter values and across data from different animals, but the detailed discrepancies are slightly different. In current simulations we are finding that the right 3-stage model does about as well as the 6-stage model shown here, and with a considerably wider range of data. The point is that the data in Figs. 7.10 and 7.11 are perfectly compatible with the right homeostatic model.

---

[13]We had to make (relatively straighforward) assumptions to translate reflex strength into response rate. Staddon & Zanutto (in press) should be consulted for these details.

DISCUSSION

It is vain to do with more what can be done with less. William of Occam

People have always asked questions about nature. But science could develop only when they ceased to assume the answers. The history of psychology is just science in miniature. We know that we learn, that we have memories and expectations. We also know we have a brain, and we have faith that thought and behavior derive entirely from it. What could be more natural, therefore, than to look to brain and mind for explanations of behavior? Natural, yes, but not necessarily correct, for the reasons we stated at the beginning. The only defense against premature assumption is parsimony, to seek not just a theory that works, but the *simplest* theory that works.

The wish to begin with the simplest possible account led us to the CINT model for the regulation of feeding behavior. The model explains a great deal, including at least some simulated-foraging phenomena that are usually thought to depend on much more complex processes. The success of this model is a vindication of Occam's razor and underlines the importance of parsimony in psychological theorizing.

The core of the CINT model is not the chain of integrators or their number, or the equations linking parameters, or the particular parameters themselves. The essential model is just the biphasic impulse response shown in Fig. 7.1, which gives precise meaning to the axiom that the satiating effects of ingested food are delayed. There are many ways of generating an impulse response like this. With an integrator chain, at least two integrators are required; but beyond that, the number in the chain seems to be relatively unimportant. There are many other ways of generating biphasic functions, and we have no reason as yet to fix on one.

In advocating the parsimonious-mechanism approach to understanding behavior, we do not mean to put other approaches out of business—just to remove their exclusive franchise. The dominant approach at this time is behavioral neuroscience, which aims to explain behavior by direct reference to neural structure and function. This approach discounts the possibility that behavioral mechanisms may be much simpler than the underlying neural mechanisms—the argument from adaptive function that we described earlier. Yet, the feeding model, as well as an equally simple model for habituation and a related model for timing, discussed elsewhere (Machado, 1997; Staddon & Higa, 1996), support the functional argument. Considerable advances have been and continue to be made in behavioral neuroscience, particularly in areas where the underlying neural structure is circumscribed and the behavior not strongly historical—for example, sensory and motor processes, as opposed to learning. More has always been learned about the involvement of particular neural structures in particular behaviors or paradigms—about brain–behavior relations—than about the actual working of the behavior itself—environment–behavior relations: The *mechanism*, in the sense we have been using the word. No approach is without limitations, so it would be absurd to suggest that behavioral neuroscience or any other promising field be abandoned. We advocate a different approach because it is one that has been neglected, because it is yielding exciting results, and because it addresses questions of mechanism that are peripheral to current neurobiology. But science follows, and should follow, many paths; different approaches work for different problems. Our argument is for inclusion and a change of emphasis, not exclusivity.

Moreover, physiology is never irrelevant. Whatever our theory of behavior, it must be put to work by real physiology. But once we have a good behavioral theory, the ball is in the other court. The theory works: Now it is up to the physiologists to figure out how the brain functions in the way the model says it does. Perhaps future physiology will tell us just how the neurophysiology of feeding generates a biphasic satiation signal and how and where it acts.

Many questions about feeding dynamics and the relation between feeding dynamics and food-reinforced operant behavior, remain to be answered. For example, will the CINT model work for open economies? Collier has repeatedly argued that open and closed economies operate according to different principles: "[W]e must question the degree to which the results obtained in the classic open-economy, operant paradigm can be generalized to natural situations" (Collier & Johnson, in press b, p. 19). "We are suggesting that the processes motivating [animals in natural environments] are different from those operating in deprived animals" (Collier & Johnson, in press a, p. 21). This cannot be right at the level of mechanism. The animal in nature is the same as the animal in the operant box. Only the animal's history and environment are different. If we know the mechanism, we should be able to predict the behavior in any environment. The behavior may well be different under the two conditions, but the mechanism must be the same. It remains to be seen whether the CINT model can bridge the gap between open and closed economies.

Theoretical behaviorism must seem a dull and colorless thing when confronted with the full richness of human behavior. What on earth can it say, or be expected to say, about uniquely human qualities such as language and consciousness, or about raising a child or comforting the disturbed—about all those things that "real" psychologists love to pronounce upon? The answer obviously is not much. But these questions deal in part with a world of social construction that goes beyond psychology—any psychology—and perhaps beyond natural science. The wider world includes psychology and biology—but also history, philosophy, political economy, economics, sociology, religion, and law—any branch of knowledge or belief that deals with human culture. Psychology has no imperial role in these areas. It is not the privileged discipline when dealing with "man in context"—indeed, there *is* no privileged discipline. Culture is a tapestry woven from many strands and no single thread can represent the whole. We should not judge a brand of psychology by what it can do for problems that are not the exclusive province of psychology, but by what it can do with those problems that natural science can expect to solve. We are happy to see theoretical behaviorism judged by that standard.

ACKNOWLEDGMENTS

We thank George Collier for much help in supplying original data and helping us navigate through the large and confusing literature on feeding behavior. Research supported by grants from NIMH and NSF to Duke University.

REFERENCES

Adolph, E. F. (1947). Urges to eat and drink in rats. *American Journal of Physiology, 151,* 110–125.

Anderson, J. R. (1978). Arguments concerning representations for mental imagery. *Psychological Review, 85,* 249–277.

Bernard, C. (1865/1927). *An introduction to the study of experimental medicine.* (English translation). New York: Macmillan.

Boakes, R. (1984). *From Darwin to behaviourism: Psychology and the minds of animals.* Cambridge, England: Cambridge University Press.

Bolles, R. C. (1980). Some functionalistic thoughts about regulation. In F. M. Toates & T. R. Halliday (Eds.), *Analysis of motivational processes.* London: Academic Press.

Booth, D. A. (Ed.). (1978). *Hunger models.* London: Academic Press.

Borgia, G. (1994). The scandals of San Marco. A review of *Understanding scientific prose.* J. Selzer (Ed.). Madison, WI: University of Wisconsin Press, 1993. *Quarterly Review of Biology, 69,* 373–375.

Brandon, R. (1990). *Adaptation and environment.* Princeton, NJ: Princeton University Press.

Churchland, P. S., & Sejnowski, T. J. (1992). *The computational brain.* Cambridge, MA: MIT Press.

Collier, G. (1986) The dialogue between the house economist and the resident physiologist. *Nutrition and Behavior, 3,* 9–16.

Collier, G., Hirsch, E., & Hamlin, P. (1972). The ecological determinants of reinforcement in the rat. *Physiology and Behavior, 9,* 705–716.

Collier, G., Hirsch, E., & Kanarek, R. (1977). The operant revisited. In W. K. Honig & J. E. R. Staddon (Eds.), *Handbook of operant behavior* (pp. 28–52). Englewood Cliffs, NJ: Prentice-Hall.

Collier, G., & Johnson, D. (in press, a). Motivation as a function of animal vs. experimenter control. In M. E. Bouton & M. S. Fanselow (Eds.), *The functional behaviorism of Robert C. Bolles: Learning, motivation and cognition.* Washington, DC: American Psychological Association.

Collier, G., & Johnson, D. (in press, b). Laboratory simulations of foraging. In G. Greenberg & M. M. Harway (Eds.), *Encyclopedia of comparative psychology*. New York: Garland.

Commons, M. L., Grossberg, S., & Staddon, J. E. R. (Eds.). (1991). *Neural networks of conditioning and action*. Hillsdale, NJ: Lawrence Erlbaum Associates.

Davis, J. D. (1989). The microstructure of ingestive behavior. In Feeding disorders. *Annals of the New York Academy of Sciences, 575*, 106–121.

Dennett, D. C. (1978). Skinner skinned. In *Brainstorms* (pp. 53–70). Montgomery, VT: Bradford.

Gould, S. J., & Lewontin, R. C. (1979). The spandrels of San Marco and the Panglossian paradigm: A critique of the adaptationist programme. *Proceedings of the Royal Society of London. Ser. B, 205*, 581–598.

Grossberg, S. (1982). *Studies of mind and brain*. Boston: Reidel.

Guillot, A., & Meyer, J.-A. (1987). A test of the Booth energy flow model (Mark 3) on feeding patterns of mice. *Appetite, 8*, 67–78.

Hirsch, E., & Collier, G. (1974). The ecological determinants of reinforcement in the guinea pig. *Physiology and Behavior, 12*, 239–249.

Johnson, D. F., Ackroff, K., Peters, J., & Collier, G. H. (1986). Changes in rats' meal patterns as a function of the caloric density of the diet. *Physiology and Behavior, 36*, 929–936.

Johnson-Laird, P. N. (1988). *The computer and the mind*. Cambridge, MA: Harvard University Press.

Kanarek, R. (1975). Availability and caloric density of the diet as determinants of meal patterns in cats. *Physiology and Behavior, 15*, 611–618.

Klopf, A. H. (1988). A neuronal model of classical conditioning. *Psychobiology, 16*, 85–125.

Le Magnen, J. (1985). *Hunger*. Cambridge, England: Cambridge University Press.

Levitsky, D. A., & Collier, G. (1968). Effects of diet and deprivation on meal eating behavior in rats. *Physiology and Behavior, 3*, 137–140.

Levitsky, D. A., Faust, I., & Glassman, M. (1976). The ingestion of food and the recovery of body weight following fasting in the naive rat. *Physiology and Behavior, 17*, 575–580.

Lucas, G. A., & Timberlake, W. (1988). Interpellet delay and meal patterns in the rat. *Physiology and Behavior, 43*, 259–264.

Machado, A. (1997). Learning the temporal dynamics of behavior. *Psychological Review, 104*, 241–265.

Mathis, C. E., Johnson, D. F., & Collier, G. H. (1995). Procurement time as a determinant of meal frequency and meal duration. *Journal of the Experimental Analysis of Behavior, 63*, 295–311.

Maynard Smith, J. (1978). Optimization theory in evolution. *Annual Review of Ecology and Systematics, 9*, 31–56.

McFarland, D. J. (Ed.). (1974). *Motivational control systems analysis*. London: Academic Press.

McNamara, J. M., & Houston, A. I. (1986). The common currency for behavioral decisions. *American Naturalist, 127*, 358–378.

McSweeney, F. K., Hinson, J. M., & Cannon, C. B. (in press). Sensitization–habituation may occur during operant conditioning. *Psychological Bulletin*.

McSweeney, F. K., Weatherly, J. N., & Swindell, S. (1996). Within-session changes in responding during variable interval schedules. *Behavioural Processes, 36*, 67–76.

Mook, D. G. (1987). *Motivation: The organization of action*. New York: Norton.

Morgan, C. L. (1894). *An introduction to comparative psychology.* London: Scott.

Panksepp, J. (1974). Hypothalamic regulation of energy balance and feeding behavior. *Federation Proceedings, 33,* 1150–1165.

Queller, D. C. (1995). The spaniels of St. Marx and the Panglossian paradox: A critique of a rhetorical programme. *Quarterly Review of Biology, 70,* 485–489.

Schilstra, A. J. (1978). Simulations of feeding behavior: Comparison of deterministic and stochastic models incorporating a minimum of presuppositions. In D. A. Booth (Ed.), *Hunger models: Computable theory of feeding control* (pp. 167–194). London: Academic Press.

Schmajuk, N. A., & DiCarlo, J. J. (1991). Neural dynamics of hippocampal modulation of classical conditioning. In M. L. Commons, S. Grossberg, & J. E. R. Staddon (Eds.), *Neural networks of conditioning and action* (pp. 149–180). Hillsdale, NJ: Lawrence Erlbaum Associates.

Sclafani, A., & Springer, D. (1976). Dietary obesity in adult rats: Similarities to hypothalamic and human obesity syndromes. *Physiology and Behavior, 17,* 461–471.

Seeley, R. J., Kaplan, J. M., & Grill, H. J. (1993). Effects of interrupting an intraoral meal on meal size and meal duration in rats. *Appetite, 20,* 13–20.

Smith, L. D. (1986). *Behaviorism and logical positivism.* Stanford University Press.

Staddon, J. E. R. (1979). Operant behavior as adaptation to constraint. *Journal of Experimental Psychology: General, 108,* 48–67.

Staddon, J. E. R. (1988). The functional properties of feeding, or why we still need the black box. *Appetite, 11,* 54–61.

Staddon, J. E. R. (1993a). The conventional wisdom of behavior analysis. *Journal of the Experimental Analysis of Behavior, 60,* 439–447.

Staddon, J. E. R. (1993b). *Behaviorism: Mind, mechanism and society.* London: Duckworth.

Staddon, J. E. R. (1993c). On rate-sensitive habituation. *Adaptive Behavior, 1,* 421–436.

Staddon, J. E. R. (1997). Why behaviorism needs internal states. In L. J. Hayes & P. M. Ghezzi (Eds.), *Investigations in behavioral epistemology* (pp. 107–119). Reno, NV: Context Press.

Staddon, J. E. R., & Higa, J. J. (1996). Multiple time scales in simple habituation. *Psychological Review, 103,* 720–733.

Staddon, J. E. R., & Zanutto, B. S. (1997). Feeding dynamics: Why rats eat in meals and what this means for foraging and feeding regulation. In M. E. Bouton & M. S. Fanselow (Eds.), *The functional behaviorism of Robert C. Bolles: Learning, motivation and cognition* (pp. 131–162). Washington, DC: American Psychological Association.

Sutton, R. S., & Barto, A. G. (1981). Toward a modern theory of adaptive networks: Expectation and prediction. *Psychological Review, 88,* 135–170.

Thomas, R. K. (in press). Lloyd Morgan's canon. In G. Greenberg and M. Haraway (Eds.), *Comparative psychology: A handbook.* New York: Garland.

Tinbergen, N. (1963). On aims and methods of ethology. *Zeitschrift für Tierpsychologie, 20,* 410–433.

Toates, F. M., & Halliday, T. R. (Eds.). (1980). *Analysis of motivational processes.* London: Academic Press.

Toates, F. M., & Rowland, N. (Eds.). (1987). *Feeding and drinking.* Amsterdam: Elsevier Science Publishers (Biomedical).

8

A Minimal Model
of Transitive Inference

C. D. L. Wynne
University of Western Australia

This chapter concerns problems of a general type exemplified by this syllogism:

Mighty Joe Young is mightier than King Kong
King Kong is mightier than Magilla Gorilla
Who is mightiest? (Sternberg, 1980)

This riddle illustrates a transitive relation. A relation is transitive if, given that it holds between items A and B and between items B and C, then it necessarily also holds between items A and C. Examples are bigger than, smaller than, and darker than. Thus, a transitive relation implies an underlying series, and so problems of this type are also known as n-term series problems (where n stands for the number of items, in this case three). Problems of this type have interested mankind, at least since Aristotle, and were incorporated into modern psychology around the beginning of this century (Burt, 1911). What has fascinated psychologists about the transitive inference problem is that it appears to demand that a subject go beyond the information given to deduce a relationship between items that have not been presented together. In the above example, the subject needs to deduce the relationship between Mighty Joe Young and Magilla Gorilla from the relationships of Mighty Joe Young with King Kong, and King Kong with Magilla Gorilla.

The study of transitive inference formation gains importance from the generality of processes of ordering information. Such processes are not only

269

omnipresent in human behavior, but must also often occur in the lives of other species. Social animals, for example, would gain adaptive fitness from being able to judge their ranking in a group from interaction with just some of their fellow group members, rather than having to enter into possibly dangerous contests with all members of the group. Even the ranking of food preferences would be greatly simplified by the ability to draw transitive inferences. Transitivity is implied in all orderings.

The fact that a transitive relation implies an underlying series has led cognitive psychologists to conclude that accurate performance on an n-term series task means that the subject has formed an ordered representation of the items involved.

> The transitive inference process . . . depends primarily on organizing the premise elements as an ordered set. The template for this ordering comes from a common ordering schema, the left-right or top-bottom arrangement. . . . Understanding a transitive inference problem entails constructing a mental model of it in this way. (Halford, 1993, p. 11)

The aim of this chapter is to show how a simple, rather old-fashioned conditioning model can account for a set of results from a complex, cognitive task. This conditioning model has no arrangement of the stimuli, and no mental model at all beyond a simple assignment of reinforcement values to the items presented. It will be seen how this simple model accounts for most of the substantiated effects in the results from a variety of studies on transitive inference in humans and other species.

COMPLEXITY AND SIMPLICITY IN MODELS

In looking at complex behavior it is quite natural to consider the need for complex models, and, as other chapters in this volume show, over the last two decades there has been an enormous growth in novel computational theories of complex behavior. Nonetheless, there are advantages to an historically conservative approach. One is that we can build on the experience of other researchers who have used these models in other contexts. We can know that we are using a model that is consistent with some range of data other than those we are considering at present. Historically conservative models are more likely to be general models of simple processes that underlie a wide range of performances, rather than specialized local models designed solely for one task and inappropriate in any other context.

An old adage in comparative psychology is Lloyd Morgan's Canon: "In no case may we interpret an action as the outcome of the exercise of a higher psychical faculty, if it can be interpreted as the outcome of one which stands lower in the psychological scale" (Morgan, 1894, p. 53). The metatheoretical expectation that theories of science be as parsimonious as possible has a long history and much to commend it. However, to follow Morgan's Canon, we need to be to be able to tell the difference between higher psychical faculty and abilities lower on the psychological scale. To use a more modern terminology, we need to be able to distinguish a cognitive from a noncognitive explanation of behavior.

Those depositories of received wisdom, the introductory textbooks, are surprisingly little help in finding useful definitions of cognition. Anderson (1975), for example, defined cognitive psychology simply as the successor title to experimental psychology for the areas of study known separately as perception, learning and thinking. Neisser (1967) offered an even broader definition: "As used here the term "cognition" refers to all the processes by which the sensory input is transformed, reduced, elaborated, stored, recovered, and used . . . it is apparent that cognition is involved in everything a human being might possibly do; that every psychological phenomenon is a cognitive phenomenon" (p. 4). This helps us little to distinguish cognitive from noncognitive.

It seems sensible to consider the complexity of a representation before awarding it the title *cognitive*. Ordering stimuli along a single dimension of, say, hedonic valance, is too simple a representation to be termed cognitive. Similarly, associative strengths are examples of representations that could not be regarded as cognitive in any useful sense. The term cognitive is reserved here for models that utilize complex representations in which various stimuli stand in some defined relation to each other. A cognitive model of the three-term syllogism with which this chapter started might state that the three individuals are stacked on top of each other. Mighty Joe Young is represented as being on top of King Kong, and King Kong is on top of Magilla Gorilla. Such a model would also need to stipulate how these three characters came to be stacked in this way, and how a subject uses this vertically structured representation to answer the question "Who is mightiest?"

THE n-TERM SERIES AND TRANSITIVE INFERENCE

When a transitive inference problem is presented as an *n*-term series task, a subject is usually presented with *n* items in *n-1* pairs. Thus, a subject may be told that Sam is taller than Len; Len is taller than Don; and Don is taller than Tom (Banks & White, 1982). These may be referred to as neighboring items on an underlying series (Sam > Len > Don > Tom). Given training on the above three pairings, a transitive subject should be able to conclude that Sam is taller than Don or that Len is taller than Tom.

This form of transitive inference was introduced into psychology by Burt (1911) as part of his development of tests of general intelligence. Research on processes of seriation and transitivity per se was stimulated by Jean Piaget when he proposed that performance on these tasks was indicative of a relatively advanced stage of cognitive development (the concrete operational stage, reached around age 8 in the Genevese; Piaget, 1928). Research on less verbal versions of the task has demonstrated transitive choice in children as young as 4 years (Bryant & Trabasso, 1971; Trabasso & Riley, 1975), and by a variety of other species (pigeons—Fersen, Wynne, Delius, & Staddon, 1991; Steirn, Weaver, & Zentall, 1995; Wynne, 1997, rats—Davis, 1991, 1992, chimpanzees—Boysen, Berntson, Shreyer, & Quigley, 1993; Gillan 1981, Squirrel Monkeys—McGonigle & Chalmers, 1977, 1985, 1992).

It was the successes of these obviously nonverbal species that prompted the development of the present model. Cognitive explanations in terms of verbal and spatial representations seemed improbable explanations of the behavior of pigeons or rats. On the other hand, the training procedures used were simply concurrent operant discriminations (situations where two stimuli are offered, a subject is rewarded if it chose the correct one, and non-rewarded if it chose incorrectly) and therefore it seemed appropriate to consider the effectiveness of traditional accounts of operant learning for the n-term series task.

TASK PRESENTATION FORMS

Over the years, a number of ways of presenting the transitive inference task have been developed. The earliest studies on transitive reasoning (Burt, 1911, 1919; Piaget, 1928) offered subjects written or verbal syllogisms like the opening example:

Mighty Joe Young is mightier than King Kong
King Kong is mightier than Magilla Gorilla
Who is mightiest?

We shall call this the fully verbal version of the task, because it relies entirely on human language for presentation of the task to the subject. Both the items and the relationship between them are presented as words, either spoken or written. The premises may be presented all together in a block of text, singly in separate written sentences, or read aloud to the subjects. It is more-or-less axiomatic that adult humans succeed on this task, though, as we shall see, there exist training conditions that can degrade their levels of transitive choice.

Bryant and Trabasso (1971) argued that young children's failure on the fully verbal version of the transitive inference task was due to the verbal complexity of the standard method of presentation, combined with the substantial memory load imposed. They developed a semi-verbal version of the task. Subjects were presented with pairs of different colored sticks of differing lengths. Thus, on some trials the subjects might be shown two sticks, say red and green and told that red was longer than green. (In some conditions the subjects were shown the actual lengths of the sticks—we ignore these conditions here because successful choice of the longer stick could follow simply from remembering the lengths of all the sticks without any need for inference formation.) On other trials the subjects might be shown green and yellow sticks together and told that green was longer than yellow. This continued until the subject had learned the relations between five sticks whose lengths formed a series. Five sticks were used, rather than the three items standard in the fully verbal version of the task, because subjects could choose the first and last sticks just because they were always longer or shorter, respectively, than the sticks with which they had been presented in training. In tests, the second and fourth sticks were presented together, and subjects were asked which was longer (or shorter).

Assuming that the different training pairs had been presented equally often, then the second stick had been a longer stick half the time, and a shorter stick half the time; and similarly for the fourth stick. Thus, choice of the second stick as the longer of the two is evidence of transitive inference formation. We term this a semiverbal form of transitive inference because the items are no longer presented verbally (they are real-life sticks, not named items), but the relation between the items is still presented verbally (the subjects are told that one stick is longer than the other—a relation they cannot see directly). Bryant and Trabasso (1971) found successful choice of the second over the fourth item, and thus transitivity, in children as young as 4 years (the youngest they tested). As we shall see, not all training procedures are equally likely to lead to successful choice in this paradigm, in children or adults.

Bryant and Trabasso's (1971) semiverbal version of transitive inference inspired McGonigle and Chalmers (1977) to develop a nonverbal version of the task suitable for use with nonhuman species (in their case, squirrel monkeys). The subjects were presented pairs of tobacco cans, each of which had been painted a different color. To provide additional feedback of correct or incorrect choice, the cans were provided either filled (*heavy*) or empty (*light*). For one group of animals the rewarded stimulus was always heavy (the incorrect choice was light), for the other group the relationship between weight and reward was reversed. However, in both cases, only two different weights were used for the whole series, and so the cans could not be ordered by weight. The difference between this version of the task and the semiverbal version is that the nonverbal subjects are not asked which object is longer (or heavier). Rather they are left to choose without verbal instruction; a correct choice is rewarded (in the squirrel monkey's case by discovery of a peanut under the chosen can); an incorrect choice is nonrewarded or punished (the squirrel monkeys were simply required to choose again if they made an error—a correction procedure). The pattern of reward and punishment is programmed so that an underlying series can be inferred.

On a nonverbal version of the n-term series task, the n terms are presented in n-1 neighboring pairs, just as for the verbal and semiverbal paradigms described above. Thus Davis (1992) for example, gave rats choices between pairs of odors. If we label the odors A through E (yet explicitly denying any natural order to the odors used), then we can say that on some trials the rats had to choose between odors A and B. Choice of A was rewarded, choice of B punished. On other trials, odors B and C were presented together, but now choice of B was rewarded, and C punished. On $C\,D$ trials, odor C was rewarded and odor D punished; and on $D\,E$ trials odor D was rewarded and E punished. Using + to denote reward, and - to denote punishment, we may summarize this training as:

A+ B-
B+ C-
C+ D-
D+ E-

(Davis' rats were also trained on A+ C-, C+ E-, and A+ E-.) In tests for transitivity, the rats were presented odors B and D together—two stimuli they had never seen together before in the same pair, and which had previously been approximately equally often rewarded and punished. The rats' choice of odor B here is equivalent to Bryant and Trabasso's children's choice of the second stick on the semiverbal paradigm, and to the transitive choices made in studies using a verbal version of the task.

THE STANDARD EFFECTS

Aside from inferential choice, certain results have been found across species and in all three forms of task presentation, and therefore form a basic set of phenomena for which any model must account.

End Anchor Effect

One of the earliest findings in this research was that subjects perform better (respond with higher levels of accuracy and/or shorter reaction times) on pairs that contain one or both of the items at the beginning and end of the series. This is not surprising since the first and last items in the series have been uniquely rewarded or punished (in the nonverbal version), or labelled bigger or smaller (in the verbal and semiverbal versions).

Serial Position Effect

A less intuitive finding was that performance on a training pair depends on the pair's position in the implied series. Pairs toward the middle of the series are less well solved than those towards the ends of the series. On a 5-term series, this effect will be confounded with the end–anchor effect—the pairs at the ends of the series contain the end-anchor items ($A+$ B-, and $D+$ E-), whereas the pairs in the middle of the series do not ($B+$ C- and C+ D-). Studies with longer series have demonstrated, however, that the serial position effect exists independently of the end–anchor effect (Fersen et al., 1991; Moyer & Bayer, 1976; Trabasso, Riley, & Wilson, 1975; Woocher, Glass, & Holyoak, 1978).

Symbolic Distance Effect

A surprising result is the finding that pairs are better solved (i.e., higher accuracy; shorter reaction time) if the items being presented together are further apart in the implied series (Moyer & Bayer, 1976). Thus, subjects typically solve the test pair AE, better than the BD, or CD pairs. This effect can also be confounded with the end–anchor effect, because the further apart two items are in the implied series, the more likely it is that one or the other of them is an end item. On a 5-term series, only the comparison of the BD test pair with the BC and CD pairs permits consideration of the symbolic distance effect without contamination from end–anchor effects. However, longer series have been trained so that more test pairs could be considered that do not contain end items. Symbolic distance effects free of any influence from the end items have been demonstrated after training 7-term or longer series by Fersen et al. (1991) with pigeons; Trabasso, Riley, and Wilson (1975) with children and adults on a semiverbal version of the task; and Moyer & Bayer (1976) and Woocher, Glass, and Holyoak (1978) with adults on the verbal task.

Effect of First Item/Congruity Effect

In verbal and semiverbal studies, it is found that pairs are better solved if the form of the question agrees with the items being tested. For example, verbal subjects may answer faster, or with greater accuracy, when asked which of two items is larger if both items are at the large end of a series, than if the items are closer to the small end of the series. A possibly analogous effect was found by Fersen et al. (1991) when testing pigeons. These authors found that the pigeons' response accuracies were highest for test pairs containing stimuli closest to the always-rewarded end of the series. Thus, on a 7-term series, pair *BD* was more accurately solved than pair *CE*, despite the fact that the items in both pairs were the same distance apart in the implied series (so no symbolic distance effect would be expected), and neither pair contained an end item.

PAST ATTEMPTS TO MODEL TRANSITIVE INFERENCE

An early attempt was made to explain transitive inference as a form of *associative chaining* by Anderson and Bower (1973). According to theories of this type, subjects integrate (chain together) the information from the neighboring pairs in order to solve test problems in which nonneighboring items are presented together. All chaining accounts predict that the farther apart two items are in the underlying series, the more training pair information must be chained together for solution, and therefore the slower and less accurate responding will be. This is exactly the opposite of the symbolic distance effect typically observed.

McGonigle and Chalmers (1977) proposed that when a subject is confronted with an unfamiliar stimulus pair (such as the *B D* test pair) it proceeds in three phases. The subject:

1. Recognizes the stimulus (or stimuli) missing in the implied series (in this case stimulus C).
2. Considers the possible pairwise combinations of these stimuli (here *B C, C D*, and *B D*).
3. Chooses one of these pairs at random and makes the appropriate response.

If either of the pairs familiar from training is chosen, then a correct response is made: If the subject chooses the *B D* pair then it selects a stimulus at random. This *binary sampling* model suffers the same problem as the *associative chaining* model in that it predicts the reverse of the symbolic distance effect. The further apart in the series the two stimuli in a test pair are, the more stimuli would have to be filled in to complete the series, the more potential pairs there would be to choose from, and an ever higher proportion of those pairs would not be training pairs and would therefore lead to random choice. Thus, increasing the distance in the series between test stimuli can only lead to more errors, the opposite of what is observed.

A third unsuccessful theoretical approach was proposed by Humphreys (1975). He pointed out that some researchers, when training a 4-term series verbally or semiverbally, had not ensured that each item had been labelled equally often with each verbal label (e.g., bigger or smaller). Some authors (e.g., Potts, 1972, 1974), in addition to training the three neighboring pairs from a 4-term series, also trained the nonneighboring pairs, *AC*, *BD*, and *AD*. With this pattern of training, item *A* was bigger than other items in three premises; item *B* was bigger than other items in two premises, and item C was bigger than another item in just one premise. Humphreys pointed out that this frequency of "bigger than" information was sufficient to ensure accurate performance on all possible pairs. However, in subsequent studies (e.g., Potts, 1977), it was shown that successful transitive inference performance can be observed even when the frequency of use of the different verbal labels is properly balanced.

The most popular models for explaining transitive inference postulate some kind of ordered mental representation. DeSoto, London, and Handel (1965) and Huttenlocher (1968) argued that performance on the transitive inference task was governed by a spatial representation. Trabasso (1975, 1977) and Trabasso and Riley (1975) proposed that the subject builds a mental representation characterized as a linear ordering of the premise terms. First, the end items are identified and placed at opposite ends of a left-to-right or top-to-bottom mental line. Then, the terms that are presented with the end items in training are placed next to them (items *B* and *D*); finally (on the 5-term series), item *C* is placed in the middle. Once constructed, all test and training pairs are solved by scanning along the series representation. A number of models differing in detail share this basic structure (Breslow, 1981; Trabasso & Riley, 1975).

The end-inward construction of the mental line provides an explanation for the end anchor and serial position effects. End items have a special status because they are the only items uniquely labelled, and they are placed on the mental line first, ensuring that performance on pairs containing them is better than on all other pairs. The serial position effect occurs because the farther into the center of the series a training pair is, the later it will be added into the mental line, and thus the less good the performance will be. The symbolic distance effect arises because the farther apart two items are in the series, the faster one or the other of them will be found in a search along the mental line from the ends. A congruency effect can be predicted if it is assumed that the search along the mental line is biased by the form of the question posed. Thus the question *Who is taller?* biases the subject to start searching the representation from the tall end, whereas the *Who is shorter?* question biases the search to start from the short end of the representation.

Clark (1969), Banks (1977), and Trabasso (1975) among others, developed models of (verbal) transitive inference in terms of a linguistic or semantic representations. The key feature of such models was that the items are coded as degrees of intensity of the verbal label used to compare them. Thus, given "red is bigger than green" for example, subjects will encode red as being *big+*, and green as *big*. For "blue is smaller than yellow", subjects encode as blue as *small+*, and yellow as *small*. Various operations were proposed for the conversion of small and big terms to a common currency, the recognition of items present in more than statement, and the solution of test questions.

There are certainly aspects of the solution of verbal transitive inference problems that depend on the form of words used; thus, the recognition of a role for verbal processing seems necessary. The congruency effect was already mentioned and it has been found that subjects respond faster to marked comparative terms—such as "bigger" and "better," than to unmarked ones ("smaller," "less good"). Sternberg (1980) combined the most effective components of prior linguistic and spatial models into a mixed linguistic–spatial model. As in a linguistic model, the premises are resolved into linguistically coded information about the items. For example, "*A* is taller than *B*" becomes "*A* is *tall+*" and "*B* is *tall*." This information is then used to create a spatial array such as *A–B*. On the assumption that processing is strictly serial, this model can account for a large proportion of the variance in reaction times in data from fully verbal versions of transitive inference tasks involving just three items.

This brief summary of extant models of transitive inference performance makes clear that only the models assuming some kind of ordered representation of the series have proven capable of predicting the standard effects observed in the literature. There are several problems associated with models of this type, however. One problem is their ontological status. How is a *mental line* to be conceived? Where do such structures come from? How are they generated and maintained? A second problem with these models is that they are not fully developed dynamic theories. As we shall see, the particular pattern of training used can have a large effect on the results obtained, and yet mental line models do not well account for this problem. Third, these models do not fulfil the criterion of parsimony being followed here—they have a complexity that, as we shall see, is quite unnecessary to successful task solution.

THE MODEL

This model was developed for animal studies where reward and punishment were used to train the subjects. Thus, it was natural to use models in which the stimuli gained and lost reinforcement value. However, nothing in these models depends on the assumption that the values are reinforcement values; they can be considered as whatever the relation is that is being trained in a particular transitive inference task. In our opening example "Mighty Joe Young is mightier than King Kong; King Kong is mightier than Magilla Gorilla; Who is mightiest?" the values could represent the *mightiness* of the three protagonists. We will start by looking at the nonverbal tasks for which this model was developed, and then consider how well it accounts for data from semiverbal and fully verbal versions of the transitive inference task.

The first suggestion that transitive inference performance could be accounted for using familiar learning rules was made by Couvillon and Bitterman (1992). These authors proposed a model that used the associative learning rule of Bush and Mosteller (1955) to update stimulus values. The increment in value of a stimulus (ΔV) is given by

$$\Delta V = Up\beta(1 - V) \qquad \text{on reward} \quad (1a)$$

and

$$\Delta V = -Dn\beta V \qquad \text{on nonreward,} \quad (1b)$$

where $Up\beta$ and $Dn\beta$ are rate parameters controlling how rapidly value is gained or lost after reward and nonreward, respectively. These values were then combined to produce a response probability for each trial using a choice rule proposed by Couvillon & Bitterman (1986):

$$r = V(A)/(V(A) + V(B)) \quad (2a)$$

$$p(A|AB) = 0.5 + s(2r - 1)^K \text{ (for } r \geq 0.5) \quad (2b)$$

$$p(A|AB) = 0.5 - s(1 - 2r)^K \text{ (for } r < 0.5) \quad (2c)$$

where r is an intermediate variable, $V(A)$ is the value of stimulus A, $p(A|AB)$ represents the probability of choosing stimulus A on an AB trial and s and K are scaling parameters. It was shown that this model with four free parameters ($Up\beta$ and $Dn\beta$, s and K), when trained on 1,000 trials in random order could fit the mean data from a group of pigeons studied by Fersen et al. (1991) quite closely. Fersen et al. trained eight pigeons on a 5-term series using the four training pairs outlined above. Brief access to food grains served as reward, after an incorrect response the trial was repeated until the correct response was made. The stimuli were white squiggles on a black background, presented on a horizontal work surface.

Figure 8.1 gives an intuitive sense of how a model of this kind works. Consider the five items, A through E trained in four overlapping pairs, $A+ B-$ through $D+ E-$, where a + indicates the rewarded (or bigger, longer, mightier) item, and - indicates the nonrewarded (or smaller, shorter, less mighty) item. Every time the pair $A+ B-$ is presented and item A is chosen its reward value (or bigness) is increased by a small amount: Every choice of B leads to a small decrement in that item's reinforcement value (or bigness). Choice on these trials depends on the difference in value between the two stimuli A and B.

After a certain number of trials, the difference in values will be sufficient that no further errors will be made, and thus all subsequent choices are of item *A*. Notice how the value of the rewarded stimulus, *A*, continues growing in a negatively accelerated fashion long after the value of the punished stimulus, *B*, stops falling—this is because, after a time, no further responses are made to the nonrewarded stimulus.

Now we switch to training *B*+ *C*-. At first the subject will make erroneous responses to *C*, until this stimulus' value falls below that of stimulus *B*. Now the value of stimulus *B* will gradually rise as the value of *C* gradually falls further. After a time, the difference in value between the two stimuli is such that no further erroneous *C* responses are made. Now *C*'s value stabilizes, and the value of *B* continues to grow. With certain assumptions about the detailed form of the model and suitable parameter values, we find that the value of stimulus *B* after *B*+ *C*- training is lower than the terminal value of stimulus *A*. This is because the value of stimulus *B* was depressed at the beginning of *B*+ *C*- training. Note also in Fig. 8.1 that stimulus *C* now has less value than stimulus *B* did at the end of *A*+ *B*-training. This fact ensures that when we train *C*+ *D*-, stimulus *C* will not reach the level of value that stimulus *B* did on *B*+ *C*-training (and that was lower than the value of stimulus *A* on *A*+ *B*-training). In this way, once all four neighboring pairs have been trained, the stimulus values are ranked so that *A* has more value than *B*; *B* has more than *C*; *C* more than *D*; and *D* more than *E* (*E* is never rewarded, so this last difference is the largest of the four). Now, whatever combination of stimuli we present (including the critical *BD* test pair), the subject will always choose the correct one, since the stimulus values are ranked in the order of the implied series.

Wynne, Fersen, and Staddon (1992) demonstrated that several different reinforcement learning models could predict transitive choice after training

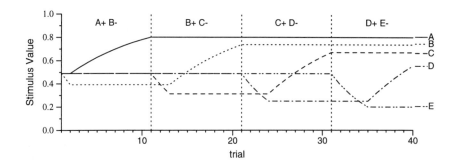

FIG. 8.1. Development of stimulus values across the 5-term series. Pair *A*+*B*- is presented for the first ten trials, then *B*+*C*-, *C*+*D*- and finally *D*+*E*-. Based on Equations 8.1 and 8.3 with Upβ = 0.1 and Dnβ = 0.2.

on a random sequence of 1,000 trials. These include a Bush–Mosteller model similar to that of Couvillon and Bitterman but with only three parameters. The update rule used was the same, but the probability of choice of a stimulus was given simply by Equation 8.2a, omitting Equations 8.2b and 8.2c and thereby reducing the number of free parameters by two:

$$p(A|AB) = V(A) / (V(A) + V(B)) \quad (3)$$

Wynne et al. (1992) also showed that models based on update rules proposed by Luce (1959) and Horner and Staddon (1987) could achieve transitive choice after this training series.

Because these are trial-by-trial dynamic models, the fact that they can predict transitive choice after training on a random series of 1,000 trials, does not necessarily mean that they can predict the results of Fersen et al. (1991), after the series of training trials that those subjects actually experienced. Fersen et al. used a rather complex sequence of training, including phases in which the pairs were presented in random orders, as well as a phase in which the pairs were presented in blocks of only one pair at a time. In a simulation in which the model was exposed to exactly the series of training trials as were the pigeons in Fersen et al., it was shown that the model based on Equations 8.1 and 8.3 can fit this study well (Wynne, 1995). The results from the best-fitting parameters are shown in Fig. 8.2.

In Phase 1 and 2 (Experiment 1 of Fersen et al., 1991) the subjects were trained on a 5-term series (four training pairs from $A+ B-$ to $D+ E-$). Phase 1 shows performance after training on the stimulus pairs in random orders; Phase 2 shows performance after a phase of blocked presentation of each stimulus pair. Thereafter (Phase 3), the 5-term series was extended to 7 terms by the addition of an new stimulus on each end of the training series (stimulus X in $X+ A-$ trials, and stimulus F on $E+ F$-trials), and more test pair performance was assessed.

In Phases 1 and 2 the end–anchor and serial position effects are clear in the pigeon data, and are simulated well by the model for the initial 5-term series. The end–anchor effect is simply that performance on $A+ B-$ and $D+ E-$ is better than on any other training pair. The serial position effect is clearest in Phase 3, where it can be seen that performance on the training pairs forms a U-shape, even if performance on the end pairs is ignored. Particularly interesting is how the serial position effect reforms in Phase 3 when the series is extended by the addition of an additional item at each end. The stimulus pairs that were previously at the ends of the series and well-solved, $A+ B-$ and $D+ E-$, are now relatively poorly solved, as their new positions in the series require. This reformation is well captured by the Bush–Mosteller model. The test pair data show clearly the symbolic distance effect, and this is also well captured

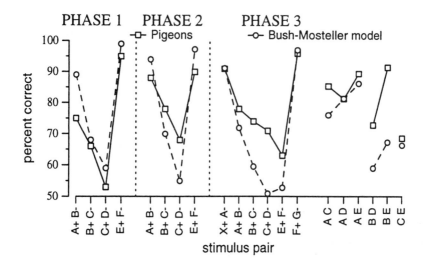

FIG. 8.2. Comparison of Fersen et al.'s (1991) pigeon subjects and the Bush–Mosteller-based model on three phases of training. Phase 1 is the original assessment of 5-term series performance; Phase 2 is 5-term series performance following a period of blocked presentation of the stimulus pairs. Phase 3 is a 7-term series formed by adding a stimulus at each end of the original 5-term series. For the test pairs, which were presented without reward or punishment, correct responding means transitive responding. Squares connected by solid lines show the pigeon performance; circles connected by dashed lines show predicted performance from the model.

by the model. The symbolic distance effect is demonstrated by the progressive improvement in performance across test pairs *AC, AD,* and *AE.*

Although the absolute levels of correct responding for each stimulus pair are not always perfectly predicted (the assumption that parameter values remained constant over the year-long course of the experiment may not be realistic), the important trends (the end–anchor, serial position, and symbolic distance effects, the reformation of the serial position effect after extending the series, as well as the effect of first item) are all well accounted for by this three-parameter model.

In keeping with the principle of parsimony, the simplest model—that based on Bush and Mosteller (1955) with three free parameters—would be preferable if it could account for all the available data. It turns out, however, that there is a relatively simple experiment that refutes this account of transitive inference.

Models of this type, unlike most other theories of transitive inference, operate on a trial-by-trial level; thus, the predictions of the models depend on the precise series of trials presented. Figure 8.3 shows that if the training pairs are presented in blocks of only one pair at a time, then these same parameters can only predict transitive choice if the training pairs are presented in one order.

If the pairs are presented first $A+B$-, then $B+C$-, then $C+D$-, and finally $D+E$- then, as Fig. 8.3 shows, these parameter values predict successful choice of B on BD test pairs. However, if the training pairs are presented in the opposite order (i.e., first $D+E$-, through to $A+B$-) this effect is not observed. At least, it is not found with the same parameter values. Because there is no basis for assuming different parameters after forward and backward training if the various experimental parameters are kept the same for these two conditions (e.g., reward magnitude, nature and duration of punishment, intertrial intervals, etc.), we can predict that if subjects are equally able to form transitive inferences after forward and backward training, this will refute the Bush–Mosteller-based model of transitive inference formation.

This experiment has been done. Steirn et al. (1995; Experiment 2) trained two groups of pigeons in standard operant chambers, in a manner broadly similar to that of Fersen et al. (1991). Stimuli were simple colors. Reward was 1.5-second access to food grains; incorrect responses led directly to the intertrial interval. One group was trained first on $A+B$- for three sessions of 96 trials each. Then they received the same amount of exposure to $B+C$-, and so on through the series. Finally they were presented the novel stimulus pair BD in tests in which reward and punishment were randomly assigned to the two stimuli. The other group was trained similarly, but first on $D+E$- for 288 trials, then on $C+D$-, and so on. The results of BD tests for these two groups are included in Fig. 8.3. It is clear that the direction of training had no impact on BD choice performance, thereby refuting the Bush–Mosteller-

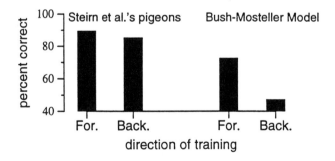

FIG. 8.3. Percentage transitive choices on the BD test pair after different directions of training. Left side: results of pigeons from Steirn, Weaver, and Zentall (1995). Right side: Results predicted from the Bush–Mosteller-based model after the same sequence of training using parameters that fit results of Fersen et al. (1991), shown in Fig. 8.2. *For.* indicates subjects trained first on $A+B$-, then $B+C$-, etc. *Back.* indicates subjects trained first on $D+E$-, then $C+D$-, etc.

based model.

In 1972, Rescorla and Wagner proposed a theory of classical conditioning that contains at its core a modified version of the Bush–Mosteller learning rule. Where Bush and Mosteller assumed that each stimulus gains and loses value entirely independently, Rescorla and Wagner proposed that stimuli compete for associative strength. Whereas in the Bush–Mosteller formulation, the increment of stimulus value on a trial depends on the discrepancy between *that* stimulus' value and the asymptote (λ - V), in the Rescorla–Wagner formulation the increment depends on the difference between asymptote and the sum of the values of *all* stimuli present (λ - ΣV_i). The next stage in attempting to find the simplest possible model of transitive inference formation was to explore the ability of a model based on the Rescorla–Wagner update rule to account for the data.

The update rule for the Rescorla–Wagner based model can be expressed:

$$\Delta V = \beta(\lambda - \Sigma V_i) \quad (4)$$

where the ΣV_i term indicates that the values are being summed over all stimuli present, and λ is 1 for reward, and 0 for nonreward. This model loses a parameter in its update rule because the assumption of different effects of reward and nonreward proved unnecessary, but it gains a scaling parameter, α, in the choice rule:

$$r = \frac{V(X)+V(Z)}{V(X)+V(Y)+2V(Z)} \quad (5)$$

$$p(X|XY) = \frac{1}{1+e^{-\alpha(2r-1)}} \quad (6)$$

where r is an intermediate variable and $V(Z)$ is the value of the background.

Figure 8.4 shows the fit of the Rescorla–Wagner model to the results of Fersen et al. (1991). The fit is good on the first two phases of training and the test pairs, but generally performance is underestimated on the training pairs in the third phase of training. However, the main effects (end–anchor, serial position and symbolic distance), and the reformation of the serial position effect from Phase 1 to Phase 2 are all fitted by the model.

As Fig. 8.5 shows, the Rescorla–Wagner model can also account for Steirn et al.'s (1995) data from forward and backward training, achieving similar levels of accuracy after both directions of training, just as the pigeons did. This simulation uses the same parameter values as that of the Fersen et al. results.

Aside from the somewhat poor fit to certain training pairs from the Fersen et al. (1991) experiment, there exists a condition of training for which the Rescorla–Wagner model cannot account. Fersen et al. went on to train together the ends of the transitive series into a closed loop. Stimuli G and X were presented together in a G+ X- pair, thereby closing the series. It is impossible

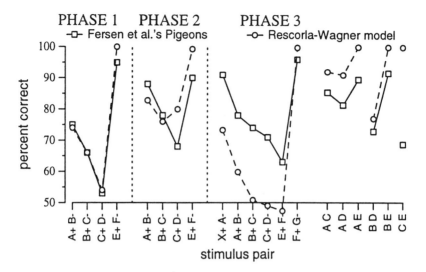

FIG. 8.4. Comparison of Fersen et al.'s (1991) pigeon subjects and the Rescorla–Wagner-based model on three phases of training. Squares connected by solid lines show the pigeon performance; circles connected by dashed lines show predicted performance from the model. Other details as Fig. 8.2.

for a circular series to be learned by assigning independent stimulus values to the stimuli (this is intuitively clear if one considers the smallest possible circular series; $A+ B-$, $B+ C-$, and $C+ A-$). Therefore, the successful performance of Fersen et al.'s subjects when trained on this sequence shows that the Rescorla–Wagner model cannot be the whole story of performance on these tasks.

A modification of the model that is able to account for performance on a

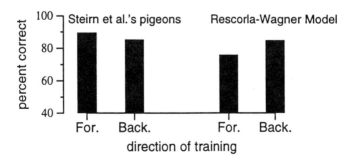

FIG. 8.5. Percentage transitive choices on the $B D$ test pair after different directions of training. Left side: Results of pigeons from Steirn, Weaver, and Zentall (1995). Right side: Results predicted from the Rescorla–Wagner-based model after the same sequence of training using parameters that fit results of Fersen et al. (1991). Further details as Fig. 8.3.

circular series also solves the problem of the poor performance on the central training pairs in Phase 3 of Fersen et al. (1991; see Fig. 8.4). The modification involves rejecting the assumption of independent stimulus values and allowing the value of a stimulus to be partially bound to its context. Thus stimulus B in the context of a $B+$ $C-$ trial has a different value (that we shall term $<B|BC>$) than it does in a $A+$ $B-$ context ($<B|AB>$). These configural values are added into the Rescorla–Wagner model, weighted by a parameter, γ. The update rules remain the same (though now also operating on the configural stimulus values as well as the independent ones). The choice rules are modified to admit the influence of the configural representations:

$$r = \frac{V(X)+V(Z)+\gamma V(<X|XY>)}{V(X)+V(Y)+2V(Z)+\gamma V(<X|XY>)+\gamma V(<Y|XY>)} \qquad (7)$$

The r values are scaled as before:

$$p(X|XY) = \frac{1}{1+e^{-\alpha(2r-1)}} \qquad (6)$$

This model has gained one parameter, γ, over the Rescorla–Wagner model. When γ is small the behavior of this model is the same as the Rescorla–Wagner model. When γ is very large the model gains the ability to solve any set of training pairs, but loses the ability to solve any test pairs.

Figures 8.6 and 8.7 show that the model can also predict the performance

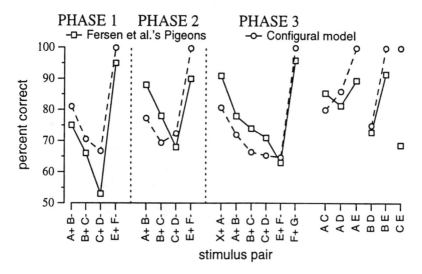

FIG. 8.6. Comparison of Fersen et al.'s (1991) pigeon subjects and the Configural model on three phases of training. Squares connected by solid lines show the pigeon performance; circles connected by dashed lines show predicted performance from the model. Other details as Fig. 8.2.

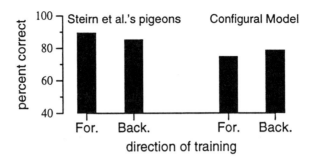

FIG. 8.7. Percentage transitive choices on the *B D* test pair after different directions of training. Left side: Results of pigeons from Steirn, Weaver, and Zentall (1995). Right side: Results predicted from the Configural model after the same sequence of training using parameters that fit results of Fersen et al. (1991). Further details as Fig. 8.3.

of Fersen et al.'s (1991) and Steirn et al.'s (1995) pigeons respectively. Performance on Phase 3 of Fersen et al.'s study was improved. All other features were very similar to the Rescorla–Wagner model.

As the preceding considerations of the studies of Fersen et al. (1991) and Steirn et al. (1995) have shown, the precise sequence of training pairs used has a major impact on the performance predicted by the model. In the Steirn et al. study, the training pairs were presented in blocks of only one at a time in the order of the series, and learning was fairly rapid (a total of 1,152 trials was presented). In the Fersen et al. study, both random and block orders of presentation were used and the subjects took much longer to learn the 5-term series (around 5,000 trials). How much difference does training order make? What would happen if the stimulus pairs were *never* presented in blocks, but only in random orders? To test this eight pigeons were trained (Wynne, 1997) with food grains as reward, and a brief time-out as punishment, in a manner similar to Fersen et al.'s and Steirn et al.'s subjects. Of these eight subjects, five were removed from the study after they failed to make substantive progress after 12,000 training trials. The three subjects that succeeded in learning all four training pairs needed over 17,000 trials to do so. The performance of these three subjects is shown in Fig. 8.8, along with the predicted levels of choice from the Configural model, based on the exact sequence of trials to which these three subjects had been exposed and using the parameters that fitted the results of Fersen et al. (Fig. 8.4) and Steirn et al. (Fig. 8.5). It can be seen that the model also fits the observed results quite well. The end–anchor and serial position effects are present in the training pair data, and the symbolic distance effect is also well modelled. Thus, the model is consistent with the pigeon results in predicting successful performance on random-order training after a far longer period of exposure to the task.

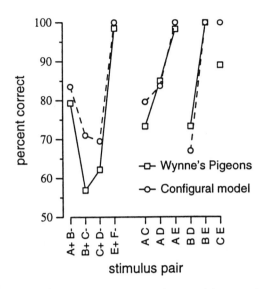

FIG. 8.8. Comparison of Wynne's (1997) pigeon subjects and the Rescorla–Wagner-based model. For the test pairs, which were presented without reward or punishment, correct responding here implies transitive choice. Squares connected by solid lines show the pigeon performance; circles connected by dashed lines show predicted performance from the model.

Overall, the model based on the Rescorla–Wagner update rule with configural stimulus values gives a good fit to the data from studies of transitive inference in pigeons. The main effects (end–anchor, serial position, and symbolic distance) are well accounted for. The model also accounts for the reformation of the serial position effect when the series trained is extended by the addition of new stimuli, and it learns the series more rapidly when the stimulus pairs are presented in blocks of one type in the order of the implied series than when they are presented randomly intermixed.

APPLYING THE MODEL TO OTHER DATA SETS

We consider next to what extent this model can be considered as an account of transitive choice on the n-term series in other species.

Other Nonhuman Studies

The only study of transitive inference in rats trained them in a manner similar to that already described for pigeons in the study of Steirn et al. (1995). Davis (1991, 1992) trained a total of 10 rats in three experiments using five different odors as stimuli. The odors were presented in pairs of tunnels that contained hinged wooden doors. Choice of the correct tunnel gave access through a door to shelled sunflower seeds. The incorrect tunnel was also baited, but its door was blocked so that the subject could not gain access to the food behind it. Subjects were trained on the four neighboring pairs, but also on the stimulus pairs $A+ E-$, $A+ C-$, and $C+ E-$. The results of Davis' Experiments 1 and 2 mirrored closely Steirn et al.'s results with pigeons—the rats made transitive test choices on BD test trials no matter the direction in which they had been trained on the series. Because Davis had presented all other possible combinations of stimuli in training, no further tests were performed.

McGonigle and Chalmers (1977, 1985, 1992) presented results from several studies using squirrel monkeys as subjects. McGonigle and Chalmers (1977) trained eight squirrel monkeys using tobacco cans as described previously. These subjects were first trained to master each stimulus pair separately, before the pairs were presented mixed together in each testing session. Seven of the squirrel monkeys succeeded in solving all four training pairs at a high level when they were randomly intermixed in each session. Subsequently, on tests with the stimulus pair BD, these seven subjects averaged 90% transitive choices. As the summary of these results in the top panel of Fig. 8.9 shows, the end–anchor and serial position effects were clearly present in this study. However, the symbolic distance effect is less clear here.

Gillan (1981) trained three chimpanzees on the 5-term series problem using methods similar to those of McGonigle and Chalmers (1977). The stimuli were colored plastic boxes, and correct choices were rewarded through the uncovering of a food item in one box of each training pair. The subjects were initially trained on each pair separately, and later the training pairs were presented randomly intermixed in each session. Of the three subjects, only one made transitive choices on first exposure to the BD test pair. The results shown in Fig. 8.9, being only from one subject, are rather noisy. Nonetheless, a symbolic distance effect can be seen in the fact that $B\ D$ performance is less consistently transitive than $B\ E$ performance in Experiment 2B (right side of figure). The end–anchor and serial position effects are less apparent here. However these results stem from just one subject, so it seems unwise to over-interpret the presence or absence of these effects here.

Boysen et al. (1993) trained three chimpanzees on the 5-term series problem, also using colored boxes and food as reward and procedures generally similar to those of McGonigle and Chalmers (1977) and Gillan (1981). All three of Boysen et al.'s subjects made transitive choices on the BD test (mean 94.3%). Insufficient data are presented to permit assessment of the end–anchor, serial position, or symbolic distance effects.

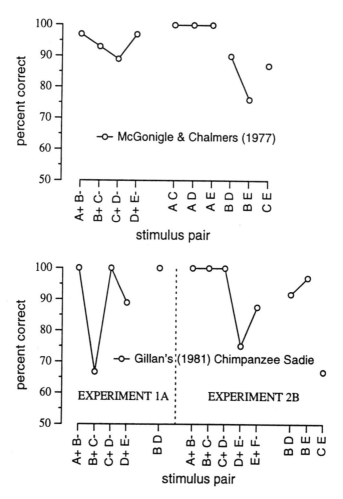

FIG. 8.9. Top panel: Performance of McGonigle and Chalmers' (1977) squirrel monkey subjects on a 5-term series. On test pairs, correct responding is defined as transitive responding. Bottom panel: Performance of Gillan's (1981) chimpanzee Sadie, after training on a 5-term series (left side), and on a 6-term series formed by adding a stimulus at the negative end of the series.

Human Transitive Inference: Nonverbal

It is not immediately obvious that a human subject would necessarily make a transitive choice if trained on a completely nonverbal version of the task. However, several studies have shown that human subjects do choose transitively after training similar to that used with nonhuman species.

Siemann and Delius (1993) trained 24 adults on a version of a six-term series presented as a video game. Subjects progressed through a labyrinth in which they were repeatedly confronted with a choice between two doors covered with different patterns (the stimuli). If the subject selected the door containing the rewarded stimulus, the door opened on to a treasure chamber where gold coin could be found. Incorrect choices led the subject into an alley in which he had to give a coin to a beggar. At first the stimulus pairs were presented in blocks of eight identical trials at a time, but over training, the size of the blocks was progressively reduced until finally the subjects were responding to random series of the five training pairs.

In tests, novel pairs of stimuli were presented on the doors and choice of either stimulus led to an empty passage.

As Fig. 8.10 shows, performance on the training pairs showed strong end–anchor and serial position effects. Fifteen of the 24 subjects were successful on the transitivity test pairs, however, their response accuracies to the three test pairs that did not contain end items failed to show a symbolic distance effect. Of particular interest in this study is the fact that, of the 15 successful subjects, 7 of the subjects were unable to report any rule governing their behavior. They were unable to form a series when given cards with the stimuli on them, and they expressed surprise when the ordering was pointed out to them. However, the results from the subjects unable to verbalize what they had done were indistinguishable from those of the subjects who could describe the series.

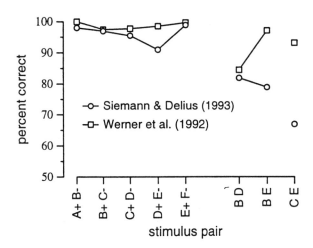

FIG. 8.10. Performance of adult human subjects in two studies on a 6-term series. On test pairs, correct responding is defined as transitive responding. Squares—Werner, Köppl, and Delius (1992). Circles—Siemann and Delius (1993).

Werner, Köppl, and Delius (1992) demonstrated similar results using a somewhat simpler video game to present a 6-term series. The pairs of stimuli were presented concurrently in the center of a computer screen, and subjects indicated their choice by pressing keys on the computer keyboard. Correct choices led to the acquisition of points, which could later be exchanged for money; incorrect choices led to the subtraction of points from the subject's total. Stimulus pairs were presented in pseudo-random orders, and once a subject had attained 70% correct responses in a session, the test pairs were incorporated. Test trials led neither to the addition nor subtraction of reward points, no matter what choice was made. As shown in Fig. 8.10, the end–anchor and serial position effects were, here again, very clear for the training pairs. In these data, a symbolic distance effect was also present in the test pair results. Again, some successful subjects could not report the underlying series of stimuli, though the ability to lay out a set of cards with the stimuli on them in the order of the implied series did correlate positively with test pair performance in this study.

In the only study of children on a nonverbal transitive inference task, Chalmers and McGonigle (1984) trained 6-year-old children under two conditions. In Experiment 1, there were 10 children trained on a 5-term series following as closely as possible the procedures that these authors used in their studies of squirrel monkeys described above (McGonigle & Chalmers, 1977, 1985, 1992). That is, the stimuli were colored cans painted in uniform colors. These cans were weighted so that the correct choice was always heavier than the incorrect choice (for half the subjects, weights were counterbalanced for the other subjects). Only two weight values were used, so choice in transitive tests could not be based on the weights of the containers. In training, correct choice was reinforced with verbal feedback and the discovery of a colored counter under the correct stimulus. These counters could later be exchanged for candy. On B D tests, both stimuli were heavy for half the group, light for the other half, and responses to either were rewarded. In Experiment 2, a similar group of children was trained on a semiverbal version of the task, modelled on those developed by Bryant and Trabasso (1971). The subjects were shown the same cans as in Experiment 1, but were asked which stimulus was the heavier (or lighter) one, and had to respond verbally.

As Fig. 8.11 shows, a ceiling effect in the response accuracies on the training pairs removes the possibility of observing end–anchor or serial position effects. Response accuracy on the crucial B D test pair is substantially lower than on the training pairs, in contradiction of the symbolic distance effect. Possibly these subjects noticed that any response on the test pairs was rewarded and therefore did not maintain a high level of attention to the stimuli presented. Comparison of the several test pairs does suggest a symbolic distance effect.

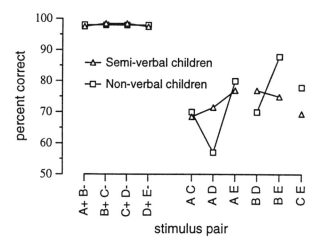

FIG. 8.11. Comparison of performance of 4-year-old children trained on a 5-term series nonverbally, and semiverbally (from Chalmers & McGonigle, 1984). On test pairs, correct responding is defined as transitive responding. Squares—nonverbal training. Triangles—verbal training.

Most marked in Fig. 8.11 is the strong similarity between the results of similar subjects tested nonverbally and semiverbally.

Some studies of transitive inference formation in human subjects have included training of the nonneighboring test pairs alongside training of the neighboring pairs. Because it is the ability to spontaneously make correct choices on the nonneighboring pairs (such as *B D*) that constitutes the test for a subject's ability to form a transitive inference, these studies cannot be said to be testing for transitive inference formation. Nevertheless, some are worthy of discussion because they include evidence about the end–anchor, serial position, and symbolic distance effects.

In one of the earliest tests of the ability of human subjects to learn a nonverbal transitive series, Humphreys (1975) presented all 10 possible pairs from a 5-term series (i.e., the nonneighboring as well as the neighboring pairs) on cards. The correct stimulus was always on the left and underlined during training. In test, the correct stimulus was no longer underlined, and its left–right position was randomized. Subjects had to indicate which member of each pair they believed was correct. Results were reported as the percentage of total errors made by a subject on each pair. Thus, a result of 50% for a pair indicated that, on average, 50% of all errors made were on this pair. A result of 0% indicated that no errors were made on a pair. All the standard effects appear in Fig. 8.12 inverted (the serial position effect as an inverted U). This figure shows an end–anchor effect only at the negative end of the series, and a correspondingly asymetrical serial position effect. The symbolic distance

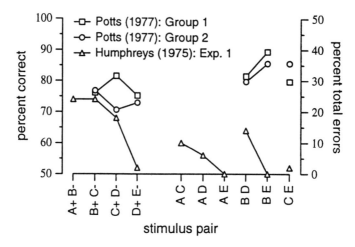

FIG. 8.12. Performance of adults on early nonverbal studies. Triangles—percentage of total errors (right axis) on each stimulus pair from Humphreys (1975, Experiment 1). Squares—percentage correct (left axis) from Potts (1977, Group 1); circles—Potts (1977, Group 2).

effect is clear in the *downward* slope of the lines linking test pairs *A C*, *A D*, and *A E*, and well as *B D* and *B E*.

In a similar study, Potts (1977) trained adults on a 6-term series where the stimuli were presented in pairs on a computer screen. The correct item was underlined. Again, the transitive test pair was included in training, making this not a test of transitive inference formation per se, and only the results from pairs not containing one of the end items were reported (included in Fig. 8.12). Thus assessment of the end–anchor and serial position effects is impossible. However, comparison of the *B D* and *B E* test pairs suggests a symbolic distance effect.

Semiverbal. As we progress to studies with an increasing verbal compo-nent, so the literature expands substantially. Semiverbal studies, that is, those with real objects as stimuli, but using language to describe the relationship between them, are very numerous. As outlined above, these are based on a study by Bryant and Trabasso (1971), in which children were shown sticks of different colors that they were told were also of different lengths. Figure 8.13 summarizes the results of several studies using this paradigm by Trabasso and co-workers. The top panel shows results from Bryant and Trabasso (1971) with 4-, 5-, and 6-year old children, as well as a replication with 4-year-olds by the same authors reported in Trabasso (1975). Although the absolute levels of performance decrease with age on the test pairs, the same patterns are apparent at all ages. On average, end–anchor and serial position effects can be

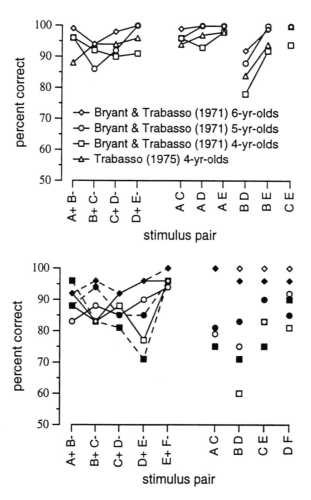

FIG. 8.13. Top panel: Performance of different age groups on semiverbal 5-term series (from Bryant & Trabasso, 1971; Trabasso, 1975). Diamonds—6-year-olds; circles—5-year-olds; squares—4-year-olds; triangles—replication of 4-year-olds.
Bottom panel: Performance of different age groups on semiverbal 6-term series from Trabasso, Riley, and Wilson (1975). In each case, filled data point connected by dashed line is *Longer?* question form; open data point connected by solid line is *Shorter?* question form. Diamonds—adults; circles—9-year-olds; squares—6-year-olds.

seen in the training pair data, and the symbolic distance effect is clear in the comparison of the *B D* and *B E* test pairs.

The lower panel shows results from Trabasso, Riley, and Wilson (1975). These authors compared performance of 6- and 9-year-old children with adults on a 6-term series presented in the same semiverbal fashion. Here also, the performance to longer and shorter question forms was assessed separately. Again, the end–anchor and serial position effects look similar across the different age groups and question forms, though the absolute levels of performance differ between age groups. The limited number of test pairs presented makes it impossible to assess the symbolic distance effect.

In the development of a reinforcement model of transitive inference performance, the comparison of groups trained of pigeons trained with different sequences of the stimulus pairs played an important role. As outlined above, the fact that groups of pigeons trained forward and backward through the implied series were equally successful in transitivity tests (Steirn et al., 1995) was important in rejecting a Bush–Mosteller-based model in favor of one based on the Rescorla–Wagner theory of conditioning. Additionally, the faster learning rate of subjects trained on the stimulus pairs in blocks in the direction of the implied series compared to those trained on random series of training pairs was also significant for the development of the model. In the literature on semiverbal transitive inference, three studies have made comparisons of different training orders, albeit with somewhat contradictory results.

De Boysson-Bardies and O'Regan (1973) trained 4-year-old children on Bryant and Trabasso's (1971) colored-stick version of the 5-term series. In one experiment the children were presented the training pairs either in forward or backward orders (the results from these two orders of training were presumably similar—results are presented collapsed across the two orders of training). In a second experiment, a similar group of children was trained with the stimulus pairs in a disordered sequence. As Fig. 8.14 shows, the results from both training methods are very similar. The end–anchor effect is only apparent at the negative end of the series, and the serial position effect is accordingly asymmetrical. The symbolic distance effect is very clear in these results. Unfortunately, it was not reported whether the two groups required similar amounts of training to reach these similar levels of performance. Perhaps the strangest feature of this study is that the authors reported that, when they compared performance of adults on these two forms of training, the adults performed worse in the condition in which the stimulus pairs were presented randomly. However, no details of the amounts of training, nor the particular pairs on which the adults failed, are presented.

A more complete report of a comparison of different training sequences in human subjects is presented by Kallio (1982). Again utilizing Bryant and Trabasso's (1971) stick task, Kallio trained a 5-term series with 4-year-old children in several different training orders. Of most interest here is the comparison of a group trained in forward or backward series order, and a group trained with blocks of one training pair at a time, but with the order of the blocks randomized. Four-year-olds learned the training pairs faster, and made

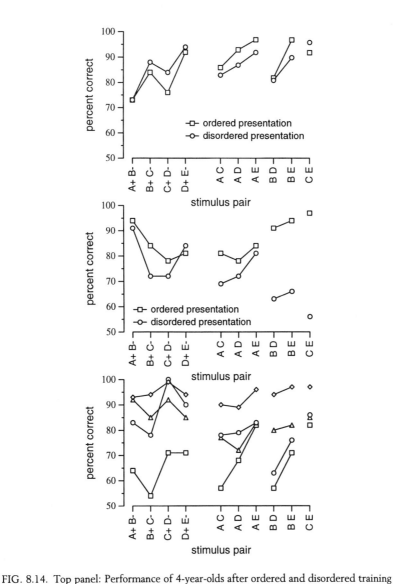

FIG. 8.14. Top panel: Performance of 4-year-olds after ordered and disordered training on semiverbal 5-term series (from De Boysson-Bardies & O'Regan, 1973; see text for further details). Squares—ordered presentation; circles—disordered presentation.

Middle panel: Performance of 4-year-olds after ordered and disordered training on semiverbal 5-term series (from Kallio, 1982, Experiment 1; see text for further details). Squares—ordered presentation; circles—disordered presentation.

Bottom panel: Performance of preschool, 2nd-grade and 4th-grade children and college students on disordered presentation of semiverbal 5-term series (from Kallio, 1982, Experiment 2; see text for further details). Diamonds—college students; circles—4th-grade children; triangles—2nd-grade children; squares—preschool children.

fewer errors on the test pairs when the pairs were presented in serial order than when they were presented in a random order. In subsequent experiments with children of different ages (4-year-olds, 7-year-olds, and 9-year-olds), using only the random order of presentation, Kallio found that rate of learning and success on the test pairs improved with increasing age (although the relationship was not entirely stable across a replication). The results of these studies are included in the middle and bottom panels of Fig. 8.14. Both sets of results show that, although the absolute levels of accuracy improve with age on the disordered form of presentation, the main effects (end–anchor, serial position, and symbolic distance) are apparent in all age groups.

A third study comparing orders of training an n-term (in this case 4-term) series semiverbally with young children as subjects is that of Halford and Kelly (1984). These authors found that success using the stick task of Bryant and Trabasso (1971) was very dependent on age, with the youngest subjects (4-year-olds) failing to reach criterion when trained on the premises in a random order. Only 1 child out of 15 in this age group reached criterion, as compared to 13 (out of 15) 7-year-olds. In their Experiment 3 these authors compared performance of 4-year-olds on a 4-term series trained in the order of the series, and randomly. Again, as Kallio found, the children were much better on the ordered than random condition. Unfortunately, details of performance on individual training pairs were not reported.

Those authors who found that the transitive inference performance of adults, but not of young children, was degraded by having the stimulus pairs presented in mixed orders, considered two possibilities, "We can conclude that children implement transitivity differently, and sometimes better than adults. Alternatively . . . we can suppose that the children in these experiments were not using transitivity at all" (De Boysson-Bardies & O'Regan, 1973, p. 533)—and chose the latter explanation. Conversely, those researchers who found the opposite pattern of results (that only the performance of younger children was negatively impacted by presenting the stimuli in mixed orders; Halford & Kelly, 1984 and Kallio, 1982), concluded that *this* result showed that adults, but not children, were forming true transitive inferences. Both studies summarized that the younger children, who only succeeded reliably when the stimulus pairs were presented in the order of the series, relied on serial order cues in the presentation order to aid transitive choice.

If one group of subjects fails to reach a criterion of success on a particular pattern of training in a given time, then, within the model under consideration here, this simply implies that they have a lower learning rate-parameter (β), than a group that reaches criterion in that time. This means that in De Boysson-Bardies and O'Regan's (1973) study, for some reason the child subjects showed a higher learning rate than the adults—perhaps the stimuli used were more appealing to children. The results of Kallio (1982) and Halford and Kelly (1984) imply that learning-rate parameters increase with age. Such an increase could be explained by many factors, including simply improved attentiveness to the stimuli.

Fully Verbal. There exists a large number of studies of fully verbal transitive inference, but many of them are of little use in attempting to model this phenomenon for a variety of reasons. Very often, subjects are given exposure to the nonneighboring transitive test pairs as part of their training, making it impossible to test for the development of transitivity. In several studies, subjects were given uncontrolled access to the premises, often as a passage of prose that they could read as often as they liked, making it impossible to reconstruct how much effective exposure to the training items the subjects had received. Even the relative exposure to different training pairs is not controlled because it is unknown how often subjects read different parts of the text. Notwithstanding these problems, it is worthwhile considering the results of some of these studies, because they address the question of whether verbal transitive inference formation is really different from the types of nonverbal and semiverbal inference formation we have considered thus far.

The question of the relative difficulty of different training orders has been addressed in three studies. Smith and Foos (1975) read out three sentences representing a 4-term series to groups of adult subjects. The sentences referred to a sequence of heights of common professions ("The doctor is taller than the farmer," etc.). The four titles were presented in various different orders and subsequently the subjects were required to write down the series they had heard. Smith and Foos recorded the proportion of correct series reconstructions. The most successful series construction followed presentation of the series in forward series order (AB, BC, CD), and the second most successful was the backward series order (CD, BC, AB). The least successful orders were those in which the order of presentation failed to correspond to the (forward or backward) order of the series, particularly if the second sentence introduced two new items in the series. Thus these results are similar to those predicted from the configural model and observed in nonverbal and semiverbal forms of the transitive inference task, that presenting the stimulus pairs in the order of the implied series leads to better learning than disordered presentation.

In a follow-up study, Foos and Sabol (1981) examined the ease of learning 5- and 6-term series, again presented aurally. This study also found a clear superiority of recall when the premises were presented in the order of the underlying series, and again performance following presentation of the premises in the reverse series order was the second most successful presentation order. The remaining disordered sets of premise presentation produced lower levels of successful recall.

In a further study addressing this issue, Mynatt and Smith (1977) trained adults on a 4-term series by presenting sentences of the same form as used by

Smith and Foos (1975) one at a time on a computer screen. Subsequently, the subjects had to note down the series they had seen. Error rates were generally low, and it is the time spent reading the premises that are the key data from this study. It was found that reading times were shortest for the premises presented in the order of the series, second fastest for the reverse series order and slowest for the disordered conditions, particularly those where the second sentence introduced two new items.

In a study with interesting data on the symbolic distance effect, and its reformation after extension of a transitive series, Woocher, Glass, and Holyoak (1978) trained adults with pairs of profession names using methods similar to those of Smith and Foos (1975). Initially, each subject was read two sets of seven pairs of names (two independent 8-term series). Thereafter subjects had to decide which of two tachistoscopically presented names was the taller in the series. Subsequently the subjects were told that the shortest person in one list was taller than the tallest person in the other list, thereby joining the two 8-term series to form a single 16-term series. Thereafter various test pairs were again presented. The results presented are the reaction times to test pairs (Fig. 8.15), and thus all the standard effects are inverted. These show clearly the end–anchor, serial position, and symbolic distance effects, and how these effects reformed after combination of the two 8-term series into a 16-term series. This reformation of the serial position effect is similar to that observed in pigeons by Fersen et al. (1991) when the 5-term series was expanded to 7 terms, which, as shown in Fig. 8.6, can be accounted for within the model.

A second study that addresses the issue of the reformation of series after initially training one length of series and subsequently adding new items is that of Banks, White, and Mermelstein (1980). Adults were given the premises of a 4-term series in written form and told to memorize them. The items were 3-letter men's names, related by the comparatives taller and shorter. Thereafter, test pairs were presented on a computer screen so that reaction times could be measured. In a subsequent phase, the series was lengthened by addition of a new term, either at one or other end or in the middle of the series. The subjects were then tested with the lengthened series, before finally being retrained and retested with the original, shorter series. The results (reaction times) are shown in Fig. 8.16. It is clear again here that the patterns of reformation of the serial position and symbolic distance effects are consistent with those found by Fersen et al. (1991) and fit by the reinforcement-based model of transitive inference performance.

CONCLUSIONS

Certain effects have recurred in the data sets considered here from a variety of species and training regimens. This legitimized considering them to be

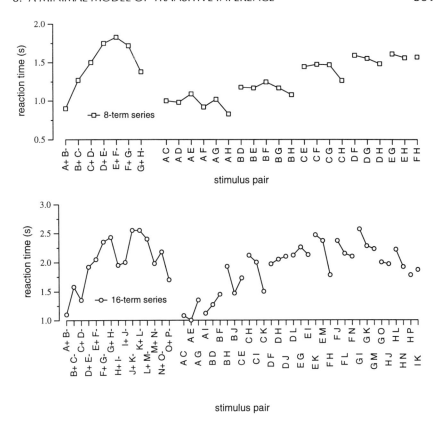

FIG. 8.15. Reaction times to stimulus pairs from 8-term (top panel) and 16-term (bottom panel) series from Woocher et al. (1978).

standard effects. End–anchor and serial position effects have been regularly seen in the training pair results (see Fig. 8.2, Figs. 8.8–8.16), and the symbolic distance effect is present in most sets of test pair results (see Fig. 8.2, Figs. 8.8–8.16). The exceptions to these generalities are not concentrated among any one species or method of task presentation. The data from pigeon subjects in Figs. 8.2 and 8.8 show some of the clearest end–anchor and serial position effects. Symbolic distance effects are less consistent, but are no more likely in human than nonhuman subjects. Figure 8.11, comparing 4-year-olds trained nonverbally and semiverbally, gives no suggestion that the main effects depend on verbal presentation.

The reinforcement model of transitive inference developed here not only predicts the standard effects (end–anchor, serial position, and symbolic distance)—because it is a dynamic model, it also makes predictions about the effects of different orders of training. The model predicts that, if a transitive

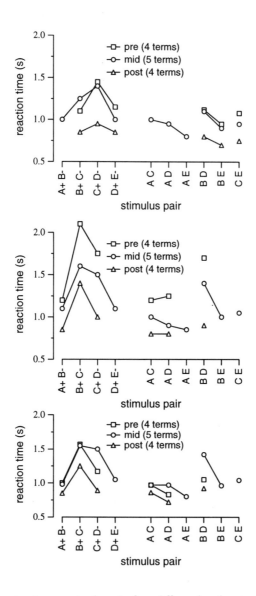

FIG. 8.16. Reaction times to stimulus pairs from different length series from Banks et al. (1980). In each panel, squares indicate results from the original 4-term series before the addition of an extra term; circles show results of five-term series with additional term; triangles show performance after removal of the extra term. Top panel: Addition of a term at the positive (long) end of the series (Banks et al., 1980, Experiment 1). Middle panel: Addition of a term at the negative (short) end of the series (Experiment 1). Bottom panel: Addition of a term in the middle of the series (Experiment 2). The new term becomes item C, and the original terms C and D are re-labelled D and E for the figure.

series is extended by the addition of more items, the serial position and symbolic distance effects will reform, even if this means that performance on some pairs must decline. This was modelled on results from pigeons of Fersen et al. (1991), but has also been observed in Gillan's (1981; see Fig. 8.9) study of a chimpanzee, and in studies on humans by Woocher et al. (1978; see Fig. 8.15) and Banks et al. (1980; see Fig. 8.16).

Modelling of the pigeon studies of Fersen et al. (1991), Steirn et al. (1995) and Wynne (1997) showed that the model predicts that ordered presentation of the training pairs is much more rapidly learned than disordered patterns of presentation. Several studies have compared orders of training in human subjects. De Boysson-Bardies and O'Regan, (1973; see Fig. 8.14) found that adults, though not 4-year-old children, were more successful on an ordered sequence of training pairs. Kallio (1982) as well as Halford and Kelly, (1984; see Fig. 8.14) found that young children were worse on disordered (than ordered) training patterns, and that the difference became smaller with increasing age. These authors have argued for developmental changes in the representation of the transitive inference task. The reinforcement model permits a simpler explanation. Subjects that fail in a given amount of training on the disordered task but succeed on the premises presented in series order are simply operating with a smaller learning rate parameter (β) than subjects that succeed on the disordered task presentation with a similar amount of training. Because the learning parameter would be influenced by such factors as the subjects' attention to the task and the stimuli, it is easy to see how younger subjects might fail on the disordered task presentation, yet succeed under ordered conditions of training pair presentation.

It could be argued that because humans are highly verbal animals, they may utilize verbal reasoning to solve transitive inference problems, even though other species can achieve similar results through simpler means. Aside from contradicting the principle of parsimony, an explanation in these terms cannot explain why Siemann and Delius (1993) found that adult subjects who could verbalize the transitive series were at no advantage to those who could not, nor the lack of any difference in the performance of Chalmers and McGonigle's (1984) 6-year-old subjects trained semiverbally and nonverbally (Fig. 8.11).

When the task is presented verbally, there are clearly some aspects of the results obtained that relate to the use of language. The marking effect is purely linguistic—the fact that test questions are more accurately and rapidly answered if they contain an unmarked comparative (bigger, taller) than a marked one (smaller, shorter). The congruency effect (that questions are better answered if the form of the question is congruent with the end of the series being asked about—for example, "Who is shorter?" is answered better at the short end of the series) may be another such, though the observation of similar asymmetries in the serial position curve in the results from pigeons of Fersen et al. (1991; see Fig. 8.2) suggests this need not be a verbally based effect. Nonetheless, even though there are aspects of performance on verbally presented transitive inference tasks that relate to the verbal form of task presentation, this need not prove that linguistic constructs are used in task solution. To use a computer metaphor—it is apparent that a linguistically presented task would have to go through a linguistic front-end, which could impress upon the pattern of results—it need not necessarily imply the use of a linguistic processor. This is an area of transitive inference research in which more experimental work on the nature of the effects of verbal and nonverbal task presentation forms is required.

There are those who would argue that nonverbal transitive inference is not real transitive inference at all (e.g., De Boysson-Bardies & O'Regan, 1973; Markovits & Dumas, 1992) because, by definition, inference involves language. These authors need to explain why it is that, in human subjects, the presence or absence of language in the task presentation (and in the subjects' abilities to report the nature of the task) makes so little difference to performance.

Many have argued that an apparently inferential process that can be explained by a simpler mechanism is ipso facto not a truly inferential process (De Boysson-Bardies & O'Regan, 1973; Markovits & Dumas, 1992; McGonigle & Chalmers, 1977). It is a strange science that would deny the existence of a phenomenon once it finds an explanation. Where once physical scientists were chastised for finding explanations for physical phenomena that did away with the need for God's continuous oversight, so in our times psychologists are criticized for finding explanations of cognitive phenomena at a more basic, mechanistic level.

No doubt, the model proposed here will in time be overtaken as novel research adds to our knowledge of transitive inference in beast and man. However, if this model flags the fact that a complex cognitive task can be solved through the application of a simple mechanistic model, then it will have served a useful function in the development of our understanding of complex performances.

ACKNOWLEDGMENTS

I thank Jim Kehoe for helpful comments on an earlier draft. Research supported by a grant from the Australian Research Council.

REFERENCES

Anderson, B. J. (1975). *Cognitive psychology: The study of knowing, learning and thinking.* New York: Academic Press.

Anderson, J. R., & Bower, G. H. (1973). *Human associative memory.* New York: Wiley.

Banks, W. P. (1977). Encoding and processing of symbolic information in comparative judgments. In G. H. Bower (Ed.), *The Psychology of Learning and Motivation, Vol. II* (pp. 101–159). New York: Academic Press.

Banks, W. P., & White, H. (1982). Single ordering as a processing limitation. *Journal of Verbal Learning and Verbal Behavior, 21*, 39–54.

Banks, W. P., White, H., & Mermelstein, R. (1980). Position effects in comparative judgments of serial order: List structure vs. differential strength. *Memory and Cognition, 8*, 623–630.

Boysen, S. T., Berntson, G. G., Shreyer, T. A., & Quigley, K. S. (1993). Processing of ordinality and transitivity by chimpanzees (Pan troglodytes). *Journal of Comparative Psychology, 107*, 1–8.

Breslow, L. (1981). Re-evaluation of the literature on the development of transitive inference. *Psychological Bulletin, 89*, 325–351.

Bryant, P. E., & Trabasso, T. (1971). Transitive inferences and memory in young children. *Nature, 232*, 456–458.

Burt, C. (1911). Experimental tests of higher mental processes and their relation to general intelligence. *Journal of Experimental Pedagogy, 1*, 93–112.

Burt, C. (1919). The development of reasoning in school children: I. *Journal of Experimental Pedagogy, 5*, 68–77, 121–127.

Bush, R. R., & Mosteller, F. (1955). *Stochastic models for learning.* New York: Wiley.

Chalmers, M., & McGonigle, B. O. (1984). Are children any more logical than monkeys on the five-term series problem? *Journal of Experimental Child Psychology, 37*, 355–377.

Clark, H. H. (1969). Influence of language on solving three-term series problems. *Journal of Experimental Psychology, 82*, 205–215.

Couvillon, P. A., & Bitterman, M. E. (1986). Performance of honeybees in reversal and ambiguous-cue problems: Tests of a choice model. *Animal Learning and Behavior, 14*, 225–231.

Couvillon, P. A., & Bitterman, M. E. (1992). A conventional conditioning analysis of "Transitive Inference" in pigeons. *Journal of Experimental Psychology: Animal Behavior Processes, 18*, 308–310.

Davis, H. (1991). Logical transitivity in animals. In W. Honig & G. Fetterman (Eds.), *Complex and Extended Stimuli* (pp. 405–429). Hillsdale, NJ: Lawrence Erlbaum Associates.

Davis, H. (1992). Transitive inference in rats (Rattus norvegicus). *Journal of Comparative Psychology, 106*, 342–349.

De Boysson-Bardies, B., & O'Regan, K. (1973). What children do in spite of adults' hypotheses. *Nature, 246*, 531–534.

DeSoto, C. B., London, M., & Handel, S., (1965). Social reasoning and spatial paralogic. *Journal of Personality and Social Psychology, 2*, 513–521.

Fersen, L. von, Wynne, C. D. L., Delius, J. D., & Staddon, J. E. R. (1991). Transitive inference formation in pigeons. *Journal of Experimental Psychology: Animal Behavior Processes, 17*, 334–341.

Foos, P. W., & Sabol, M. A. (1981). The role of memory in the construction of linear orderings. *Memory and Cognition, 9*, 371–377.

Gillan, D. J. (1981). Reasoning in the chimpanzee: II. Transitive inference. *Journal of Experimental Psychology: Animal Behavior Processes, 7,* 150–164.

Halford, G., & Kelly, M. E. (1984). On the basis of early transitivity judgments. *Journal of Experimental Child Psychology, 38,* 42–63.

Halford, G. (1993). *Children's understanding: The development of mental models.* Hillsdale, NJ: Lawrence Erlbaum Associates.

Horner, J. M., & Staddon, J. E. R. (1987). Probabilistic choice: A simple invariance. *Behavioral Processes, 15,* 59–92.

Humphreys, M. S. (1975). The derivation of endpoint and distance effects in linear orderings from frequency information. *Journal of Verbal Learning and Verbal Behavior, 14,* 496–505.

Huttenlocher, J. (1968). Constructing spatial images: A strategy in reasoning. *Psychological Review, 75,* 550–560.

Kallio, K. D. (1982). Developmental change on a five-term transitive inference. *Journal of Experimental Child Psychology, 33,* 142–164.

Luce, R. D. (1959). *Individual choice behavior: A theoretical analysis.* New York: Wiley.

Markovits, H., & Dumas, C. (1992). Can pigeons really make transitive inferences? *Journal of Experimental Psychology: Animal Behavior Processes, 18,* 311–312.

McGonigle, B. O., & Chalmers, M. (1977). Are monkeys logical? *Nature, 267,* 694–696.

McGonigle, B. O., & Chalmers, M. (1985). Representation and strategies during inference. In T. Myers, K. Brown, & B. McGonigle (Eds.), *Reasoning and discourse stratagies processes* (pp. 141–164). London: Academic Press.

McGonigle, B. O., & Chalmers, M. (1992). Monkeys are rational! *Quarterly Journal of Experimental Psychology, 45B,* 189–228.

Morgan, C. L. (1894). *Introduction to comparative psychology.* London: Scott.

Moyer, R. S., & Bayer, R. H. (1976). Mental comparison and the symbolic distance effect. *Cognitive Psychology, 8,* 228–246.

Mynatt, B. T., & Smith, K. H. (1977). Constructive processes in linear order problems revealed by sentence study times. *Journal of Experimental Psychology: Human Learning & Memory, 3,* 357–374.

Neisser, U. (1967). *Cognitive psychology.* New York: Appleton-Century-Crofts.

Piaget, J. (1928). *Judgement and reasoning in the child.* New York: Harcourt Brace.

Potts, G. R. (1972). Information processing strategies used in the encoding of linear orderings. *Journal of Verbal Learning and Verbal Behavior, 11,* 727–740.

Potts, G. R. (1974). Sorting and retrieving information about ordered relationships. *Journal of Experimental Psychology, 103,* 431–439.

Potts, G. R. (1977). Frequency Information and distance effects: A reply to Humphreys. *Journal of Verbal Learning and Verbal Behavior, 16,* 479–487.

Rescorla, R. A., & Wagner, A. R. (1972). A theory of Pavlovian conditioning: Variations in the effectiveness of reinforcement and nonreinforcement. In A. H. Black & W. A. Prokasy (Eds.), *Classical conditioning II: Current theory and research.* New York: Appleton-Century-Crofts.

Siemann, M., & Delius, J. D. (1993). Implicit deductive reasoning in humans. *Naturwissenschaften, 80,* 364–366.

Smith, K. H., & Foos, P. W. (1975). Effect of presentation order on the construction of linear orders. *Memory and Cognition, 3,* 614–618.

Steirn, J. N., Weaver, J. E., & Zentall, T. R. (1995). Transitive inference in pigeons: Simplified procedures and a test of value transfer theory. *Animal Learning and Behavior, 23,* 76–82.

Sternberg, R. J. (1980). Representation and process in linear syllogistic reasoning. *Journal of Experimental Psychology: General, 109,* 119–159.

Trabasso, T. (1975). Representation, memory, and reasoning: How do we make transitive inferences? In A. D. Pick (Ed.), *Minnesota symposia on child psychology* (Vol. 9, pp. 135–172). Minneapolis: University of Minnesota Press.

Trabasso, T. (1977). The role of memory as a system in making transitive inferences. In R. V. Kail & J. W. Hagen (Eds.), *Perspectives on the development of memory and cognition* (pp. 333–366). Hillsdale, NJ: Lawrence Erlbaum Associates.

Trabasso, T., & Riley, C. A. (1975). On the construction and use of representations involving linear order. In R. L. Solso (Ed.), *Information processing and cognition. The Loyola symposium* (pp. 381–410). Hillsdale, NJ: Lawrence Erlbaum Associates.

Trabasso, T., Riley, C. A., & Wilson, E. G. (1975). The representation of linear order and spatial strategies in reasoning. In R. Falmagne (Ed.), *Reasoning, representation and process* (pp. 201–229). Hillsdale, NJ: Lawrence Erlbaum Associates.

Werner, U. B., Köppl, U., & Delius, J. D. (1992). Transitive Inferenz bei nicht-verbaler Aufgabendarbietung [Transitive Inference under non-verbal conditions], *Zeitschrift für experimentelle und angewandte Psychologie, 39,* 662–683.

Woocher, F. D., Glass, A. L., & Holyoak, K. J. (1978). Position discriminability in linear orderings. *Memory and Cognition, 6,* 165–173.

Wynne, C. D. L. (1995). Reinforcement accounts for transitive inference performance. *Animal Learning and Behavior, 23,* 207–217.

Wynne, C. D. L. (1997). Pigeon transitive inference: Tests of simple accounts of a complex performance. *Behavioural Processes, 39,* 95–112.

Wynne, C. D. L., Fersen, L. von, & Staddon, J. E. R. (1992). Pigeons' inferences are transitive and are the outcome of elementary conditioning principles: A response. *Journal of Experimental Psychology: Animal Behavior Principles, 18,* 313–315.

Author Index

Subject Index

A

Adaptive behavior, 2, 18.
Adaptive resonance theory, 33–57
Adaptive timing, 32, 50–55, 71–73
Amygdala, 30, 31, 33
Arousal, 129–142, 153
Attention, 33–34
Automatically defined function, 158–160, 171–174, 183, 195–197
Avoidance and escape learning, 135, 136, 142, 143, 201–235
    acquisition of, 219–223
    blocking the avoidance response, 216, 217
    different escape and avoidance responses, 224, 225
    discriminative avoidance, 225, 226
    extinction of, 213–215
    shocking all responses, 217, 218
    shocking the avoidance response, 217–219
    Sidman avoidance, 226–228
    trace avoidance, 223, 224

B

Baker map, 6–9
Blocking, 31, 32, 48

C

Cerebellum, 30–32, 48, 54–56
Conditioning, see also Reinforcement learning, 129, 130
    aversive, 136

classical, 31, 32, 49, 51, 53, 58–61, 64, 65, 67–73, 88, 92, 93, 106, 116, 118, 121
compound, 94, 95, 107–114, 117, 118, 557
Configuration,
    spontaneous, 91, 92
Contextual discrimination, 121
Crossover, 157, 161–163, 173, 174, 182–186, 191, 194

D

Delayed nonmatch to sample, 50, 52, 54
Doubly gated signal spectrum, 65, 66
Drive, 57, 58, 61, 62, 72, 73

E

Exponentially weighted moving average, 149, 150
Extinction, 129, 130, 141–152
    partial reinforcement extinction effect, 144–148, 150

F

Feeding, 242–265
    eating in meals, 245–247, 251–253, 260–262
    in closed economy, 243, 258, 260–262
    in open economy, 243, 258, 264
    model for, 247–265
Fitness algorithm, 162
Fixed interval schedule of reinforcement, 146, 152